Industrial Electricity and Electronics

Robert G. Seippel

RESTON PUBLISHING COMPANY, INC.
A Prentice-Hall Company
Reston, Virginia

Library of Congress Cataloging in Publication Data

Seippel, Robert G.
Industrial electricity and electronics.

Includes index.
1. Electric engineering. 2. Electronics. I. Title.
TK146.S35 1983 621.319'24 83-23114
ISBN 0-8359-3073-4

A Prentice-Hall Company
Reston, Virginia 22090

10 9 8 7 6 5 4 3 2 1

Printed in the United States of America

Contents

Preface

Industrial electricity/electronics has been for some time a leader in our field. Without this giant, other components are simply not manufactured. It would seem that we in electricity/electronics should dedicate more of our time and energy to the advancement of the discipline. The purpose of this book is to provide a survey of the industrial electricity/electronics field.

The subject material is directed toward the function and operation of industrial devices. Descriptive terms are applied, rather than mathematical. The book is intended to satisfy the needs of the technician, but it may indeed provide resource material for the engineer. Readers in community colleges, technical colleges and institutes, industrial high schools, and industrial training programs may benefit equally.

Chapter 1 is a dissertation on electrical/electronic safety, a must for everyone in the field. Chapters 2 and 3 provide general component descriptions. Chapters 4 through 7 are related to electrical/electronic circuitry. Chapter 8 covers a major area usually forgotten by writers, that is, transducers. Machine monitoring is dealt with in Chapter 9. Chapter 10 covers an extremely broad industrial field—optoelectronics. Chapter 11 is directed toward motion monitoring and control—optical encoders.

All the areas covered in this book are extremely large in scope. The author would hope that this cursory look at industrial electricity/electronics will stimulate the reader to further study.

I would like to thank all the companies that I have dealt with in the past years for providing me with the background to write this work. I would also like to thank my artist Lara Snider and my editor Patricia Rayner. Final accolades are due my partner and wife, Hazel Seippel, for her typing, proofing, and patience.

ROBERT G. SEIPPEL

1

Industrial Safety

INTRODUCTION

Safety in electrical systems concerns three different areas: protection of life, protection of property, and protection of uninterrupted productive output. The required investment to accomplish improved safety often consists merely of additional planning effort without any extra equipment investment. The protection of human life is paramount. Electrical plant property can be replaced and lost production can be made up, but human life can never be recovered nor human suffering compensated for. To achieve improved safety to personnel, special attention should be directed to energized equipment, adequate short-circuit protective devices, a good maintenance program, simplicity of the electrical system design, and proper training of personnel who work around electricity. Many of the items necessary to give improved protection to life will also secure improved protection to plant property and minimize breakdown of electrical system equipment. This chapter deals with electrical safety, switching practices, precautions, and accident prevention.

This chapter is reprinted with permission from A. S. Gill, *Electrical Equipment*, Reston Publishing Co., Reston, Va., 1982.

The following rules are basic to electrical accident prevention:

- Know the work to be done and *how* to do it.
- Review working area for hazards of environment or facility design that may exist, in addition to those directly associated with the assigned work objective.
- Wear flame-retardant coveralls and safety glasses, plus other recommended protective devices/equipment.
- Isolate (de-energize) the circuits and/or equipment to be worked on.
- Lock out and tag open all power sources and circuits to and from the equipment/circuit to be worked on.
- Test with two pretested testing devices for the presence of electrical energy on circuits and/or equipment (both primary and secondary) while wearing electrical protective gloves.
- Ground all sides of the work area with protective grounds applied with "hot sticks." All grounds must be visible at all times to those in the work area.
- Enclose the work area with tape barrier.

ON-SITE ELECTRICAL SAFETY

Prior to going on an on-site electrical assignment, each worker should receive the following rules and should review and abide by them while on the assignment.

A foreman or qualified employee should be designated for each on-site assignment to provide on-site work direction and safety coordination. All personnel assigned to on-site electrical work should comply with the following directions:

- Know the work content and work sequence, especially all safety measures.
- Know the proper tools and instruments required for the work, that they have the full capability of safely performing the work, and that they are in good repair and/or are calibrated.
- Check to determine that all de-energized circuits/equipment are locked out and that grounds are placed on all sides of the work area prior to beginning work.
- Segregate all work areas with barriers or tapes, confine all your activities to these areas, and prevent unauthorized access to the area.

- Insure that all energized circuits/equipment adjacent to the work area are isolated, protected, or marked by at least two methods (e.g., rubber mats, tapes, signs, etc.) for personnel protection.
- Do not perform work on energized circuits/equipment without the direct authorization of your unit manager. When work on energized circuits/equipment has been authorized, use complete, safety-tested equipment (i.e., rubber gloves, sleeves, mats, insulated tools, etc.).
- Your foreman/qualified employee must inform you of all changes in work conditions. You then must repeat this information to your foreman/qualified employee to ensure your recognition and understanding of the condition.
- Do *not* work alone; work with another worker or employee at all times.
- Do *not* enter an energized area without direct permission from your foreman/qualified employee.
- Discuss each step of your work with your foreman/qualified employee before it is begun.
- Do *not* directly touch an unconscious fellow worker since he or she may be in contact with an energized circuit/equipment. Use an insulated device to remove him or her from the suspect area.
- Do *not* perform, or continue to perform, any work when you are in doubt about the safety procedure to be followed, the condition of the equipment, or any potential hazards. Perform this work only after you have obtained directions from your foreman/qualified employee.
- Do *not* work on, or adjacent to, any energized circuits/equipment unless you feel alert and are in good health.

ON-SITE SAFETY KIT

The following are recommended protective tools to be used in preparation for and in the performance of on-site electrical work:

- Red safety tape (300 ft)
- Red flashing hazard lights (6)
- Safety cones (6)
- Red "Do Not Operate" tags (15)
- Padlocks, keys, and lock shackle (6)
- Ground fault circuit interrupter, 15 ampere, 125 volt (1)
- Fire extinguishers (2)

- Personal protective equipment
  - Flame-retardant coveralls
  - Safety glasses
  - Face shields
  - Hard hats
  - Other items required for protection on the job
- Combustible gas/oxygen detectors
- Portable ventilation blower
- Ground loop impedance tester, ohmmeter (1)
- Voltage detectors
  - Statiscope
    - (1) Station type (1)
    - (2) Overhead extension type (1)
  - Audio
    - (1) Tic Tracer
    - (2) ESP
- Voltage/ampere meter (1)
  - Amprobe
  - Simpson
- Rubber gloves and protectors of appropriate class
- Grounding clamps, cables
- Hot sticks

WORK AREA CONTROL

When workers are setting up the control area, it should be standard procedure that the safety coordinator be present and provide the required information.

Tape—Solid Red

A red tape barrier (with safety cones and red flashing lights) must be used to enclose an area in which personnel will be working. Other persons may not enter the isolated area unless they are actively working in conjunction with the personnel on the assigned work.

The purpose of the solid red tape barrier is to enclose and isolate an area in which a hazard might exist for individuals unfamiliar with the equipment enclosed. The only persons permitted within the solid red barrier are individuals knowledgeable in the use and operation of the enclosed equipment.

For their safety, workers shall not interest themselves in nor enter any area not enclosed by the red tape barrier except for a defined route to enter and leave the site.

It is important that the tape barrier is strictly controlled and the restrictions regarding its use are enforced.

When any workers are using solid red tape to enclose an area, the following requirements must be satisfied:

- Place the tape so that it completely encloses the area or equipment where the hazard exists.
- Place the tape so that it is readily visible from all avenues of approach and at such a level that it forms an effective barrier.
- Be certain that the area enclosed by the tape is large enough to give adequate clearance between the hazard and any personnel working in the tape enclosed area.
- Arrange the tape so any test equipment for the setup can be operated safely from outside the enclosed area.
- Use the tape to prevent the area from being entered by persons unfamiliar with the work and associated hazards. Do not use the tape for any other purpose.
- Remove the tape when the hazard no longer exists and the work is completed.

It shall be standard procedure that the workers should consider all areas outside the red barrier work area as energized and undertake no investigation unless accompanied by a knowledgeable plant employee.

Tape—White with Red Stripe

White tape with a red stripe is used to enclose and isolate a temporary hazard (mechanical or electrical). No one is to enter this enclosed area. Obviously, if the enclosing of a hazardous area with tape is to be protective, the use of the tape barrier should be controlled and the restrictions on entering the area strictly enforced.

When any personnel are using white tape with a red stripe, the following requirements must be satisfied:

- Place the tape so that it completely encloses the area or equipment where the hazard exists.

- Place the tape so that it is readily visible from all avenues of approach and at such a level that it forms an effective barrier.
- Be certain that the area enclosed is large enough to give adequate clearance between the hazard and any personnel outside the enclosed area.
- Arrange the tape so that the test equipment for the setup can be operated outside the enclosed area.
- Use the tape only to isolate a temporary mechanical or electrical hazard: do *not* use the tape for any other purpose.
- Consider a striped tape area similar to an interlocked enclosure and treat as such.
- Remove the tape when the hazard no longer exists.

LOCK-OUT AND/OR TAGGING

For the protection of personnel working on electrical conductor and/or equipment, locks must be placed on all open isolation devices designed to receive them. "DANGER" tags signed by the foreman or qualified employee must *also* be placed on the open isolation device.

Danger Tags

Danger tags may be applied only by authorized personnel, and the tags must be dated and signed by the person applying the tag. The following requirements must be satisfied when danger tags are used:

- Danger tags are to be used only for personnel protection when the personnel are required to work on or near equipment that, if operated, might cause injury.
- Danger tags are attached to primary disconnecting devices as a means of "locking out" equipment. Tag each source of power to the equipment and associated feeds (instrumentation circuits, PT's, CT's, etc.) to the equipment which is to be locked out.
- Danger tags should be left on the equipment only while the personnel are working on the equipment or when a hazard to the personnel exists.
- A device bearing a danger tag must not be operated at any time (see Figure 1-1).

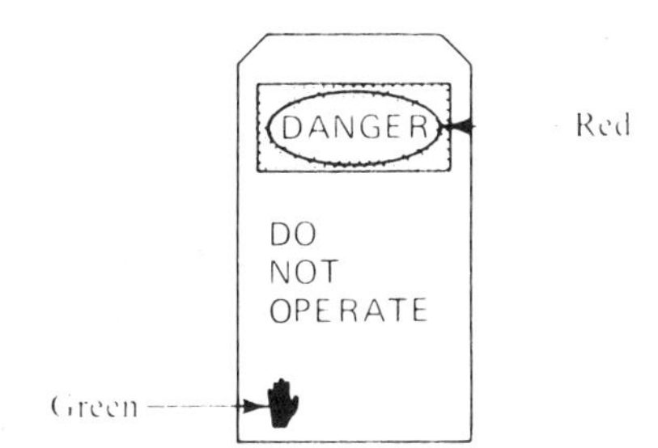

Figure 1-1 Danger Tag

Out of Order Tags

Out of order tags are used to restrict the operation of equipment which has a mechanical defect or for other reasons that are not related to the safety of personnel. Complete information concerning the reasons for the tag and a list of all persons authorized to operate the tagged device must be written on the tag (see Figure 1-2).

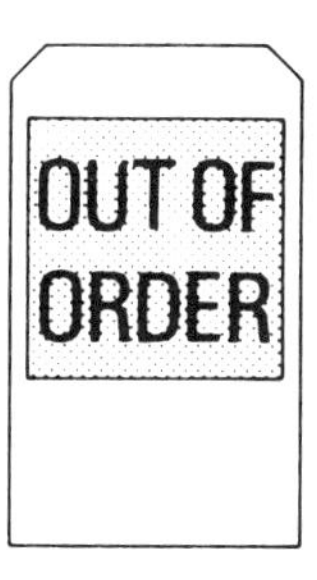

Figure 1-2 Out-of-Service Tag

Out of Service Tags

Out of service tags are used to indicate equipment that has been taken out of service. It is a white tag with white letters on a black background.

Caution Tags

Caution tags are used to indicate potential hazard or unsafe conditions. These are yellow tags with yellow letters on a black background.

Use of Danger Tags

Danger tags are authorized for use on any isolation device as a method of "locking out" equipment. These tags must be hung so there is no doubt as to which device they control. The tags may be used and signed only by authorized personnel, who are designated by the plant manager.

When more than one crew or trade is working on the same equipment, each crew must attach its own tag and place its own lock on the device. Gang lock clips can be used as shown in Figure 1-3 to provide maximum protection to a number of crews working on the same equipment or conductors.

Figure 1-3 Danger Tag

Danger tags may be removed only by the person who originally placed and signed the tag. If that person is absolutely unable to remove the tag, a committee selected by the plant manager will fully investigate the situation. This committee will remove the tag only when they are satisfied that they have full knowledge of the intention of the original "tagger" and that the tag may be removed without endangering anyone.

PROTECTIVE APPAREL—OPERATING ELECTRICAL EQUIPMENT

All personnel must wear the following protective apparel when working on electrical equipment which is, or might be considered, energized, or become energized as a result of the work:

- Fire-resistant coveralls buttoned fully at the throat and wrists.
- Electrical lineman's safety gloves with protectors.

- A face shield which also provides forehead and hair protection, or a face shield which can be attached to a hard hat.
- All personnel shall wear protective apparel when withdrawing and inserting circuit breakers, connecting and disconnecting ground connections, and testing for energized circuits and/or equipment.
- Only qualified personnel shall be allowed to operate switching equipment.

TESTING OF ELECTRICAL CIRCUITS AND/OR EQUIPMENT

General

- *All circuits and equipment are to be considered as energized* until proven de-energized by testing with voltage detectors, and grounding cables are connected. The voltage detectors selected should be for the class of voltage supplied to the circuits and equipment to be serviced.
- *Personnel* assigned to on-site electrical service work should be *supplied* with at least two (2) electrical *voltage detectors.* The *voltage detectors* provided shall be capable of *safely detecting* the *voltage* present in the circuits and/or equipment to be serviced. The assigned *personnel* shall be instructed in the correct operation of each detector *before each* on-site electrical *job.*
- Each electrical circuit and/or piece of equipment to be serviced should be tested by an assigned craftsman with two detectors and then tested by one other person who has been trained in the correct operation of the voltage detectors. This testing shall be performed in the assigned craftsman's presence to ensure that the electrical circuit and/or equipment is de-energized.
- The voltage detectors *should be checked* for proper operation immediately *prior to* and *immediately after* testing the electrical circuits and/or equipment to be serviced. These checks should be made on a known source of energized voltage, such as on the spark plug of a running automobile engine if a "glow stick," or with a specifically designed tester supplied by the detector vendor.
- While testing circuits and/or equipment, the craftsman performing the tests shall wear lineman's safety rubber gloves designed for the class of voltage in the circuits and/or equipment to be serviced and other protective equipment for this work.

Capacitors

A capacitor to be serviced must be removed from operation in the following sequence:

- Isolate the capacitors by opening the breakers or isolation devices connecting them to the electrical system.
- Permit the capacitors to drain off the accumulated charge for five to ten minutes. (There is generally a built-in device which accomplishes this drain.)
- Short circuit and ground the capacitors in the manner and with the protective equipment noted in "Grounds—Personnel Protection." While performing these procedures, be very careful that sufficient distance is maintained from the capacitors with a hot stick in the event the drain-off device is not properly functioning.

Vacuum Circuit Breaker Hipotting—Cautions

Although the procedure for hipotting a vacuum circuit breaker is similar to that used for any other electrical device, there are two areas that require the exercise of extra caution.

During any hipotting operation, the main shield inside the interrupter can acquire an electrical charge that usually will be retained even after the hipot voltage is removed. This shield is attached to the midband ring of the insulating envelope. *A grounding stick should always be used* to discharge the ring as well as the other metal parts of the assembly before touching the interrupter, connections, or breaker studs.

High voltage applied across open gaps in a vacuum can produce hazardous X-radiation if the voltage across the contacts exceeds a certain level for a given contact gap. Therefore, *do not make hipot tests on an open breaker at voltages higher than the recommended 36 kV ac* across each interrupter. During the hipot test, the steel front panel and partial side panels should be assembled to the breaker. Personnel should stand in front of the breaker to take advantage of the shielding afforded by the panels. If this position is not practical, equivalent protection can be provided by limiting personnel exposure to testing four 3-phase breakers per hour with the personnel not closer than three meters (9 ft 10 in.) to the interrupters. During equipment operation in the normal current-carrying mode, there is no X-radiation because there are no open contacts.

Electrostatic Coupling

When personnel are working on a de-energized circuit that is adjacent to an energized circuit, it is important to be certain that solid grounds are attached to the de-energized circuit at all times. *A substantial voltage charge can be generated* in a de-energized circuit by electromagnetic coupling with the energized circuit. The solid grounds will drain off this voltage charge.

RUBBER GLOVES FOR ELECTRICAL WORK—USE AND CARE

Rubber gloves with leather protectors that have been tested to at least 10,000 volts must be worn when work is performed on or within reach of energized conductors and/or equipment. The rubber gloves and protectors of the appropriate class should be available to all trained personnel as part of the safety kit for on-site electrical work.

The rubber gloves and protectors are of two types:

- *Low-voltage* rubber gloves and protectors (Class 0). These gloves are tested and approved for work on equipment *energized at 750 volts or less.* (Permission should be given by the foreman for the use of low-voltage gloves when working on conductors and/or equipment *energized below 750 volts.*)
- *High-voltage* rubber gloves and protectors. The gloves are tested at 10,000 volts (Class 1) for use on 5 kV or less, tested for 15,000 volts for use on 10 kV or less (Class 2), and at 20,000 volts (Class 3) for use on 15 kV or less voltage ratings.

Both high- and low-voltage rubber gloves are of the gauntlet type and are available in various sizes. To get the best possible protection from rubber gloves, and to keep them in a serviceable condition as long as possible, here are a few general rules that apply whenever they are used in electrical work:

- Always wear leather protectors over your gloves. Any direct contact of a rubber glove with sharp or pointed objects may cut, snag, or puncture the glove and rob you of the protection you are depending on.
- Always wear rubber gloves right side out (serial number and size to the outside). Turning gloves inside out places a stress on the preformed rubber.

- Always keep the gauntlets up. Rolling them down sacrifices a valuable area of protection.
- Always inspect and give a field air test (described later) to your gloves before using them. Check the inside of the protectors for any bit of metal or short pieces of wire that may have fallen in them.
- Always store gloves where they cannot come into contact with sharp or pointed tools that may cut or puncture them.
- All gloves are to be inspected before use.

High-Voltage Rubber Gloves

- These gloves must be tested before they are issued. All gloves should be issued in matched pairs in a sealed carton. If received with the seal broken, return them for testing.
- When high-voltage gloves are issued to individuals for use over a three-month period, they shall be inspected and tested at least every three months by a certified testing laboratory. All gloves must be tested when returned to the tool crib after the job is completed.

Low-Voltage Rubber Gloves

- Low-voltage rubber gloves must be inspected (see "Inspection of Rubber Gloves") before each use.
- Defective gloves, or gloves in a questionable condition, must be immediately replaced.

Leather Protector Gloves

- Approved leather protectors must be worn over rubber gloves to protect them from mechanical injury.
- Protectors that have been soaked with oil should never be used over rubber gloves.
- Protectors that are serviceable for use over rubber gloves are not to be used as work gloves.
- Protectors should be replaced if they have faulty or worn stitching, holes, cuts, abrasions, or if for any other reason they no longer protect the rubber gloves.

Inspection of Rubber Gloves (All Classes)

Before rubber gloves are used, a visual inspection and an air test should be made at least once every day and at any other time deemed necessary during the progress of the job.

Visual Inspection When inspecting rubber gloves in the field, stretch a small area at a time (see Figure 1-4a), checking to be sure that no defects exist, such as (a) embedded foreign material, (b) deep scratches, (c) pin holes or punctures, (d) snags, or (e) cuts. In addition, look for signs of deterioration caused by oil, tar, grease, insulating compounds, or any other substance which may be injurious to rubber. Inspect the entire glove thoroughly, including the gauntlet.

Gloves that are found to be defective should not be mutilated in the field but should be tagged with a yellow tag and turned in for proper disposal.

Air Test After visually inspecting the glove, other defects may be observed by applying the air test as follows:

- Hold the glove with thumbs and forefingers as illustrated in Figure 1-4b.
- Twirl the glove around quickly to fill with air (Figure 1-4c).
- Trap the air by squeezing the gauntlet with one hand. Use the other hand to squeeze the palm, fingers, and thumb in looking for weaknesses and defects (Figure 1-4d).
- Hold the glove to the face to detect air leakage or hold it to the ear and listen for escaping air.

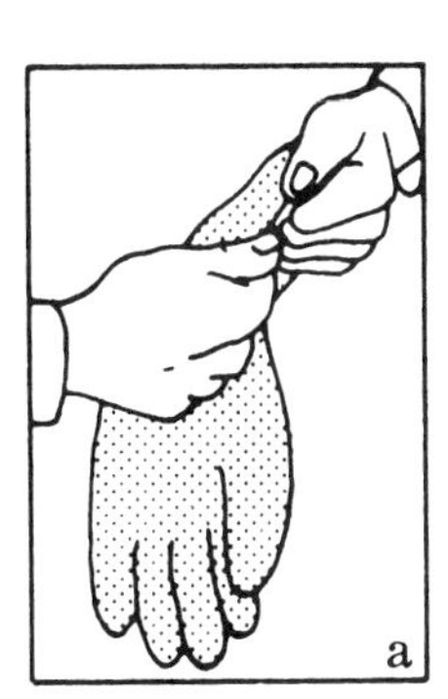

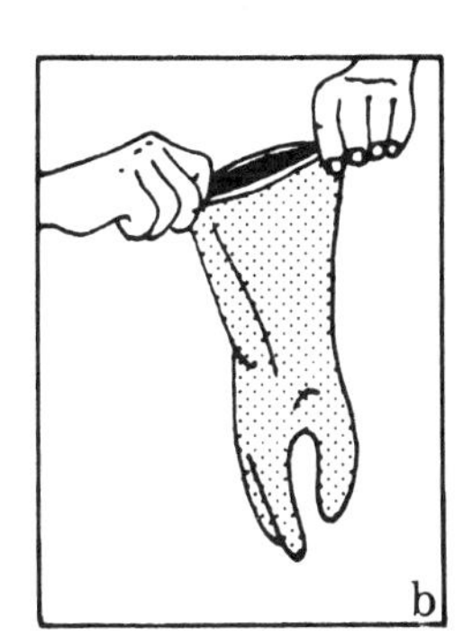

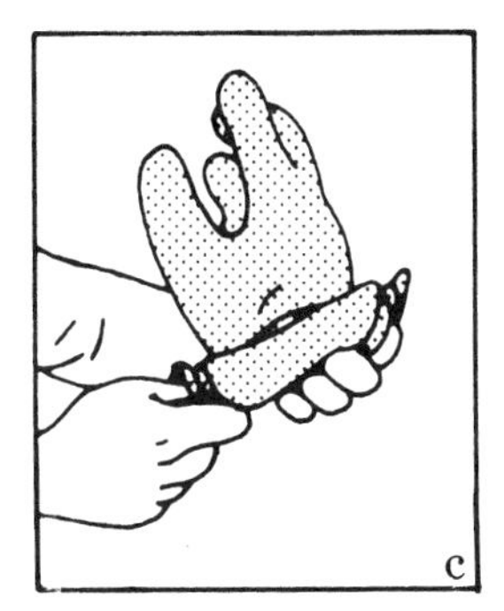

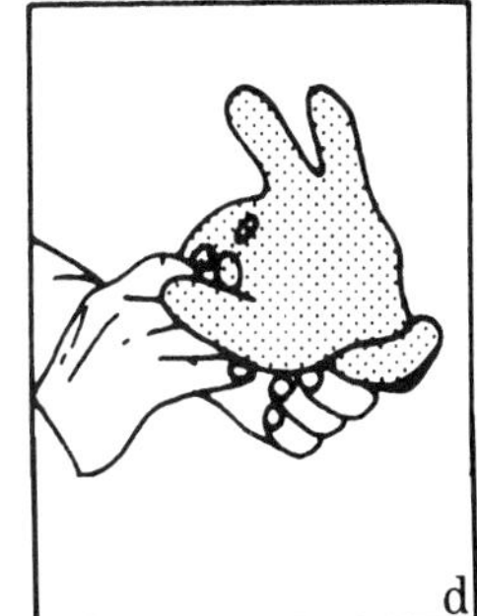

Figure 1-4 Inspection of Rubber Gloves

LOW-VOLTAGE TESTER

This tester may be used for measuring ac or dc voltage from 110 to 600 V when accuracy is not required. It can be used to test for continuity, blown fuses, grounded side of a circuit or a motor, and polarity. This tester operates on the principle that the current passed through the solenoid of the instrument is proportional to the voltage under test and will cause the tester solenoid plunger to move in the same proportion. A pointer attached to the plunger indicates the voltage on the tester scale. This instrument has no internal protection; therefore, extreme caution must be used at all times.

Some models have a two-part neon bulb. Both parts glow when energized by alternating current. Only the part that is connected to the negative side of a circuit will glow when energized by direct current.

Use of Low-Voltage Tester When the low-voltage tester is used,

- Wear rubber gloves with protectors.
- Check the operation of the tester by testing a known energized circuit.
- Assure good contact with the tester probes across the circuit being tested.
- Read the voltage on the tester.
- Because the low-voltage tester is designed for intermittent use only, continuous operation might burn out the solenoid, especially on the higher voltage.

Tic Tracer (An Audio Voltage Detector)

The Tic Tracer is an audio voltage detector, which detects the electrostatic field surrounding an energized alternating current circuit and/or equipment.

Use of the Tic Tracer This detector will operate only on unshielded ac circuits and/or equipment. It will detect the presence of voltages ranging from approximately 40 up to 600 V when hand held (higher voltages when used with approved hot sticks).

- Wear high-voltage rubber gloves and leather protectors when you are using the Tic Tracer to test circuits and/or equipment to be serviced.
- Turn on the actuating switch on the side of the Tic Tracer.
- Check the tracer by bringing it close to a conductor known to be energized. Check for proper tracer operation by placing it near a lighted fluorescent bulb or at any known energized conductor of ac voltage.

MEDIUM- AND HIGH-VOLTAGE DETECTORS

Proximity Type

The *proximity type* high-voltage (HV) tester is an instrument intended for use in detecting the electrostatic field surrounding an electrical conductor that is energized with alternating current at high potential. *It is used only on alternating current circuits and/or equipment.* The lowest voltage that can be reliably detected by this device is about 2000 V. Most detectors of this type have hard rubber or plastic tubular cases with one end for testing and a handle at the other end. A neon tube is used for voltage detection. There are several designs in various lengths for use in different situations. When the test end is brought near an uninsulated conductor that has been energized with alternating current at high potential, the neon tube will light with a red glow. A conductor that is surrounded by grounded metal has its electrostatic field effectively limited by the grounded metal; therefore, care should be taken that a conductor under test is not shielded in such a way as to interfere with the operation of the detector.

Use of a Proximity HV Detector

- Wear high-voltage rubber gloves of the appropriate class with protectors.
- Wipe detector clean and dry.
- Check detector by bringing the test end close to an uninsulated conductor known to be energized at high voltage. Or you can use a portable tester (Figure 1-5) by placing the lamp-end metal terminal of the detector against the testing point of the tester and pulling the trigger. If the neon bulb of the detector glows, it is in good condition.

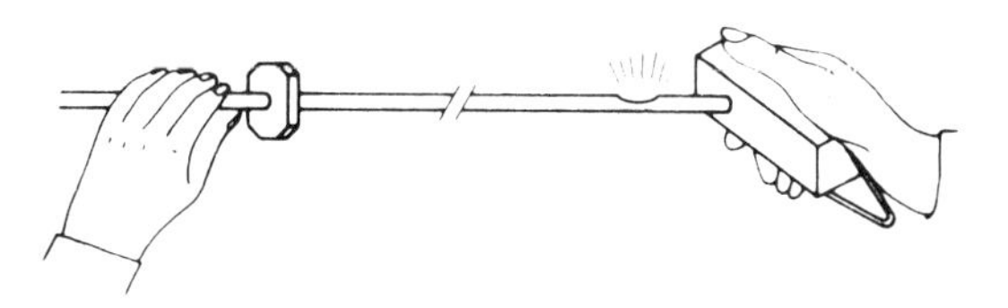

Figure 1-5 High-Voltage Detectors, Proximity

- To test an uninsulated conductor to determine whether or not it is energized, bring the test end of the detector close to the conductor. A red glow from the neon tube indicates that the conductor is energized. If there is no red glow from the neon tube, recheck the detector as explained above to make certain that the instrument is properly functioning.

- In using the detector, turn the neon tube away from the direct rays of strong light in order to make the red glow from the tube more visible.
- Wear rubber gloves and leather protectors when making tests with HV detectors and keep hands in back of guards on handles.
- The proper type of detector, having sufficient length or extension, should be used to maintain proper body clearances for the particular voltage being tested. Some detectors are equipped with a periscope (Figure 1-6) so that the glow from the neon tubes will be visible from a greater distance. Care must be exercised in the use of the periscope type so the guard or hand does not block the line of sight.

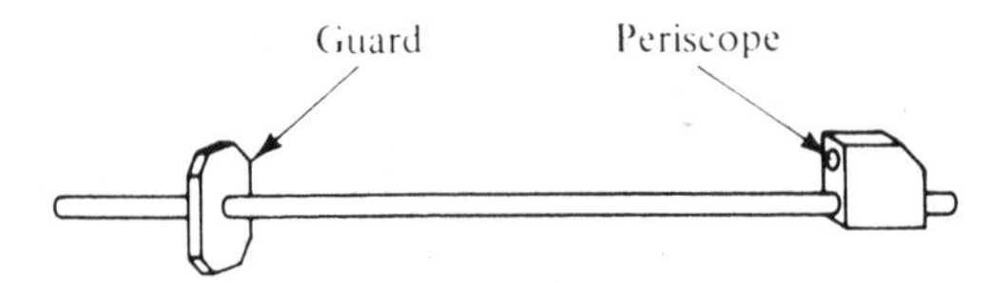

Figure 1-6 Direct Contact High-Voltage Detector

Direct Contact Type

The *direct contact type* high-voltage detector is an instrument intended for use in detecting the presence of an ac voltage with respect to ground by direct contact between the detector and the energized conductor. *It is used only on alternating current circuits and/or equipment.* The lowest voltage that can be reliably detected by this instrument is about 2400 V to ground. The actual detection is normally done by a neon tube connected to one side of a forked contact terminal. The bulb is illuminated by a very small current flow due to the capacitance between an internal electrode and ground. *This type of detector should not be used on ungrounded systems.*

A special type of phasing detector with two high-voltage wands should be used if the system is ungrounded. These two wands are each touched to a different phase and the neon indicator detects phase-to-phase voltage. This device can also be used on grounded systems. The lowest voltage that can be reliably detected by this device is about 2000 V.

Important Points to Remember

- Assume all circuits to be energized until proven otherwise.
- All protective equipment is to be proof tested for the voltage being worked.
- Maintain all leads, probes, clips, and terminals in good condition. Repair or replace defective leads.
- Wear high-voltage rubber gloves of the appropriate class with protectors.

GROUNDS—PERSONNEL PROTECTION

Grounding is installed to provide a metallic connection from ground to de-energized circuit and/or equipment to be serviced. This is for the purpose of draining off static and induced electricity but, most importantly, to *protect the worker* in the event that the equipment becomes accidentally energized. Before grounds are attached, the cable, bus, or equipment must be de-energized, isolated, locked out, and tagged. It must then be definitely established that the equipment to be grounded is de-energized by testing the circuits and/or equipment with voltage detectors.

The selection of the ground clamps is based on both the configuration and the electrical capacity according to the type of equipment to be grounded. The ground cable is to be a flexible insulated copper conductor. The rules for sizing the ground cables to be used as protective grounds are:

- The minimum size of cable to be used is a No. 1/0 conductor.
- The size of ground cables to be used must be at least equal to the size of the conductors feeding the circuit and/or equipment to be serviced. When the size of the ground cables or clusters is too large due to the system capacity, then bus sections or similar conducting materials must be used.
- The cross-sectional area of the shorting paths and to ground must be sufficient to carry the short-circuit current. One or several conductors in a cluster may function as the grounding cable to carry the current. (A 4/0 Geoprene insulated welding cable will pass 30,000 amperes for 1/2 second without melting the insulation.)

When installing ground clamps on electrical circuits and/or equipment, all workers shall use hot sticks rated for the voltage being worked. Several types of hot sticks, from 6 to 8 feet in length, are as follows:

- Rotary blade—universal end
- Two-prong—universal end
- Fixed blade—universal end
- Rotary prong—universal end

While grounds are applied, the following protective equipment must be worn:

- Flame retardant coveralls
- Safety glasses
- Electrical lineman's gloves with leather protectors
- Hard hat
- Face shield

In applying grounds, perform the following steps in sequence:

- Attach the protective ground cable to the station or building ground grid. If a ground system is not available, drive ground rods of sufficient cross section and number to carry the fault current. Be positive that a *solid* ground connection is made.
- Test the value of the impedance of the ground cable and clamps with an impedance meter. The value should be much less than 1 ohm. Unless this value is extremely low, the ground connection is *not* adequate for personnel protection.
- Connect one ground cable to the closest phase of the system and connect each succeeding phase in order of closeness. When removing the grounds, reverse the order so the application or removal of a ground will not require the crossing of an ungrounded system phase. The grounds are applied to phases A, B and C, in that order, and removed in the reverse order.
- Connect grounds to each phase of the circuit and/or equipment.
- Check to determine that all connected ground cables are visible at all times when work is being performed.
- Install grounds on *all sides of the work area.*

Switchgear Ground and Test (G & T) Device

A ground and test device is an auxiliary device, used with metal-clad switchgear, to ground equipment, or to permit various tests, when equipment is out of service. *The G & T device resembles a circuit breaker but is not designed to interrupt a circuit.*

Switchgear Dummy Element

A dummy element is a device used to provide a current path through a breaker compartment. The element frame resembles a breaker. *The element is not designed to interrupt a circuit.*

ON-SITE CIRCUIT BREAKER MAINTENANCE SAFETY CHECKLISTS

Low-Voltage (600 V and below) Checklist

Preparation

- Telephone channels must be made available to summon emergency personnel when needed. The telephone must be close to the work site and functional throughout the period during which the work is to be done.
- The light level in the work area must be sufficient to perform the work safely. (M-G sets, high-powered self-contained lighting systems, and/or emergency feeders will be used/supplied if the electrical shutdown is complete.)
- No employee should work on site for more than 12 continuous hours, and the work period should be preceded and followed by a minimum of 8 hours off (rest).
- The qualified craftsman should provide technical direction on site.
- *No one shall work alone.*
- General workers shall not energize equipment or systems. These activities are to be performed by an assigned qualified person.
- Damp or recently flooded areas shall be worked completely de-energized. Control power also shall be de-energized.
- *All* stationary (bolted in) and plug-in-type circuit breakers (nondrawout) shall only be worked on when both the line and load sources are de-energized.
- When primary power circuits are energized, no conducting materials, including hardware or tools, shall be inserted in the cubicle.

Examination of Equipment/Breakers Prior to working on equipment/breakers, the following precautions must be observed:

- On solenoid operating mechanisms, trip the breaker "open."
- On stored energy mechanisms, trip the breaker "open" and completely discharge "stored energy springs." (See the appropriate breaker instruction book and determine the exact procedure.)
- Check for proper operation of mechanical or electrical interlocks. (All low-voltage drawout power circuit breakers have either mechanical or electrical interlocks to protect both personnel and equipment while the breaker is being inserted or withdrawn from its cubicle.) Always check these devices to confirm their proper operation. *In all cases,* consult the manufacturer's instruction to obtain interlock adjustment data (dimensions and tolerances).
- Check for defeated or bypassed interlocks. (This condition enables the circuit breaker to be withdrawn or inserted in a closed position. *This is an extremely hazardous condition.*) If defeated interlocks are found, the following steps are to be performed:

  Reactivate the interlock (or remove the bypass) and test to verify proper performance.

  If equipment or materials are not available to make the repair, notify the appropriate person and do *not* reinsert the circuit breaker. The responsible person informed of the risk should decide whether to have his personnel reinsert a breaker with defective interlocks.

  To minimize personnel and equipment exposure, de-energize the equipment (including control power). If this is not possible and the responsible person plans to reinsert the breaker, all other personnel should remove themselves from the immediate area.

- Place the keys to interlocks on the equipment being worked in the possession of the qualified craftsman or foreman providing technical and safety direction for the on-site job.
- Check carefully that the spring-loaded contacts of the primary disconnect assemblies on the low-voltage drawout breakers are mounted properly, that the hardware is tight, and none is missing. Also check to determine that springs are in good condition and exert the proper pressure to ensure good contact.
- Check the hinge pins and spring clips on the primary disconnect assemblies. [The primary disconnect assemblies on competitive breakers may employ hinge pins mounted horizontally and passing through each disconnect cluster. The pins are retained on both ends by spring (or cee) clips. There is one pin per cluster and a total of six clusters, three line and three load. The length of these pins is sufficient to bridge the spacing

between the phases, and the pin will still be retained in its original primary cluster. If not properly clipped, these pins will travel with vibration or other external forces. If the pin movement is extreme, a phase-to-phase fault can result.]

- Check each low-voltage breaker with a 1000-V megger phase-to-phase-to-ground to assure adequate dielectric resistance between phases and to ground. These tests will prevent reapplying a breaker, which could cause a serious flashover due to the effects of aging and environment.
- Prior to operation of circuit breaker mechanisms, remove all tools, parts, and equipment from the breaker proper. All personnel are to stand clear of the breaker while it is being energized by a qualified person.

Racking in Precautions

- First ensure that the breaker is open.
- Inspect the cubicle for foreign objects (such as tools, rags, hi-pot wire, loose hardware, etc.). Adequate lighting is necessary to thoroughly inspect the cubicle.
- Exercise care when cleaning and inspecting cubicles. Use only insulated nonconducting tools (brushes, vacuum hoses, screwdrivers, etc.) to clean or adjust elements within the cubicle. Handles or grips are to be sufficiently long to avoid the necessity of major extensions of the arm into the cubicle. Long sleeves and rubber gloves/gauntlets are mandatory for all interior cubicle adjustments when the system is energized. Technicians are to wear long-sleeved shirts and remove all jewelry such as watches and rings. If the stationary line side or load side stabs or bars require maintenance which involves other than vacuuming, the system is to be de-energized.
- Check to determine that the control circuits (24V to 250V dc, 120V to 550V ac) are de-energized. Pulling the fuses on control circuits will ensure that these circuits are de-energized because they are not necessarily de-energized by opening the circuit breaker.
- Inspect the circuit breaker on the lift table or overhead crane just prior to insertion. This is to ensure that all parts are tight and in their proper positions. It is also intended to ensure that all foreign materials (such as rags, tools, hi-pot wire, or loose hardware) are removed.
- Perform a 1000-V megger test with the circuit breaker in the open position (the last step prior to racking in). Megger phase-to-phase and phase-to-ground on all the circuit breaker primary disconnects.

- The steps above shall be verified by the qualified craftsman. On those competitive units using cee clip retainers, a count is to be made to ensure that all clips required are in position and all hardware is properly mounted.
- Before racking in (or racking out) the circuit breaker, communicate audibly your intention to the other members of the work crew.
- Be certain to wear the required protective equipment and position yourself to either side of the cubicle.

Medium-Voltage (601 through 15,000 V) Checklist

Preparation

- Telephone channels must be made available to summon emergency personnel when needed. The telephone must be close to the work site and functional throughout the period during which the work is to be done.
- The light level in the work area must be sufficient to perform the work safely. (M-G sets, high-powered self-contained lighting systems, and/or emergency feeders will be used/supplied if the electrical shut-down is complete.)
- No employee should work on site for more than 12 continuous hours, and the work period should be preceded and followed by a minimum of 8 hours off (rest).
- The qualified craftsman should provide technical direction on site.
- *No one shall work alone.*
- General workers shall not de-energize and/or energize equipment or systems. These activities are to be performed by an assigned qualified person.
- Damp or recently flooded areas shall be worked completely de-energized. Control power also shall be de-energized.
- *All* stationary (bolted in) and plug-in-type circuit breakers (nondraw-out) shall only be worked on when both the primary and secondary (control power) sources are de-energized.
- When primary power circuits are energized, no conducting materials, including hardware and tools, shall be inserted into the cubicle.

Examination of Equipment/Breakers Prior to working on equipment/breakers, the following precautions must be observed:

- On solenoid operating mechanisms, trip the breaker "open."
- On stored energy mechanisms, trip the breaker "open" and completely discharge "stored energy springs." (See the appropriate breaker instruction book and determine the exact procedure.)
- Check for proper operation of mechanical or electrical interlocks. (All medium-voltage drawout power circuit breakers have either mechanical or electrical interlocks to protect both personnel and equipment while the breaker is being inserted or withdrawn from its cubicle.) *Always* check these devices to confirm their proper operation. *In all cases*, consult the manufacturer's instruction to obtain interlock adjustment data (dimensions and tolerances).
- Check for defeated or bypassed interlocks. (This condition enables the circuit breaker to be withdrawn or inserted in a closed position. *This is an extremely hazardous condition.*) When defeated interlocks are found, the following steps are to be performed:

  Reactivate the interlock (or remove the bypass) and test to verify proper performance.
  If equipment or materials are not available to make the repair, notify the appropriate person and do *not* reinsert the circuit breaker. The responsible person informed of the risk should decide whether to have his personnel reinsert a breaker with defective interlocks.
  To minimize personnel and equipment exposure, the equipment (including control power) should be de-energized. If this is not possible and the responsible person plans to reinsert the breaker, all other personnel shall remove themselves from the immediate area.

- Place the keys to interlocks on the equipment being worked on in the possession of the qualified craftsman or foreman providing technical and safety direction for the on-site job.
- Examine the condition of the ball-type contacts on the medium drawout breakers. This examination is to be performed with the breaker removed from the cubicle. Spring-loaded clusters form the mating contact to the primary bus. These clusters are protected by a sliding safety shutter that moves (open or closed) with the breaker elevating mechanism, which is part of the cubicle construction. The clusters must not be exposed by

sliding open the shutter when the breaker is not in the cubicle or until the cubicle is *completely* de-energized, tested, and grounded.

- Check the hinge pins and spring clips on the primary disconnect assemblies. [Competitive breakers may employ hinge pins mounted horizontally and passing through each primary disconnect cluster. The pins are retained on both ends by spring (or cee) clips. If not properly clipped, these pins will travel with vibration or other external forces. If the pin movement is extreme, a *phase-to-phase fault* can result.]
- Check each medium-voltage breaker with a 2500-V megger phase-to-phase-to-ground to assure adequate dielectric resistance between phases and to ground. These tests will prevent reapplying a breaker which could cause a serious flashover due to the effects of aging and environment.
- Prior to operation of circuit breaker mechanisms, remove all tools, parts, and equipment from the breaker proper. All personnel are to be away from the breaker and *to keep hands off the breaker.*

Racking in Precautions

- First ensure that the breaker is open.
- Inspect the cubicle for foreign objects (such as tools, rags, hi-pot wire, loose hardware, etc.). Adequate lighting is necessary to thoroughly inspect the cubicle.
- Exercise care when cleaning and inspecting cubicles. Cubicle heaters, powered from a CPT source, are an example type of this potential problem. The stationary secondary coupler (control power connections) is mounted vertically and is recessed. Care must be exercised when working close to the bottom of the coupler since potentially dangerous voltages could exist on several of the contact points.
- Check that the control circuits are de-energized (24V to 250V dc; 120V to 550V ac). Pulling the fuses on control circuits will ensure that these circuits are de-energized because they are not necessarily de-energized by opening the circuit breaker.
- Inspect the circuit breaker just prior to inspection. This is to ensure that all parts are tight and in their proper positions. It is also intended to ensure that all foreign materials (such as rags, tools, hi-pot wire, or loose hardware) are removed.
- Perform a 2500-V megger test with the circuit breaker in the open position (the last step prior to racking in). Megger phase-to-phase and phase-to-ground on all the circuit breaker primary disconnects.

- The steps above shall be verified by the qualified craftsman. On competitive units using cee clip retainers, a count is to be made to ensure that all clips required are in position and all hardware is properly mounted.
- Before racking in or racking out the circuit breaker, communicate audibly your intention to the other members of the work crew.
- Close the cubicle door *prior* to closing the circuit breaker.

CONFINED SPACES—PROCEDURE FOR ENTERING

General A confined space is an enclosed structure or space with restricted means of entry (such as a manhole, transformer vault, transformer tank, elevator pits, motor basements, etc.). The confined space is so enclosed and of such volume that natural ventilation through openings provided does *not* prevent the accumulation of dangerous air contaminants nor supply sufficient oxygen to protect the life, health, and safety of any person occupying such structure or space.

General workers are *not* to enter confined spaces (see definition above) where dangerous air contaminants have been present, are present, or could be introduced from potential sources. Workers may enter these confined spaces *only* after the atmosphere has been tested and found free of dangerous air contaminants.

Any such confined space shall be continuously maintained free of dangerous air contaminants by mechanical ventilation or equivalent means during any period of occupancy. If, however, due to emergency conditions, any such confined space cannot be cleared of dangerous air contaminants by mechanical ventilation or equivalent means, any person entering such confined space shall be provided with and shall use an approved air line respirator or approved self-contained breathing apparatus.

Dangerous contaminants that may be found in confined spaces may be grouped as follows:

- Fuel gases (e.g., manufactured gas, natural gas, or liquefied petroleum gases).
- Vapors of liquid fuels and solvents (e.g., gasoline, kerosene, naphtha, benzene, and other hydrocarbons).
- Products of combustion (e.g., carbon monoxide–engine exhaust or carbon dioxide).

- Nitrogen and/or carbon dioxide used for testing or burning gases and volatile substances within industrial drainage.
- Gases from fermentation of organic matter (e.g., hydrogen, hydrogen sulfide, methane, carbon dioxide, and mixtures deficient in oxygen).
- Gases generated by the customers' processes.

The hazards of explosion, fire, and asphyxiation may all be encountered in the preceding contaminants because mixtures of these classes of contaminants are not uncommon.

Preparing to Enter a Confined Space *All confined spaces shall be considered hazardous until proven safe by tests.* General workers shall not enter a confined space, even momentarily, until it has been tested for oxygen and combustible gas content and then power ventilated for a minimum of 5 minutes or four complete air volume changes, whichever is the greater.

Smoking, or any device which produces a spark, shall not be allowed in a confined space. In addition, smoking is not permitted within 10 feet of an open confined space.

Every employee that is to enter a confined work area should be properly trained in the procedures for detecting hazardous conditions and must be provided with the proper equipment to make this determination. Before a confined space is entered, the foreman/qualified employee must also review with the general workers the work to be performed and the hazards that may be encountered.

Testing a Confined Space Every confined space that has been closed for any period of time should be tested to determine if sufficient life-supporting oxygen is present and if combustible gases are present. In addition, any instruments that are used to sample a confined space environment must first be tested for proper working operation before they are used. Periodic calibration of the test instrumentation as recommended by the manufacturer of the instrument is a mandatory requirement, and such calibrations must utilize the type of equipment suitable for the air contaminants involved.

- If doors or covers contain vents, the preentry test is made with the doors or covers in place in order to test conditions of confined space before it has been disturbed. If the cover or door is unvented, it is opened enough only to admit the test hose or the instrument to be inserted.
- When a test indicates hazardous gases, the cover or door must be very carefully opened or removed in order not to create sparks. If the test

indicates that air contaminants are in excess of safe concentrations or that explosive hazards are present in the confined space, the space must be purged by forced ventilation until another test indicates that the air contaminant concentration is safe (see "Ventilation Procedures").

- If initial tests indicate that the atmosphere is safe, the confined space must be force ventilated for a minimum of 5 minutes, using a blower of 500 CFM (cubic feet per minute) capacity or more, or four complete air volume changes before it is entered. The blower must be operated during the entire time period when an employee is occupying the confined space.
- After the confined space is entered and the blower hose is positioned, initial testing for gas is accomplished by sampling in the areas of possible gas entrance and then generally throughout the confined space. If the test results are satisfactory, the work may be performed.
- If an unsatisfactory atmosphere is found as the result of the preceding test, employees must immediately leave the confined space. The blower must be operated for 10 additional minutes and then a second test is performed. The second test must be performed away from the direct output of the blower.
- If the second test indicates that the atmosphere is safe, the confined space may be entered. The blower shall be operated during the entire time period that personnel are occupying the confined space. Again, a test is made for gas by sampling or testing in the area of possible gas entrance and then generally throughout the confined space.
- If the confined space is still contaminated and cannot be cleared, after venting and retesting, the cause must be determined. If the area where the contaminant is entering can be plugged, sealed, or capped to render it safe, these procedures shall be performed by personnel wearing an approved air line respirator or approved self-contained breathing apparatus. The area is then to be retested, continuously vented, and monitored.

Ventilation Procedures Confined spaces containing air contaminants and/or explosion hazards must be purged by mechanical ventilation until tests indicate that the concentration of air contaminants in the confined space is not more than 10% of the lower explosive level of such air contaminants and that there is sufficient oxygen to support life available in the confined space.

Personnel performing the ventilation procedure must be familiar with the operating instructions for the particular equipment being used, and must also perform the following general procedures:

- Place the blower so it will not be subject to damage, obstruct traffic, or present a hazard to pedestrians.
- On sloping surfaces, avoid placing the blower on the upgrade side of the confined space opening. If it is necessary to place the blower on the upgrade side of a manhole or vault, block the unit so that vibration will not cause it to move toward the manhole opening.
- Do *not* operate or store blower in a confined space.
- Always remove the blower hose from a confined space before the blower is turned off.
- Place blower on a firm level base at least 10 feet from a manhole or vault opening and in accordance with above.
- Attach the blower hose to the air outlet of the blower by slipping the end of the hose, which is equipped with a strap type clamp, over the air outlet and then pull the strap tight to hold the hose in place.
- Connect the power cord to the power source to start the blower. (Only grounded electrical equipment shall be used.)
- Let the blower run for 1 minute with the hose out of the confined space. Check the end of the hose to see that the hose is securely attached to the air outlet.
- Place the blower hose in the confined space and adjust the position of the blower so the hose will run directly into the confined space without unnecessary bends. The optimum position of the output end of the blower hose is with the hose opening directed toward an end wall.
- If the ventilating blower stops, leave the confined space immediately. Remove the hose from the confined space. Do not replace the hose in the confined space until the blower is operating. When the blower is again operating, purge the hose and test the atmosphere before replacing the hose in confined space.

Equipment Necessary for Entering a Confined Space Any person entering a confined space should be provided with and should use the following additional safety equipment:

- Either an approved life belt, approved safety harness, approved wrist straps, or approved noose-type wristlets should be worn.
- A lifeline should be attached to such life belt, approved safety harness, approved wrist straps, or approved noose-type wristlets with the other end securely anchored outside the confined space.
- A safe means of entering and leaving the confined space (such as a portable ladder) should be provided. Such means must not obstruct the access opening.

- An explosion-proof battery-operated portable light in good working order.
- Nonsparking striking, chipping, hammering, or cutting tools and equipment where the confined space may contain explosive or flammable air contaminants.

Safety Monitors If the confined space is found to be contaminated, a person designated as a safety monitor should be stationed at the access opening of any confined space while such space is occupied for any reason. The safety monitor is responsible for performing the following:

- Maintaining visual contact with every person in the confined space where the construction of the space permits.
- Having continuous knowledge of the activities and well-being of every person in the confined space either through verbal communication or other positive means at all times.
- Assisting a person in a confined space with such tasks as handling tools or supplies or removing containers of refuse or debris, provided that these tasks do not interfere with his or her primary duty as a safety monitor.

The safety monitor selected should have the following characteristics:

- Be an alert, competent person, and fully capable of quickly summoning assistance for the administration of emergency first aid when required.
- Be physically able to assist in the removal of a person from a confined space under emergency conditions.

The following should be available to the safety or rescue personnel for use if required.

- Approved air line respirator, approved hose mask, or approved self-contained breathing apparatus.
- Explosion-proof battery-operated portable light in good working order.

The emergency equipment should be located at the access opening of the confined space or not more than 15 feet from such opening. In the case of a manhole or in-the-ground enclosure, a universal tripod should be set up before the confined space is entered.

Emergency Conditions *The safety monitor should not enter a confined space until he or she is relieved at his or her post.* An additional employee or another person should be available to summon aid immediately. The monitor will attempt to remove the victim by the use of the lifeline and to perform all other necessary rescue functions from the outside. Upon arrival of help, the monitor may enter the confined space for rescue work only when assured that outside assistance is adequate. Rescuers entering confined space should be protected with the approved safety equipment required by the situation, such as lifeline and harness and proper personal protection equipment.

ELECTRICAL SWITCHING PRACTICES

There are many types and designs of disconnecting switches, commonly known as disconnects, which are used to sectionalize a line or feeder, make connections, and isolate equipment on electrical systems. The type used depends upon the kind of service, voltage, current-carrying capacity, and the equipment design. This section discusses only the most common types, which use air as an insulating medium, and the hazards involved and what measures should be taken to avoid them.

General Precautions

- Management should thoroughly define who has the authority to disconnect switches or operate electrical controls or apparatus that will in any manner affect the safety of personnel or interrupt electrical service. Switching should be done by persons who are fully qualified and authorized to do this work and by other individuals only when they are under the direct supervision of such qualified and authorized persons.
- All apparatus should be legibly marked for easy identification. This marking should not be placed on a removable part.
- Switching orders should be in written form, with every step in the switching sequence spelled out in detail. Telephone or radio orders should be written down and then repeated. These procedures are particularly important for long or complicated operations.
- Every manual switching operation exposes the operator to some degree of hazard. Therefore, for his own safety he must understand the switching job to be done and be completely familiar with every detail of his part of the operation. An operator should not start a switching sequence until he has carefully checked the written order and is satisfied that it is correct in every respect. Once he has begun the operation, he must keep his mind on what he is doing, ignoring distractions, until the job

is completed. If his attention is diverted to another task while he is executing a switching operation, he should not continue the operation before carefully checking what has already been done.

Loads and Currents

- Ordinary disconnects should not be used to interrupt loads and magnetizing currents or to energize lines, cables, or equipment unless all the following conditions are met:

  The amount of current should be small.
  The kVA (kilovolt-ampere) capacity of the equipment being interrupted should be relatively low.
  The location and design of the disconnect assure that it can be operated without danger of flashover.
  Experience has shown that the disconnect can be used successfully for the particular purpose. Therefore, disconnects should be properly connected and installed before proceeding with an operation.

- Disconnecting switches are frequently used to break parallel circuits. As the blade leaves the clip, a relatively light or weak arc is drawn, which is quickly broken as the arc resistance increases. This operation is safe provided that the impedance of the circuit is low enough to permit the arc to break. Here, again, experience is the best guide as to which parallels can be broken. Energizing or magnetizing current is the most difficult to break because of its low power factor.
- Disconnects should never be used to de-energize lines, cables, capacitors, transformers, and other equipment unless specific approval is given and then with full knowledge that the disconnects will interrupt the current.
- Underhung disconnects are mounted horizontally, and careful consideration must be given before they are used to break a parallel or to interrupt load current. The heat and ionized gas of even a small arc may be enough to cause a flashover.

Switch Sticks

- Switch sticks or hook sticks are insulated tools designed for the manual operation of disconnecting switches and should be used for no other purpose. A switch stick is made up of several parts. The head or hook is either metal or plastic. The insulating section may be wood, plastic, laminated wood, or other effective insulating material, or a combination of several such materials. Glass fiber and epoxy resin materials are being used instead of wood by some manufacturers, and although they may cost somewhat more than switch sticks made of wood, the extra expense can be justified by longer life and reduced maintenance. Some manufacturers make a switch stick with a thin extruded plastic coating. This type of stick requires less maintenance than other types because the coating is tough and can be easily repaired.

- A stick of the correct type and size for the application should be selected. Standard switch sticks are made in lengths up to 24 feet with proportional diameters. Special or telescoping sticks are available in longer lengths.
- Switch sticks with insulated heads should be used to operate disconnects mounted indoors or on structures where the metal head of the stick might be shorted out when inserted into the eye of the switch.
- The parts of a switch stick are pinned together and are therefore subject to wear. Consequently, they should be examined frequently. Varnish or a similar nonconductive coating used to seal the wooden parts and prevent their absorbing moisture should be in good condition at all times.
- It is recommended that personnel do not approach electrical conductors any closer than indicated below unless you are certain that the conductors are de-energized.

Voltage range (phase to phase) kilovolt	*Minimum working and clear hot stick distance*
.3 to .75	1 ft 0 in.
2.1 to 15	2 ft 0 in.
15.1 to 35	2 ft 4 in.
35.1 to 46	2 ft 6 in.
46.1 to 72.5	3 ft 0 in.
72.6 to 121	3 ft 4 in.
138 to 145	3 ft 6 in.
161 to 169	3 ft 8 in.
230 to 242	5 ft 0 in.
345 to 362	7 ft 0 in.
500 to 552	11 ft 0 in.
700 to 765	15 ft 0 in.

- Storage of switch sticks is important. When stored indoors, a stick should be hung vertically on a wall to minimize the accumulation of dust. (It should be located in a convenient place but not where it might be subject to damage.)

- If a switch stick must be stored outdoors, it should be protected from sun and moisture. The varnish or insulating coating on a stick exposed to direct sunlight or excessive heat may soften and run. A long pipe capped at both ends, ventilated, and shielded or insulated from direct sun rays makes a good storage place.

Opening Disconnects by Using the Inching Method

The *inching* method of opening manually operated disconnects should be used wherever the opening operation can be controlled. The inching method should *never* be used for loadbreak disconnects, airbreak switches, or other switching devices designed to break load or magnetizing currents.

In the inching method, the operator opens the disconnect gradually until he is sure that there is no load current. He then opens the disconnect fully. If a small static arc develops, but no more than is expected, the disconnect may be opened further, with caution, until the arc breaks. The opening can then be completed. If an arc develops that is greater than the normal charging current warrants or, in the case of breaking a parallel, greater than expected, the disconnect should be quickly closed.

Using these techniques, an operator can open disconnects by the inching method nearly as fast as he can by other methods and with maximum safety.

Selector Disconnects

- A selector disconnect has three phases with a double blade in each phase. The blade may be placed in either of two positions. Each blade is operated separately. The three operations—to open, to open and close, and to transfer from one position to the other under load conditions—must be done in the right sequence for proper functioning.

  To open a set of selector disconnects, first one blade of each phase is fully opened. Then the second blade of each phase is fully opened.

  To open a set of selector disconnects from one position *and close* them to the other position, first one blade of each phase is fully opened, and then the second blade of each phase is fully opened. All six blades are then open. One blade of each phase is then closed to its selected position; and to complete the operation, the second blade of each phase is closed to this position.

To transfer selector disconnects *from one position to the other under load conditions*, first one blade of each phase is fully opened. After these blades have been opened, they are then closed one at a time to the selected position. The two sources of power for the circuit are thus paralleled. In a like manner, the second blade of each phase is opened from the original position, breaking the parallel, and then closed to the selected position. Now all six blades are closed to the selected position, and the transfer is completed without interruption to load.

- When operating selector disconnecting switches, the operator should never open a blade of one phase from one position and swing it closed to the selected position in one operation. The corresponding blades in each of the phases should be opened successively, and only one step should be taken at a time.

Circuit Breaker Disconnects

- The bus and line side disconnects of a circuit breaker must not be operated until the operator has made certain by observation of the circuit breaker indicating target or mechanism that the breaker is in the open position. (The exception to this rule is that the line side disconnects may be operated to make or break a parallel, for example, to shunt out feeder voltage regulators.)
- Checking the position of the breaker is a routine part of operation that must never be neglected. Sometimes a breaker operated by remote control may not open because the control contact has failed or the operator has not held the opening control long enough.
- Even if a mechanical failure has occurred or a control fuse has blown, it is still possible for a breaker to operate partially, reaching a semiclosed position. Wherever possible, all three phases of a breaker should be checked for failure of a lift rod or other mechanical failure that could cause a phase to remain closed. (An operator should always be on the alert for such conditions.)
- Before the circuit breaker is restored to service after maintenance work has been completed, the operator must check to make sure that it has been left in the open position.

Interrupter Switches

- The need to interrupt load currents and to de-energize regulators and similar equipment has led to the development of the interrupter switch.

There are many different designs, and the type used depends upon the voltage and the current-interrupting capacity required.

- Generally, there is an auxiliary blade or contact in addition to the regular load contacts. Before the switch is opened, it is important to check this auxiliary contact, where possible, to make sure that it is fully engaged. When the switch is opened, the load contact breaks first and then the auxiliary contact is opened. The arc is extinguished by an arc chamber, arcing horns, or other means. *No attempt should ever be made to inch an interrupter switch.*

Closing Disconnects under Load Conditions

- While disconnects are not designed to be used as load-pickup devices or to energize lines, cables, or apparatus, they may be used for these purposes where all the following conditions are met:

  The length of line or cable should be limited;
  The load and the capacity of the apparatus should be small; and
  The voltage should be low.

  Approval for these operations should be obtained from the person in charge only after all conditions have been studied.
- In all cases, the procedure to be followed is the same. An operator must be aware that he is closing a disconnect under load conditions; he should select a switch stick of the correct length and then take a comfortable stance in direct line with the disconnect; and he should first move the disconnect to about the three-fourths closed position. After checking to see that the blade is in line with the clip, he should then use a firm direct stroke to seat the disconnect completely.
- An operator should never reopen the disconnect to make a second attempt at closing it. If it is not seated completely, he should use added pressure to finish the closing. If the alignment is wrong, the lines or apparatus should be de-energized before the disconnect is opened.

Airbreak Switches

- An airbreak switch is a gang-operated disconnect designed with arcing horns and with sufficient clearance to energize and de-energize load currents, magnetizing current of power transformers, charging current of transmission lines, and to make and break transmission line parallels.
- Airbreaks, like other types of gang-operated disconnects, are connected so that operation of all three phases is controlled by means of a hand

lever. Some of the procedures governing stick-operated disconnects apply also to airbreaks. The inching procedure should not be applied to airbreaks designed to interrupt load or magnetizing currents.

- When airbreak switches are installed, their use should be specifically stated, and any limitation must be made known to all operating personnel concerned.
- Weather conditions can affect the successful breaking of current by an airbreak—a strong wind can blow the arc across phases or a heavy rain can change the normal insulating air clearances.
- Airbreaks should be firmly opened and closed. When closing an airbreak to pick up a load, an operator should be careful not to open it after the load circuit has been completed by either the arcing horns or the main contacts, regardless of whether or not the airbreak has been closed properly.
- Before airbreaks are used to break transmission line parallels, the load current should be checked to assure that it is within the capacity of the airbreak to interrupt.
- The operator should never depend upon the position of the operating level to determine whether the airbreak is open or closed. He should check the airbreak visually before and after operation and make sure that each operating blade is in the selected position. One blade may fail to operate and remain either closed or open.
- To prevent inadvertent operation when maintenance, repair, or other work is to be done, all airbreaks should be locked in an open position and tagged.

Protection Against Airbreak Flashover

- During the operation of airbreaks, a flashover causing a flow of fault current may occur. Several measures have been commonly used to protect the operator against electric shock. They include using an insulating section in the handle of the operating rod, grounding the handle of the rod, and providing ground mats or insulating stools at the operating position. (Opinions vary concerning the effectiveness of these measures and there are no widely accepted standards.) Rubber gloves should always be worn by the operator.
- The handle of the operating rod may be insulated against possible contact with energized parts of the airbreak by a section of nonconductive wood or porcelain, in order to effectively protect an operator against the hazards of an airbreak failure. However, this does not protect against

the shunting effect of the pole mounting and attachments. The ground voltage gradient is reduced but not eliminated by an insulating section.

- Whether or not an insulating section is provided, the handle of the operating rod should be grounded with the lowest possible resistance. (A large majority of airbreak installations have a continuous metal operating rod that is grounded.)
- Ground mats should be provided for operators to stand on; they will give maximum protection against touch voltage and ground gradient voltage and prevent any potential differences from occurring across the body in case of an insulation failure or flashover. Some companies provide portable ground mats of small iron mesh. Other companies install fixed ground mats, and specifications for installation vary widely. (Whether a ground mat is portable or fixed, however, it must be electrically connected to the operating rod and to ground to equalize the ground gradient in the area where an operator stands.)
- Operators should keep both feet on ground mats. Regardless of the type of installation or the protection provided, they should always stand with their feet as close together as is comfortable.

Motor-Controlled Disconnects and Airbreaks

- Many of the routine procedures and practices previously discussed for manually operated disconnects and airbreaks also apply to motor-controlled disconnects and airbreaks. In addition, the practices and precautions outlined in the following paragraphs should be observed.
- The operators should hold the remote control contact long enough for the operating relay to seal in and ensure operation of the disconnect.
- To determine whether the switch is open or closed, operators must never rely upon the indicating lights or upon the position of the operating handle. Instead, they must always visually check the position of the blades at the switch, making sure that each blade is in the selected position. (This check is particularly important for switches with high pressure contacts.)
- When a motor-operated switch is used as a tagging point for work clearance, the motor drive should be uncoupled from the operating rod. If this precaution cannot be taken, a heavy pin, lock, or blocking device must be used on the switch to prevent inadvertent operation. In addition, the switch for the motor control circuit should be tagged and locked in an open position to prevent operation of the motor in case of an accidental ground.

ELECTRICAL FIRE EMERGENCIES

This section is written as a guide for fire-fighting personnel for handling electrical fire emergencies. Electrical personnel are not usually fire-fighting experts but, because of their knowledge of electricity, they can provide vital and helpful information to others who are involved in fighting fires. Therefore, a cooperative effort is needed among the various groups when dealing with electrical emergencies. The various safety considerations dealing with such emergencies are as follows:

Never Make Direct Contact with Any Energized Object

Electricity, whether from a powerline or from a thundercloud, is always trying to get to the earth, which is at ground voltage—also called zero voltage. Voltage is a measure of the "pressure" that pushes electric charge through a conductor. An object with *any* voltage above zero is called *energized.* Any energized object will produce a flow of electric charge through a conductor placed between it and the earth or any other object at ground voltage, such as a grounded wire. Since nearly all common materials—including the human body—are conductors to some extent, the only way to keep the electricity where it belongs is to place some sort of insulator (nonconductor) between the energized object and the earth.

One can get just as "shocked" from a 120-V house current as one can from a 500,000-V powerline! In fact, a high-voltage shock, because of the clamping action it has on the heart (cardiac arrest), may prevent the deadly irregular beating of the heart (fibrillation) often associated with lower-voltage shocks. Cardiac-arrest victims often respond readily to artificial respiration and external heart massage, whereas a fibrillation victim may only respond to an electrical defibrillator device. Also with the lower-voltage shock, instead of enough current to knock you out, you may get just enough to set your muscles so you can't let go.

Stay Clear of Vicinity of Any Faulty Energized Object

One can be injured *without* touching an energized object. When an energized object is sparking, it emits excessive heat and ultraviolet rays. Such sparking occurs while trying to interrupt the flow of electric charge, such as when an energized wire is cut or when a fallen energized wire is lifted away from the earth. The electric charge tries to maintain its flow through the air—this results in a flash, an electric arc. The excessive heat from such a flash can burn human flesh several feet away.

The heat of an electric arc has been known to fuse contact lenses to the cornea of a man's eyes. Ultraviolet rays emitted from an electric flash may also damage unprotected eyes. Eye injuries may not be immediately apparent; there may be no noticeable eye irritation for several hours after exposure. If your eyes are exposed to an electric arc, consult a doctor for proper treatment without delay. Electrical employees should wear specially treated goggles to prevent ultraviolet-ray damage whenever an electric arc may occur.

Be Alert in Vicinity of Any Energized Object

We have already emphasized the danger from contacting an energized object or even getting in the vicinity of a faulty energized object, such as a fallen wire. It is just as important to be cautious in the vicinity of energized facilities that are operating properly. Most electrical emergency work is performed without de-energizing all electric facilities in the vicinity. In many cases, it is even advantageous to leave power on as long as possible. However, all personnel must continuously be alert. Don't let the quiet, "harmless" appearance lull you into a false sense of security.

Beware Covered Wires! Many overhead wires are covered. *But*, that covering is often designed to protect the wire from the weather or tree contact, not to protect you from the wire. Never consider a covered wire any safer than a bare wire. And remember, most wires on utility poles are bare, even though they may appear to be covered when viewed from the ground.

Beware "Telephone" Cables! Telephone cables are rarely dangerous when accidentally contacted. *But*, are you so sure you can tell the difference between telephone cable and electric power cable that you'd stake your life on it—and the lives of others? Although higher-voltage facilities are generally installed higher up on utility poles, this is *not* true always—electric power cables operating at 34,000 V may be attached *below* telephone cables on the same pole. And a fallen telephone cable may be contacting a powerline!

- Never rest ladders on wires or on any other electric equipment.
- Never drag hose over wires.
- Never even come too close to wires—brushing against one can be fatal.

You may have had some experience where you were able to contact energized facilities without incident. *But*, just because you "got away with it" before doesn't guarantee you will get away with it again. And remember, higher-voltage facilities have much greater "pressure" behind the electricity—something you "got away with" on 120-V facilities can bring disaster if attempted on 34,000-V facilities. And since normal water is a conductor of electricity, even slightly damp objects become much more hazardous.

Assume Every Fallen Wire Is Energized and Dangerous

Wire on Ground Some fallen wires snap and twist—bursting warning sparks. Others lie quietly—no sparks, no warning rattles like a snake. Both types are equally deadly. It is impossible to determine from the appearance of a wire whether or not it is energized. Also, automatic switching equipment may re-energize fallen wires. Always stay clear and keep everyone else clear until an electric company employee arrives and clears the wire or de-energizes it.

Wire on Object If a wire is in contact with any object—fence, tree, car, or person—that object in turn may be energized and deadly. Keep yourself and others away from metal highway dividers and metal fences that may be in contact with fallen wires. A fallen wire draped over such dividers and fences can energize them for their entire length.

Wire on Vehicle If anyone is in a vehicle which is in contact with a wire, the safest thing he can do is stay inside. If possible, he should drive the vehicle away from the contact. If the vehicle is on fire, tell him to jump free with both hands and feet clear of the vehicle when he hits the ground. At no time can the person simultaneously touch both the vehicle and the ground or any other object that is touching the ground, such as yourself. If he does, he will become a path for the electricity to flow to ground. Never board a vehicle that may be energized. A spray or fog nozzle should be used to direct water onto a burning vehicle—even then, stay back as far as practicable (at least 6 feet) whenever the wire on a vehicle may be energized.

Never Cut Wires Except to Protect Life

And even then, only thoroughly trained persons, such as electric company employees, using approved procedures and equipment, can cut wires. Otherwise, cutting wires can create more hazards than leaving them alone. When taut wires are cut, the change in tension may cause utility poles to fall or wires to slack off and sag to the ground some distance from where the wires are cut. Wire which retains some of its original "reel-curl" may coil up when cut and get out of control with resultant hazards.

Take Care After Cutting Cutting a wire at one place does not necessarily ensure that the wire on *either* side of the cut is de-energized because:

- Wires are frequently energized from both directions,
- Wires may be in accidental contact with other energized wires,
- Wires may be energized from a privately owned generator within a building.

Cutting Service Wires When protection of life requires de-energizing a building, cutting service wires should be considered *only* when it is not practicable to remove fuses, open circuit breakers, open the main switch, or wait for an electric company representative. *Specialized equipment* must be used to cut each wire individually and then bend each one back, to prevent short-circuiting the wires together. *All* wires must be cut. Never assume that one wire is a ground wire and is therefore safe. Even a "ground" wire may be contacting an energized wire at some unseen location. If the service wires can be cut on the supply side of where they connect to the building's wires, it will be possible to restore the service more quickly when required. However, far more important, service wires should always be cut on the building side of where they are first attached to the building—this avoids having wires fall on the ground.

Use Approved Procedures and Equipment If You Must Work *Near* Energized Facilities

This rule certainly applies whether or not there is any victim to be rescued. However, the presence of a victim requires you to be even more conscientious.

Notify Electric Company If you see no safe way of separating a victim from an energized object, request electric company assistance. Your first consideration must be your own protection—you cannot help by becoming a victim yourself.

Moving the Victim Electric company employees have *specialized equipment* that they can use to drag a victim clear of electric equipment. They can use other specialized equipment to keep the wire in contact with the ground while the victim is being dragged clear—this reduces the amount of electricity flowing through the victim and minimizes further injury from additional burns.

Moving the Wire Electric company employees have *specialized equipment* that they can use to remove a wire from a victim. They can control the wire to prevent it from recontacting the victim. Electric company employees will pull the wire toward themselves while walking away—rather than pushing and walking toward it—to reduce the danger to themselves in case the wire gets out of control. And, again, they can use other *specialized equipment* to keep the wire in contact with the ground while moving it—this reduces the amount of electricity flowing through the victim and minimizes further injury from additional burns.

Cutting the Wire If a victim is entangled with an electric wire, the wire on *both* sides of the victim must be cut to be certain that no source of electricity remains. Wires should only be cut by an individual who is thoroughly trained to cut wires safely and who uses *specialized equipment.*

First Aid A victim who has been separated from energized electric facilities *does not* retain an electric charge—so there is no danger in handling the victim, administering first aid, or applying artificial respiration. Electric burns, even if insignificant on the surface, may involve serious destruction of tissues and must receive expert medical treatment as soon as possible.

Avoid Using Hose Streams on Energized Facilities

The application of water on electric facilities by hand-held hoses may carry the electricity back to the nozzle. This electricity might be sufficient to cause serious injury. Tabularized "safe distances" can be misleading, since water conductivity and nozzle design vary widely. The National Board of Fire Underwriters' Special Interest Bulletin No. 91 advises that for 120-V facilities there is no danger unless the nozzle is brought within a few inches. However, fire fighters should consider all electric facilities to be high voltage, because even low-voltage wires may inadvertently be crossed with high-voltage wires.

Spray or Fog Preferred For maximum safety to the fire fighter, when either intentional or unintentional application of water on energized facilities may occur, a spray or fog nozzle should be used.

Beware Run-off Water! A dangerously energized puddle of water may be formed by water running off energized electric facilities.

Beware Adjacent Equipment Take care not to damage uninvolved electric facilities nearby. A porcelain insulator supporting energized facilities may flashover (arc), and even explode, if hit by a straight stream (even spray or fog) directed onto it. Wires may swing together, short-circuit, and burn down if hit by the force of a straight stream.

Other Extinguishing Agents Dry chemical and carbon dioxide are nonconductive and may be used around energized facilities. These may be used to extinguish a surface-type utility pole fire. Foam, soda acid, and the loaded-stream type are conductive and should not be used on fires around energized facilities.

Be Equally Alert Indoors and Outdoors

Medium-Voltage Installations Medium-voltage services do exist in many larger buildings—commercial, institutional, and industrial. Do not enter any transformer room or open any electric switch without advice of an authorized individual. Besides the obvious electric hazard, privately owned transformers may be filled with flammable oil or with nonflammable liquids. Such equipment is not required to be isolated outdoors or in a fire-resistant room and therefore may be located anywhere on the premises. The nonflammable liquid, while safe from a fire standpoint, may be caustic and may generate poisonous fumes. Call the plant electrician to identify specific hazards and to de-energize facilities as needed.

Low-Voltage Installations Low-voltage services exist in practically every building and can be as dangerous as medium-voltage facilities.

Leave Power On as Long as Possible The power may be needed to operate pumps or other equipment which, if stopped, would cause additional damage to the building or to any materials being produced in it.

Remove Fuses or Open Circuit Breakers
To shut off an affected section, remove fuses or open circuit breakers.

Open Main Switch
Open the main switch to shut off entire building when electric service is no longer useful. If you must stand in water or if the switch is wet, do not grasp the switch handle in the palm of your hand. Use dry equipment such as a piece of rope, pike pole, or handle of fire axe to open the switch. Then attach a warning tag indicating that the power has been intentionally shut off.

Cut Wires Only to Protect Life
Cut wires only when life would be endangered by leaving a building energized or when a victim must be rescued. However, cutting electric wires should only be considered when it is not practicable to remove fuses, open circuit breakers, open the main switch, or wait for an electric company representative.

Pull Electric Meter

Pull the electric meter only to protect life when no other method is practicable. Wear gloves and face shield or goggles to protect against electric arcing. Meters at most large buildings, as well as many house meters, can produce extensive arcing when removed—especially if the interior wiring is faulty. In addition, removing some meters does *not* interrupt the power. Such meters should be identified by a small label reading "CAUTION: Apply Jumpers Before Removing Meter." If a meter is removed, cover panel to protect public, and notify electric company.

Flammable Fumes

Whenever flammable fumes may be present, avoid operating any electric switch within the area—even a simple light switch—because even a small spark can cause an explosion.

Palms Inward

When walking through a building or any enclosure where visibility is poor, proceed with arms outstretched and the palms of the hands turned toward the face. In this way, if contact is made with an energized object the tendency of the muscles to contract may assist in getting free from the contact.

Protect People and Property in Surrounding Area and Don't Fight Fires on Electric Equipment until an Electric Company Representative Arrives

Where electric power equipment is involved, wait for the electric company representative and coordinate the fire-fighting operation with him to ensure maximum effectiveness and safety. Cooperate with his requests because he knows what is necessary to fight fires on his equipment.

Danger from Switches Never operate electric company switches that are mounted on utility poles or located in manholes or within substation properties. Many of these switches are not intended to open and drop the electric load, and attempting such an operation could damage the switch and even cause it to explode.

Danger from Oil Oil may be present in any pole-mounted, underground, or surface equipment, such as transformers. This oil will burn. Under the intense heat of fire, the equipment may even rupture and spray its burning oil. This may be followed by subsequent explosions caused by ignition of the mixture of air with hot oil vapor or with burning insulation vapor.

Danger from Water Water greatly increases the danger of electrocution from energized facilities. Until it is confirmed that electric facilities are de-energized, use only dry chemical, carbon dioxide, water spray, or fog—and even then, take extreme care to avoid physical contact with energized facilities. Also, take care not to direct a straight stream onto uninvolved electric facilities nearby.

Contain Liquids Leaked from Equipment Any liquid leaked from electrical equipment may be flammable oil or a nonflammable liquid such as askarel. Avoid contact with these liquids—they may be caustic and fumes may be irritating. Both types of liquids must be thoroughly cleaned up by the utility company to prevent environmental damage. After extinguishing any fire, try to contain any leaked liquid—use absorbent granules, dry sand, ashes, or sawdust. *Do not wash it away with a hose stream.*

Hose Streams May Be More Hazardous Than Helpful Until Any Underground Fault Is De-energized

Electric facilities are installed underground in many urban areas and new residential developments. Switching equipment and transformers are installed in manholes or in metal cabinets on the surface, and they supply electricity through an interconnected network of electric cables. The cables may be directly buried beneath only 2 or 3 feet of earth, or they may be installed in duct. Voltages are both low and high. The two major causes of fires are:

- Cable faults that ignite the cable insulation, or the fiber duct, or both.
- Oil-filled manhole equipment which overheats and spills oil that ignites.

Notify Electric Company Specify location of all manholes involved. A cable fault usually clears itself, or it can be cleared manually by opening appropriate switches. Until the fault which caused the fire is de-energized, no attempt should be made to extinguish the fire. An electric arc cannot be extinguished by fire-fighting techniques, and the arc is sustaining the fire.

Clear the Area Under normal conditions, the insulation and jacketing of underground cables provide adequate protection. However, an explosion or fire can remove these protective coverings and expose the energized conductors. Such a condition is a major hazard, and fire-fighting personnel are cautioned to stay clear.

Beware of Toxic or Explosive Gases! Flammable vapors, which are not always detectable by sense of smell, may be coming from (a) nearby sewers,

(b) gas mains, or (c) buried gasoline or oil storage tanks, as well as from (d) smoldering insulation and fiber duct. Inside a duct, the vapor–air mixture may be too rich to ignite. Upon reaching a source of fresh air, such as a manhole, the vapor–air mixture may fall within the explosive limits. The resulting explosions may be intermittent, with their frequency depending on how fast the vapors are coming out and mixing with the air. They may vary in intensity from a slight "puff" to an explosion of sufficient violence to blow a manhole cover high in the air. If the mixture becomes too rich to ignite within the confined space of a manhole, an explosion may occur when the manhole cover is removed.

Prepare to Assist Electric Company Employees The electric company may discharge water into a duct line to cool it after the circuit has been de-energized. If a hose line is supplied by the local fire company, let the electric company employees handle the nozzle, using their approved rubber gloves for protection.

Leave Manhole Covers as Found Only electric company employees should remove manhole covers using hooks or long-handled tools and standing safely to one side. And everyone must be kept a reasonable distance back to avoid injury. Removing manhole covers may help to ventilate the conduit system and pin down the location of the fault. However, removing a manhole cover may reignite flammable vapors, or even cause low-order explosions, if the atmosphere was too rich to burn before removing it.

Never Direct Water into a Manhole Until requested by the electric company representative, never direct water into a manhole. The source of the fire and any other facilities that might be damaged must be de-energized before water can be used safely and effectively.

EFFECTS OF ELECTRICAL SHOCK

Current is the killing factor in electrical shock. Voltage is important only in that it determines how much current will flow through a given body resistance. The current necessary to operate a 10-W lightbulb has eight to ten times more current than the amount that would kill a lineman, that is, if it actually breaks through skin and body resistance and flows at this amperage. A voltage of 120 V is enough to cause a current to flow which is many times greater than that necessary to kill. Currents of 100 to 200 milliamperes (mA) cause a fatal heart condition known as ventricular fibrillation for which there is no known remedy. The following figures are given for human resistance to electrical current:

Type of Resistance	*Resistance Values (ohms)*
Dry skin	100,000 to 600,000
Wet skin	1,000
Internal Body	
Hand to foot	400 to 600
Ear to ear	About 100

With 120 V and a skin resistance plus internal resistance totaling 1200 ohms, we would have $\frac{1}{10}$-A electric current, that is 100 mA. If skin contact in the circuit is maintained while the current flows through the skin, the skin resistance gradually decreases. The following is a brief summary of the effects of current values as shown in Table 1-1.

FIRST AID

First aid kits for the treatment of minor injuries should be available. Except for minor injuries, the services of a physician should be obtained. A person qualified to administer first aid should be present on each shift on on-site jobs.

Prior to starting on-site jobs, telephone communications should be available and tested to summon medical assistance if required. Each on-site job should have the telephone number of the closest hospital and medical personnel available.

Shock

Shock occurs when there is a severe injury to any part of the body from any cause. Every injured person is potentially a patient of shock and should be regarded and treated as such, whether symptoms of shock are present or not.

Proper Treatment for Shock Is As Follows

- Keep the patient warm and comfortable, but not hot. In many cases, the only first aid measure necessary and possible is to wrap the patient underneath as well as on top to prevent loss of body heat.
- Keep the patient's body horizontal or, if possible, position him so that

his feet are 12 to 18 in. higher than his head. In any case, always keep the patient's head low. The single exception to this positioning is the case of a patient who obviously has an injury to his chest, and who has difficulty breathing. This patient should be kept horizontal with head slightly raised to make his breathing easier.

- Do not let the patient sit up, except as indicated in chest injury or where there is a nose bleed. If there is a head injury and perhaps a fracture of the skull, keep the patient level and do not elevate his feet.
- If the patient is conscious, you may give him hot tea, coffee, or broth in small quantities since the warmth is valuable in combating shock.
- Proper transportation practice is never more *imperative* than in the case of a person who may develop shock. It is the most important single measure in the prevention and treatment of shock. Use an ambulance, if possible. If other means must be used, follow the above points as closely as possible.

TABLE 1-1
EFFECTS OF 60-HZ CURRENT ON AN AVERAGE HUMAN

Current Values through Body Trunk (mA)		*Effect*
Safe current values	1 or less	Causes no sensation; not felt. Is at threshold of perception.
	1 to 8	Sensation of shock. Not painful. Individual can let go at will, as muscular control is not lost (5 mA is accepted as maximum harmless current intensity).
Unsafe current values	8 to 15	Painful shock. Individual can let go at will, as muscular control is not lost.
	15 to 20	Painful shock. Muscular control of adjacent muscles lost. Cannot let go.
	20 to 50	Painful. Severe muscular contractions. Breathing is difficult.
	100 to 200	Ventricular fibrillation (a heart condition that results in death; no known remedy).
	200 and over	Severe burns. Severe muscular contractions, so severe that chest muscles clamp heart and stop it during duration of shock (this prevents ventricular fibrillation).

Resuscitation

- Seconds count. Begin artificial respiration as soon as possible. *In electric shock cases, do not rush and become a casualty yourself.* Safely remove victim from electrical contacts before starting artificial respiration. Do not move victim unless necessary to remove him from danger or to place him in the proper position for artificial respiration.
- Attempt to stop any hazardous flow of blood.
- Clear victim's mouth of false teeth or any foreign objects or fluids with your fingers or a cloth wrapped around your finger. Watch victim closely to see that mucous or stomach contents do not clog air passages.
- If help is available, have the following taken care of while applying artificial respiration:

  Call a doctor and ambulance.
  Loosen victim's clothing about neck, chest, and waist.
  Keep victim warm during and after resuscitation. Use ammonia inhalants. Do not give liquids while victim is unconscious.

- Continue uninterrupted rescue breathing until victim is breathing without help or until pronounced dead.
- The change of operators, when necessary, shall be done as smoothly as possible without breaking the rhythm. If necessary to move victim, continue resuscitation without interruption.
- Watch victim carefully after he revives. Do not permit him to exert himself.

Resuscitation: Mouth-to-Mouth (Nose) Method

- Place victim on his back. Place his head slightly downhill, if possible. A folded coat, blanket, or similar object under the victim's shoulders will help maintain proper position. Tilt the head back so chin points straight upward.
- Grasp the victim's jaw and raise it upward until the lower teeth are higher than the upper teeth; or place fingers on both sides of the jaw near the earlobes and pull upward. Maintain jaw position throughout resuscitation period to prevent tongue from blocking air passage.
- Pinch victim's nose shut with thumb and forefinger, take a deep breath and place your mouth over victim's mouth making air-tight contact; or close victim's mouth, take a deep breath and place your mouth over

victim's nose making air-tight contact. If you hesitate at direct contact, place a porous cloth between you and victim.

- Blow into the victim's mouth (nose) until his chest rises. Remove your mouth to let him exhale, turning your head to hear outrush of air. The first 8 to 10 breaths should be as rapid as the victim will respond; thereafter, the rate should be slowed to about 12 times a minute.

Important Points to Remember

- If air cannot be blown in, check position of victim's head and jaw and recheck mouth for obstructions; then try again more forcefully. If chest still does not rise, turn victim face down and strike his back sharply to dislodge obstruction. Then repeat rescue breathing procedure.
- Sometimes air enters victim's stomach, evidenced by swelling of stomach. Expel air by gently pressing down on stomach during exhalation period.

Two-Victim Method of Resuscitation: Mouth-to-Mouth (Nose)

In those rare instances where two men working together are in shock, both require resuscitation, and only one worker is available to rescue them, the following method may be used:

- Place two victims on their backs, with their heads almost touching and their feet extended in a straight line away from each other.
- Perform the mouth-to-mouth resuscitation method previously described. Apply alternately to each victim. The cycle of inflation and exhalation does not change so it will be necessary for rescuer to work quickly in order to apply rescue breathing to both victims.

External Heart Compression

PERFORM HEART COMPRESSION ONLY WHEN INDICATED. After rescue breathing has been performed for about half a minute, if bluish or gray skin color remains and *no pulse can be felt*, or if *pupils of the eyes are dilated*, heart compression should be started. Heart compression is always accompanied by rescue breathing. If only one rescuer is present, interrupt compression about every 10 to 15 compression cycles and give victim 3 or 4 breaths.

- Place victim on his back on a firm surface.
- Put hands on breastbone. Place heel of one hand on lower third of breastbone with other hand on top of first.

- Press downward. Apply pressure until breastbone moves 1-1/2 to 2 inches.
- Lift hands and permit chest to return to normal.
- Repeat compression 60 times per minute.

Heart compression should not be performed in the following instances:

- When victim has a pulse.
- When his pupils do not remain widely dilated.
- When his ribs are broken.

2

Industrial Electrical Components

Industrial electrical components are classified as *active* or *passive*. An active component must generate a voltage or amplify it. A passive component does not have these characteristics. However, the characteristics of the passive component may be changed by some external physical force and, in turn, modify circuit behavior. The differences between an electrical and an electronic component are debatable. For this book, electrical components will be those that are generally passive. Electronic components will be generally active. The author realizes that there will be arguments as to which is which.

Industrial electrical components may be strictly mechanical, such as a switch. They may also be mechanical and electrical, such as a relay or a solenoid. Then again they may be all electrical, such as a transformer.

Some of the more sophisticated electrical industrial components are driven or sensed by motion components such as synchros, potentiometers, and linear, variable differential transformers (LVDTs). These devices will be discussed in Chapter 8.

Other active and passive industrial components sense a physical happening and are modified or modify because of it. These devices are called transducers, sensors, and detectors. These complex devices will be discussed in Chapter 8.

This chapter will deal with mechanical and electrical components. We shall begin with the basic electrical components: the resistor, the capacitor, and the inductor.

FUNDAMENTAL ELECTRICAL COMPONENTS

The fundamental electrical components in electricity are the resistor, the capacitor, and the inductor. The basic purpose of the resistor is to oppose current flow. The capacitor opposes a change in voltage. The inductor opposes a change in current. Let's consider each of these components in some detail.

The Resistor

The resistor opposes current flow. Dependent on the size of the resistance, it will develop heat, cause a voltage drop, and/or limit the current flow in a circuit. The resistor is the primary component in all electricity.

Resistors are either composition or wire wound. Composition resistors are usually made from carbon and clay, while wire-wound resistors are made from wire. Composition resistors are used for high resistance and wire wound for low resistance. With low resistance, current and power increase; therefore, current and power ratings for wire-wound resistors are large.

Each resistor has a built-in tolerance. For instance, you may purchase a 1000-ohm (Ω) resistor with a 20% tolerance. This would be a guarantee that the resistor will range in resistance somewhere between 800 and 1200 Ω. On the other hand, you may purchase a 1000-Ω resistor with a 10% tolerance. This would be a guarantee that the resistor will range in resistance somewhere between 900 and 1100 Ω. Resistors with 1% or better tolerance are called *precision* resistors.

Each resistor also has a temperature coefficient, which tells us, basically, that the resistor changes in ohmic value as temperature changes. A positive temperature change for a resistor with a positive coefficient would note an increase in resistance with a temperature increase. The reverse is true for a decrease in temperature. A positive temperature change for a resistor with a negative coefficient would note a decrease in resistance with a temperature increase. The reverse is then true for decreases in temperature coefficients. Carbon resistors have a negative temperature coefficient. Each resistor has a manufacturer's power rating.

Resistors are built to withstand a specified amount of power without being destroyed. For instance, standard carbon-type resistors are built with power ratings of $\frac{1}{8}$, $\frac{1}{4}$, $\frac{1}{2}$ and 1 watt (W). In application, these resistors would be used in circuits that are subject to 50% of their power ratings. Resistors are constructed with either a fixed resistance or a variable resistance. Resistor types are discussed in the next several paragraphs.

Fixed Resistors Fixed resistors (Figure 2-1) are constructed to have a fixed value of resistance with a set tolerance. They are made of wire, metal, or a composition such as carbon. They attach by lugs or have wire-soldered ends. In Figure 2-1 are several types of fixed resistors. Each fixed resistor is recognized on a schematic by the same symbol. See Figure 2-2 for these schematic symbols. Some of the various types of fixed resistors are as follows.

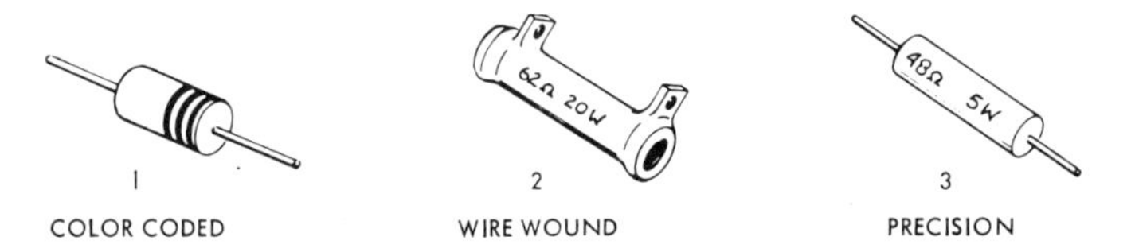

Figure 2-1 Fixed Resistors

Carbon and Clay Resistors. These are the standard resistors used most commonly in electronic equipment. They are made of a composition of carbon and clay and can be used in many applications. The carbon resistor has a negative temperature coefficient. That is, when temperature increases, the carbon resistor's ohmic value decreases, and vice versa.

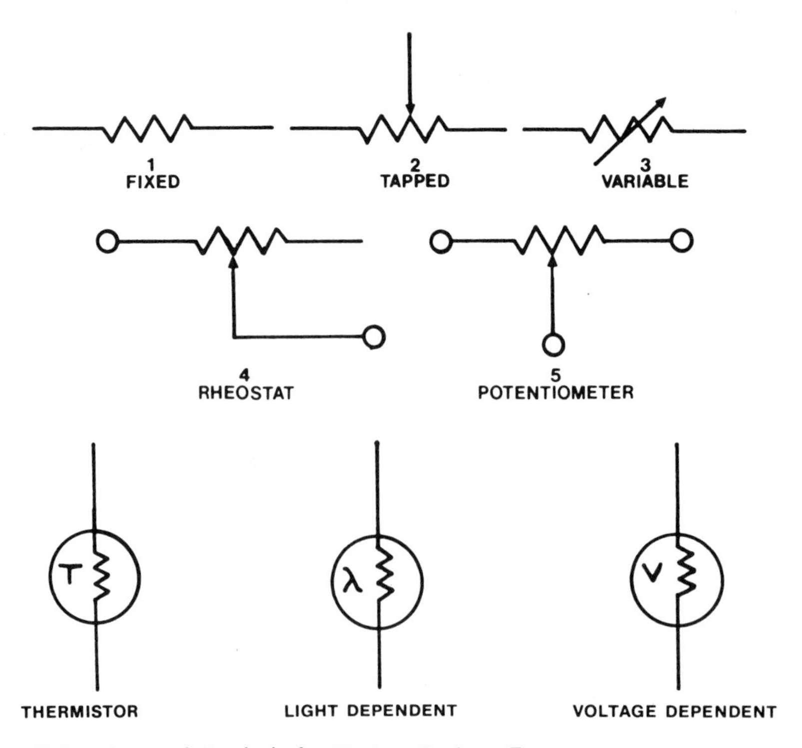

Figure 2-2 Electrical Symbols for Various Resistor Types

Wire-wound Resistors. Wire-wound resistors are made with wire wrapped around an insulator such as glass or ceramic. They are usually able to withstand rugged duty because they are large and have a hard coating of ceramic around them. Ohmic values of wire-wound resistors are usually within 1% tolerance. Therefore, they are called precision resistors. The disadvantage of wire-wound resistors is that they are expensive to make and may effect inductive reactance in high-frequency circuits because of the wire coils.

Metal-film Resistors. Metal-film resistors are made by heating metal alloys, vaporizing them on a film or a cylinder of glass, and then hermetically sealing with glass. Such resistors are manufactured with an ohmic tolerance such as 0.1% and a zero temperature coefficient. Not surprisingly, these resistors are very expensive.

Tapped Resistors. Tapped resistors are variable slide resistors that are fixed by the manufacturer from a slide or by other variable techniques. They are not normally changed from this position and often have a ceramic covering.

Variable Resistors Variable resistors (Figure 2-3) are resistors that can change their ohmic value by some mechanical method. They come in various forms and have tolerances and power ratings just as fixed resistors do. Some variable resistors are as follows.

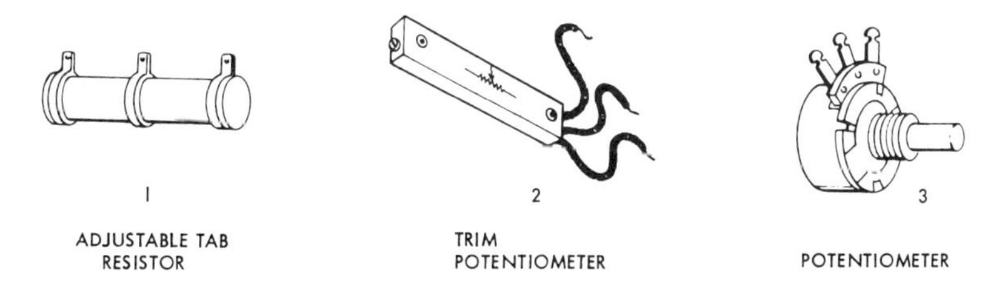

Figure 2-3 Variable Resistors

Slide Resistor. The slide resistor contains a metal sleeve that is wrapped around a wire-wound resistor so that a specific value of resistance may be picked off. The slide may be fixed at any position along the wire-wound resistor and may be adjusted at random.

Potentiometer. The potentiometer is a fixed resistor with a movable contact that picks off a specific resistance. A potentiometer has three terminals and may be of any resistance value such as 0 to 1 kilohm (kΩ) or 0 to 10 megohms (MΩ). The potentiometer may have a single turn or many turns. Many-turn potentiometers are called *helipots.* Potentiometers may rotate through turns up to 360°. Therefore, a half-turn would provide a resistance value of half the total resistance.

Rheostat. The rheostat is recognized in a circuit as having two terminals. It is built in the same manner as a potentiometer, but it is usually capable of handling higher current and is used to vary the amount of current in a circuit.

Resistor Color Coding Figure 2-4 illustrates resistor coding. You will note that colored bands on the resistor begin near one end. Reading from this end, the first band represents the first significant figure, the second band, the second significant figure, and the third band the multiplier. In the example shown, the red band is first, representing the number 2. The second band is yellow, representing the number 4. The third band is orange, representing the multiplier 1000. Therefore, the value of this particular resistor is 24,000 Ω.The silver band indicates the tolerance of 10%. This tolerance would provide the user with a resistor value of 24,000 (±2400) Ω. If there is no fourth band, the tolerance of the resistor is 20%.

Resistor Values below 10 Ω Some resistors have values of less than 10 Ω. These resistors have a third band that is gold or silver. The gold band means

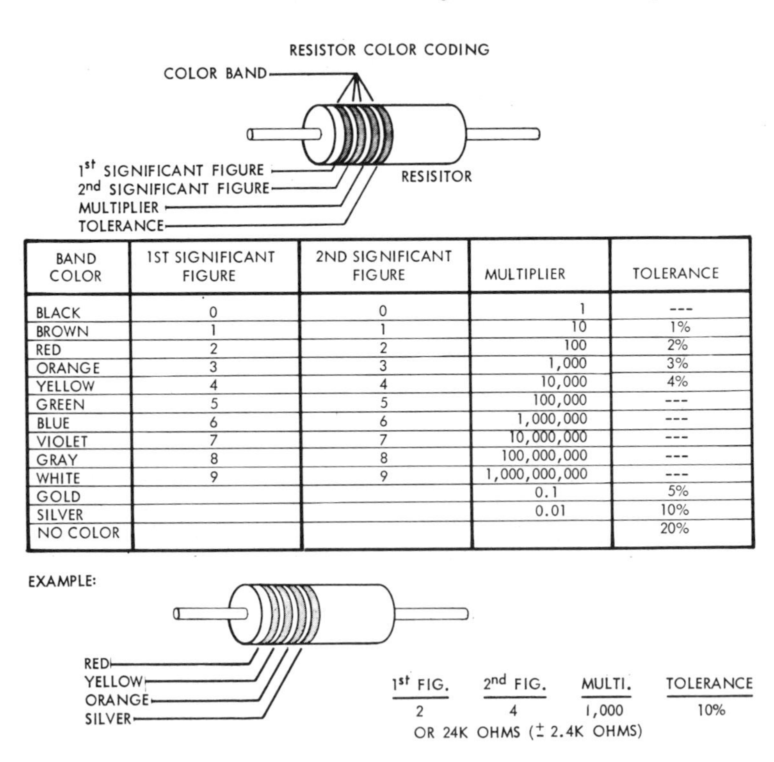

BAND COLOR	1ST SIGNIFICANT FIGURE	2ND SIGNIFICANT FIGURE	MULTIPLIER	TOLERANCE
BLACK	0	0	1	---
BROWN	1	1	10	1%
RED	2	2	100	2%
ORANGE	3	3	1,000	3%
YELLOW	4	4	10,000	4%
GREEN	5	5	100,000	---
BLUE	6	6	1,000,000	---
VIOLET	7	7	10,000,000	---
GRAY	8	8	100,000,000	---
WHITE	9	9	1,000,000,000	---
GOLD			0.1	5%
SILVER			0.01	10%
NO COLOR				20%

Figure 2-4 Resistor Color Coding

to multiply the first two digits by 0.1. Therefore, red (2), yellow (4), and gold (multiply by 0.1) bands on a resistor would represent a resistor with a value of 2.4 Ω. If this same resistor had a silver band in place of the gold band, the first two band values would be multiplied by 0.01. The value of this resistor would then be 0.24 Ω.

Standard Sizes of Resistance It is not practical to manufacture and store all different sizes of resistors. Therefore, standard values are generally used throughout the industry. These standards are readily recognized by electrical and electronic technicians. For instance, the 4.7 Ω, 47 Ω, 470 Ω, 4.7 kΩ, and 4.7 MΩ resistors are standards. Also the 2.2 Ω, 22 Ω, 220 Ω, 2.2 kΩ, 22 kΩ, 220 kΩ, and 2.2 MΩ resistors are standards. Resistor supply vendors' catalogs should be consulted for other standard size resistors.

Thermistor The thermistor is a device constructed to change resistance value as temperature changes. The word "thermistor" combines the words "thermal" and "resistance." In an actual circuit, if the resistance of the total circuit increases in value, the thermistor resistance would increase in value, thereby balancing the total resistance level of the circuit. The thermistor has a negative temperature coefficient. That is, for an increase in temperature, there is a decrease in resistance.

Resistance Temperature Detector (RTD) The RTD is similar to the thermistor in that it changes resistance as temperature changes. The RTD is opposite in function to the thermistor. The RTD has a positive temperature coefficient. That is, for a positive temperature change, there is an increase in the resistance of the RTD.

Light-Dependent Resistors Certain resistors change resistance when exposed to light. These may be used for light sensing and, in turn, switching circuits.

Voltage-Dependent Resistors Some resistors change their value when exposed to a voltage. A device called a *varistor* varies in resistance with applied voltage. A device of this type protects circuitry from being damaged by high voltages and electrical noise or transient. Varistors are often used in voltage-regulation circuits.

Some Symbols Figure 2-2 illustrates some electrical symbols that are utilized for various models of resistance. The thermistor and the voltage-dependent resistance are labeled T and V, respectively. The light-dependent resistor is labeled with the Greek letter lambda (λ). This is an electrical symbol used to represent wavelength.

The Capacitor

The capacitor has the basic purpose of opposing a change in voltage. In alternating-current circuits, the capacitor is called a *reactor*. By this reaction, it can be used for electronic sine-wave propagation and signal development.

The capacitor is a device for storing an electrical charge. It is made of two conductors (plates) that are separated by a dielectric insulator. When an electrical source is applied to the capacitor, an excess of electrons produces a negative charge on one side of the capacitor. An equal positive charge on the other side of the capacitor denotes a lack of electrons. Note that the capacitor as a whole does not have an excess or a deficiency of electrons. Thus, when we say that a capacitor has a charge of 1 coulomb (C), our meaning is that 1 C of negative charge has entered one side of the capacitor and 1 C of electronic charge has left the other side. The charge causes an electrostatic field (F) between the plates, which is measured in volts per centimeter. The charge remains on the capacitor after the power source is disconnected.

To discharge the capacitor, a short circuit is connected across the capacitor. The excess electrons on its negative side flow into and through the external circuit back to its positive side, where a lack of electrons allows the two sides to be neutralized. This action is tricky because, in reality, current is not flowing through the capacitor as it appears to. It is merely stored as a static charge on the capacitor plates and then released to go through the external circuit when the voltage drops toward zero on that alternation. Now the static charge is on the opposite plate, where it remains until the next alternation. The capacitor is also used in a pure, smooth dc circuit where the voltage is required to remain constant. In this case the capacitor builds up a static charge that cannot be discharged. The effect is equivalent to an open circuit. Any change in applied voltage across the capacitor is passed on to the circuit until the capacitor charges or discharges to an amount equal to the change.

Capacitor Details and the Farad as a Unit of Measure Plate area, dielectric thickness, dielectric strength, and dielectric constant determine the capacitance of a capacitor (in farads).

$$C = k \times \frac{A}{d} \times 22.4 \times 10^{-14}$$

is the mathematical formula for capacitance (where C = capacitance in farads (F), k is the dielectric constant, A is the area of either plate in square inches, d is the distance between plates in inches, and 22.4×10^{-14} is the conversion to farads).

In actual use, a charge (Q) of 1 coulomb is equal to 1-volt (V) potential (E) across a 1-F capacitor (C):$Q = CE$. The charge factor (Q) is discussed later. In practical use, the capacitor is made in sizes that range between microfarads (μF) and picofarads (pF). The word microfarad means one-millionth farad (10^{-6} F), and the word picofarad means one-trillionth farad (10^{-12} F). These very small submultiples of the farad are required for electronics, where the farad is much too large a unit for practical use.

Capacitor Types Capacitor dielectrics (Figure 2-5) are made from many types of material. The most commonly used are air, ceramic, mica, paper, and electrolytics. Different types are chosen because of their size, the job to be done, breakdown volts, ability to withstand heat or vibration, cost, and service life. Reference to vendor manuals will provide construction data and allow the reader to choose the right capacitor for the job.

The *air capacitor* is usually a variable capacitor. The most common is the tuning capacitor for radio receivers. A fixed set of plates and a movable set of plates are used to change the capacitance. Air is the dielectric. The plates must not touch one another.

Ceramic capacitors are made from baked earth. They are usually the disk or tubular type. Silver plates are fused on opposite sides of the ceramic dielectric to be used as the plates.

Mica capacitors are made by placing mica sheets between tinfoil and then baking the sandwich in a phenolic case. The tinfoil provides the conductor plates and the mica the dielectric.

Paper capacitors are made by wrapping tinfoil and paper into a roll. The roll is then placed in a cardboard cylinder and sealed with wax. The tinfoil provides the conductor plates, and the paper is the dielectric.

Tantalum capacitors are an electrolytic type used in low-voltage applications. Electrodes (plates) are separated by a gauze saturated with an electrolyte. An oxide film is formed on the plates and provides a layer between the positive plates and the gauze. The oxide is formed by applying direct current to the capacitor in the correct polarity, and this oxide film is an insulator. If current of enough magnitude is applied in the opposite direction, the oxide film will break down and the capacitor will be destroyed. The materials described are all placed in a compact cylinder. Aluminum electrolytic capacitors are made in the

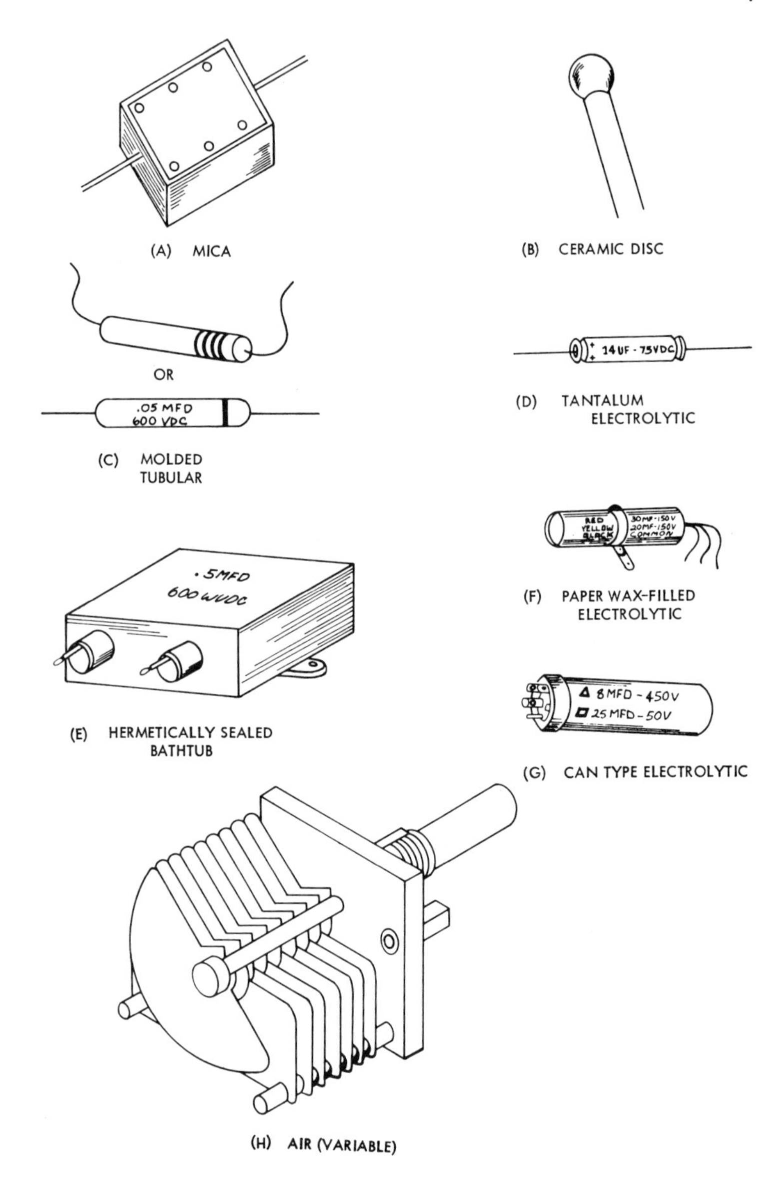

Figure 2-5 Capacitor Types

same manner as tantalum capacitors. In fact, aluminum was used before tantalum in the construction of electrolytic capacitors.

Charge Factor Q of a Capacitor The charge (Q) of a capacitor is the charge stored on the capacitor's plates. The greater the charging voltage (E), the greater the charge is. Also, the greater the capacitance (C) with a set voltage, the greater the charge amount is. These relationships of voltage and capacitance can be analyzed in the formula $Q = CE$, where Q is the charge in coulombs, C is the capacitance in farads, and E is the applied voltage in volts.

When 1 V is applied to a capacitor with a value of 1 F, the charge on the capacitor is 1 C ($Q = 1 \times 1$). If a larger capacitor, say of 5 F, is substituted, the charge would be 5 C ($Q = 5 \times 1$). Since the formula $Q = CE$ is a direct proportion, it is obvious that the size of the capacitor or applied voltage directly affects the charge Q.

Capacitor Symbols Capacitor symbols vary with manufacturer. However, the standard capacitor symbols according to military specifications are shown in Figure 2-6. These are also recognized throughout industry.

Capacitor Color Coding Tubular and mica capacitors are sometimes color coded. Other capacitors have their farad values and power ratings written on their outer coverings. Color coding for mica capacitors is slowly being replaced

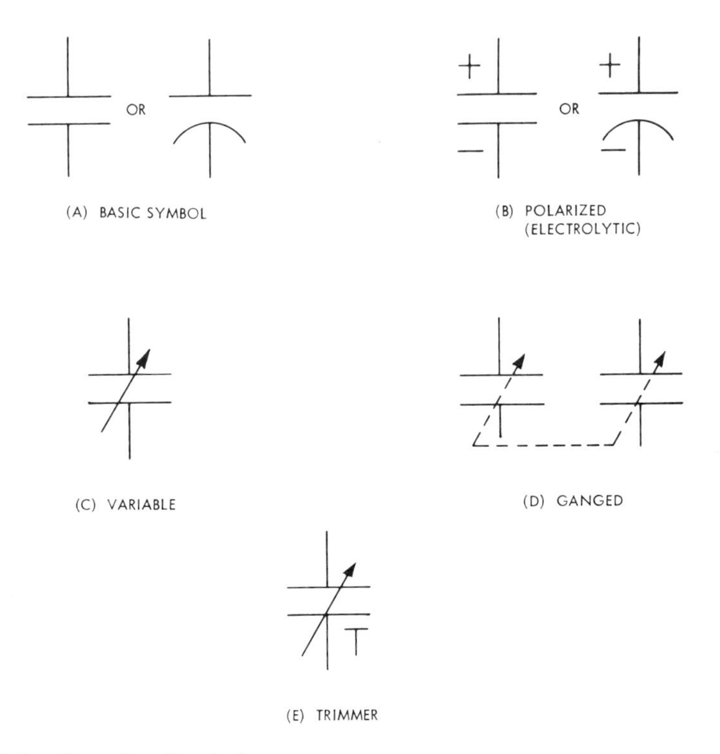

Figure 2-6 Capacitor Symbols

by printed values on capacitor covers. The reason for this is that there is confusion between the RMA, EIA, MIL, JAN, and AWS methods for color coding. Since confusion exists, it may be difficult for the user to determine which color code to use. See Figure 2-7 for color coding of mica capacitors and Figure 2-8 for color coding of tubular capacitors.

Some critics might object to including an illustration of obsolete color coding here. This is a case where merely declaring that a design, product, or method is obsolete does not make it so. It still exists until all uses of it have ceased. All conditions that a reader is likely to encounter are not ideal, and for that reason a practical book must prepare the reader to deal with present situations.

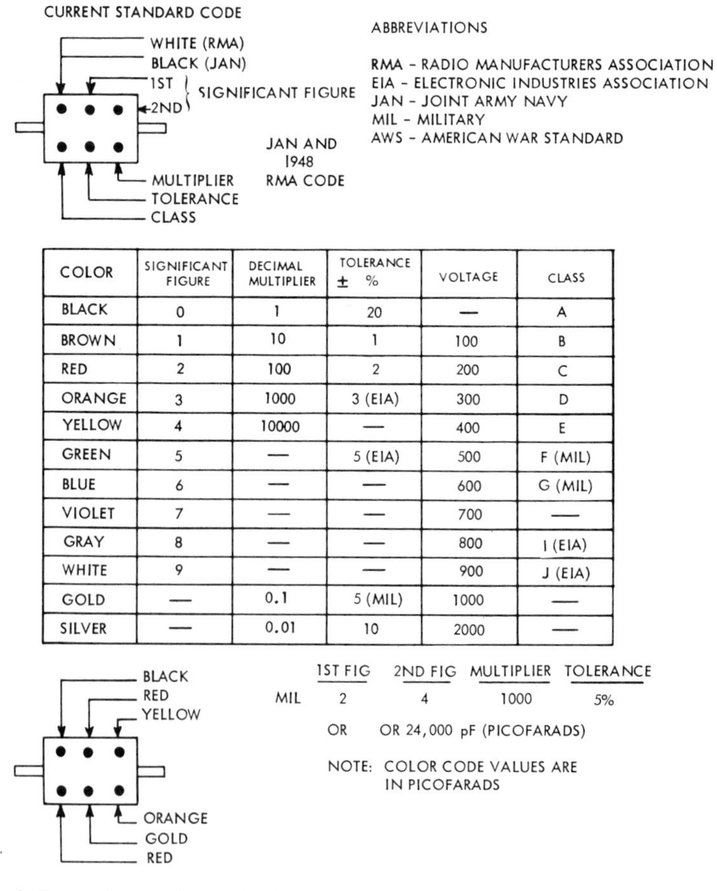

COLOR	SIGNIFICANT FIGURE	DECIMAL MULTIPLIER	TOLERANCE ± %	VOLTAGE	CLASS
BLACK	0	1	20	—	A
BROWN	1	10	1	100	B
RED	2	100	2	200	C
ORANGE	3	1000	3 (EIA)	300	D
YELLOW	4	10000	—	400	E
GREEN	5	—	5 (EIA)	500	F (MIL)
BLUE	6	—	—	600	G (MIL)
VIOLET	7	—	—	700	—
GRAY	8	—	—	800	I (EIA)
WHITE	9	—	—	900	J (EIA)
GOLD	—	0.1	5 (MIL)	1000	—
SILVER	—	0.01	10	2000	—

Figure 2-7 Color Coding of Mica Capacitors

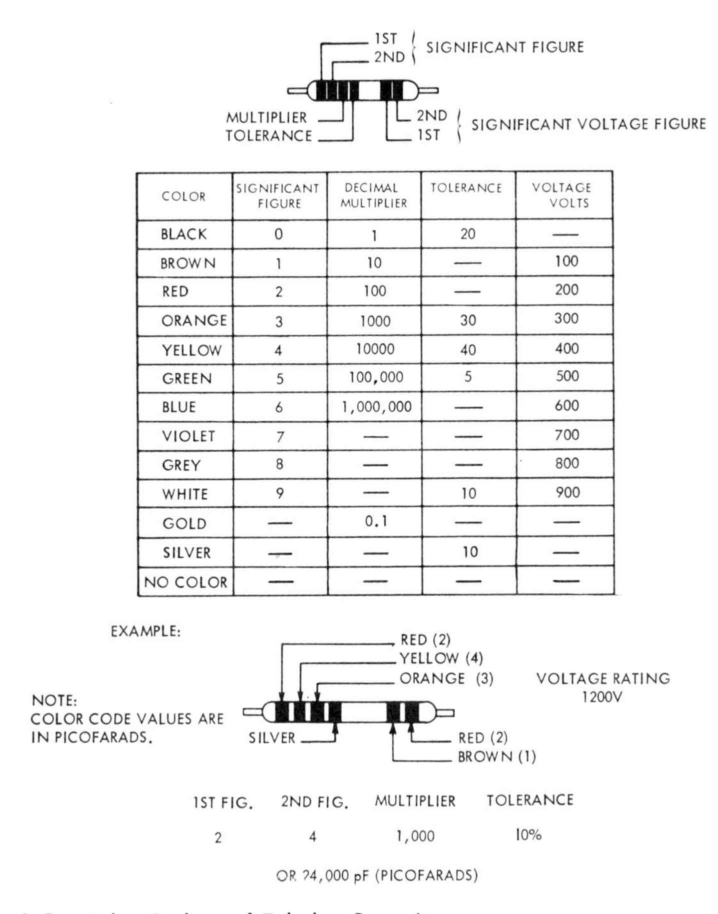

COLOR	SIGNIFICANT FIGURE	DECIMAL MULTIPLIER	TOLERANCE	VOLTAGE VOLTS
BLACK	0	1	20	—
BROWN	1	10	—	100
RED	2	100	—	200
ORANGE	3	1000	30	300
YELLOW	4	10000	40	400
GREEN	5	100,000	5	500
BLUE	6	1,000,000	—	600
VIOLET	7	—	—	700
GREY	8	—	—	800
WHITE	9	—	10	900
GOLD	—	0.1	—	—
SILVER	—	—	10	—
NO COLOR	—	—	—	—

Figure 2-8 Color Coding of Tubular Capacitors

The Inductor

The inductor's basic purpose is to oppose a change in current. In alternating current circuits, the inductor is called a *reactor*. By this reaction, it can be used for electronic sine-wave propagation and signal development.

An inductor is a coil of wire used to induce voltage at a known rate when current in the wire varies. Inductance is the ability of the inductor to provide this induced voltage. The inductor, like the capacitor, is called a reactor because it alternately stores energy and then delivers the energy back to the circuit. In the next several paragraphs we shall look at the inductor and determine what special attributes it has and what some of its problems are (see Figure 2-9).

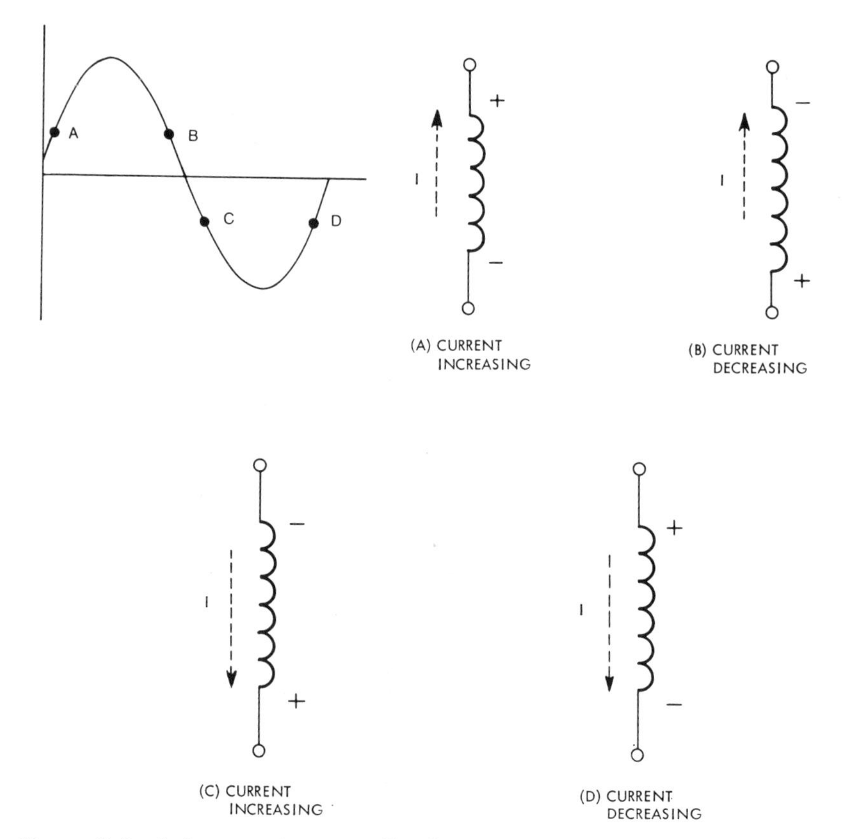

Figure 2-9 Inductance in an ac Circuit

When current increases through a conductor, the magnetic field around the conductor expands. If the current is alternating and the wire is coiled, the current produces alternately expanding and collapsing magnetic fields around the coil. This field induces a voltage in the inductor that is opposite in direction to the applied voltage. The self-induced voltage tends to keep the current moving in the same direction when the applied voltage decreases and opposes the current when the applied voltage increases. Opposition to current flow is called back- or counter-emf.

The property of the coil that opposes a change in current is called *inductance*. Since the inductance is self-imposed, this inductance is called *self-inductance*. The entire phenomenon is stated in *Lenz's law: When a voltage is induced in a coil as a result of a variation of the magnetic field with respect to the coil, the induced voltage is in such a direction as to oppose the current charge that caused the magnetic variation.*

Mutual Inductance When two coils are placed side by side so that the magnetic lines of force of one coil cut the turns of the other coil, the two coils are said to have *mutual* inductance (see Figure 2-10a).

Factors Affecting Inductance Four factors affect inductance. The first is the diameter-to-length ratio of the coil. If everything else remains constant, *the inductance of the coil varies directly with an increase in the diameter of the coil and inversely with an increase in the length of the coil.*

Second is the number of turns in the coil. The greater the number of turns in the coil, the stronger the magnetic field is and the greater the inductance. *Inductance is directly proportional to the square of the number of turns in the coil.*

Third is the type of core within the coil (see Figures 2-10b and c). *Inductance is directly proportional to the core permeability.* Coils that have non-magnetic cores have less inductance than magnetic-core coils, with all other inductance factors being equal.

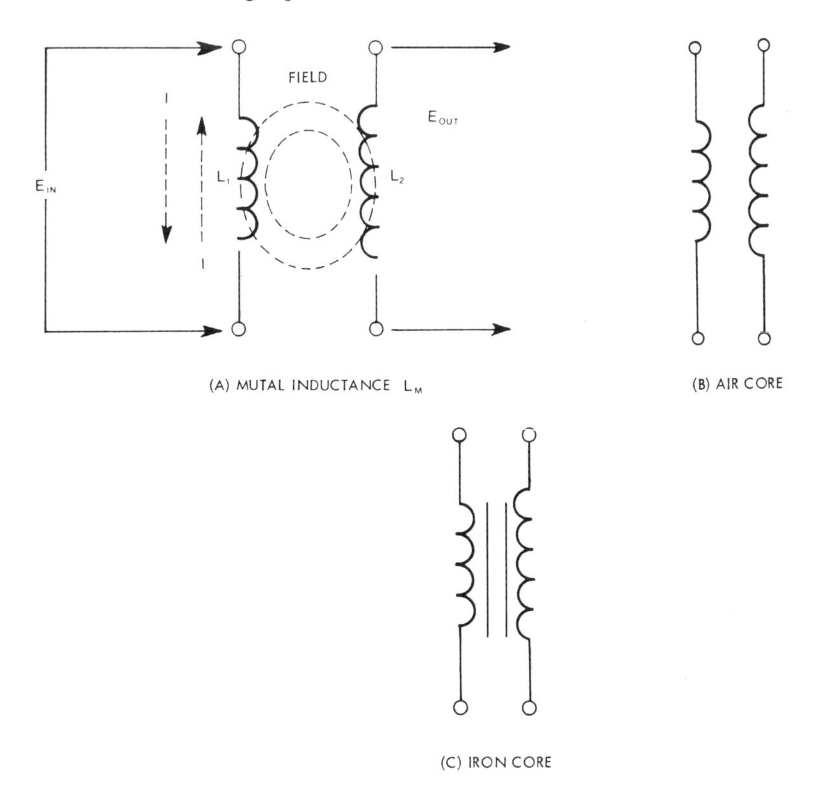

Figure 2-10 Mutual Inductance of Two Coils

Fourth is the method of winding. Inductance can be varied by changing the spaces between the windings. Coils can be wound in layers, have spaces between windings, or be wound close together. *Inductance is greatest when the windings are close and numerous.*

Flux Linkage Flux linkage is the sum of all the lines of force (flux lines) that surround a conductor, whether it is a single wire or a coil. In Figure 2-11a the flux linkage encircles the single wire. In Figure 2-11b the flux lines encircle all the loops of the coil. *Flux linkage is the product of the number of flux lines times the number of turns of the coil.* Inductance may also be said to be the number of flux linkages per unit of current.

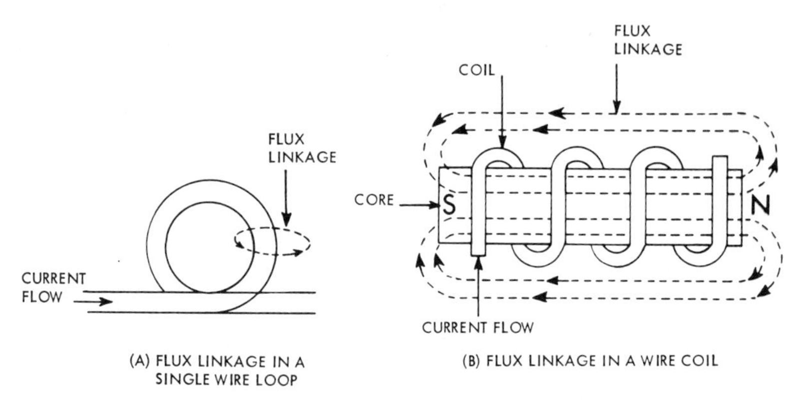

Figure 2-11 Flux Linkage

Inductor Types Inductors are manufactured in many shapes and forms (see Figure 2-12). The basic form is a coil of wire with no core (air core). Other types have iron cores and ferrite cores. The core is very important because that is where the magnetic lines of force are the most dense. The shape of an inductor is dependent on the size and packing required for the inductor to fit in an allotted space.

Inductor types are named by the job they do, such as intermediate and interstage transformers, audio- and radio-frequency transformers (chokes), and power transformers. Inductors are often called *transformers, chokes,* or *reactors.*

If an inductor is needed to change variable inductance, it may be accomplished in one of two ways. Varying the permeability of an inductor may be achieved by attaching a cylindrical metal slug that may be screwed into or out of the coil core. The second method, varying the number of turns, may be achieved by selecting the number of turns on the coil in a manner similar to that of a rheostat. A tap is then connected to the selected point.

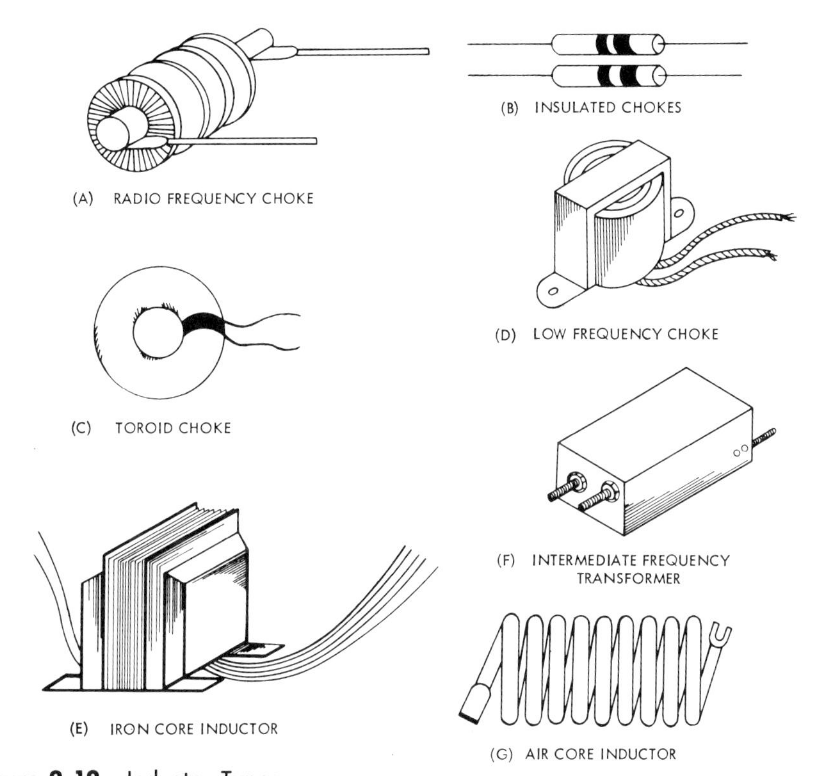

Figure 2-12 Inductor Types

Inductor Symbols Inductor symbols vary with manufacturer. However, the standard symbols according to military specifications are illustrated in Figure 2-13.

Quality of a Coil The quality (Q) of a coil is the coil's ability to produce self-induced voltage. The internal resistance of the coil reduces the current and therefore its voltage-producing ability. At low frequencies, the resistance is merely the dc resistance of the wire. At higher frequencies, the inductive reactance of the coil increases ($X_L = 2\pi fL$). Since the Q of a coil is the ratio of reactive power to real power ($Q = P_{\text{reactive}}/P_{\text{real}}$), the formula for the Q of a coil can be reduced as follows:

$$Q = \frac{P_{\text{reactive}}}{P_{\text{real}}}$$

or

$$Q = \frac{I^2 X_L}{I^2 R} \quad \text{or} \quad \frac{X_L}{R}$$

As an example, a coil with an inductive reactance of 200 Ω and a dc resistance of 10 Ω would have a Q of 20 (this value is numerical only and reflects the ratio X_L/R).

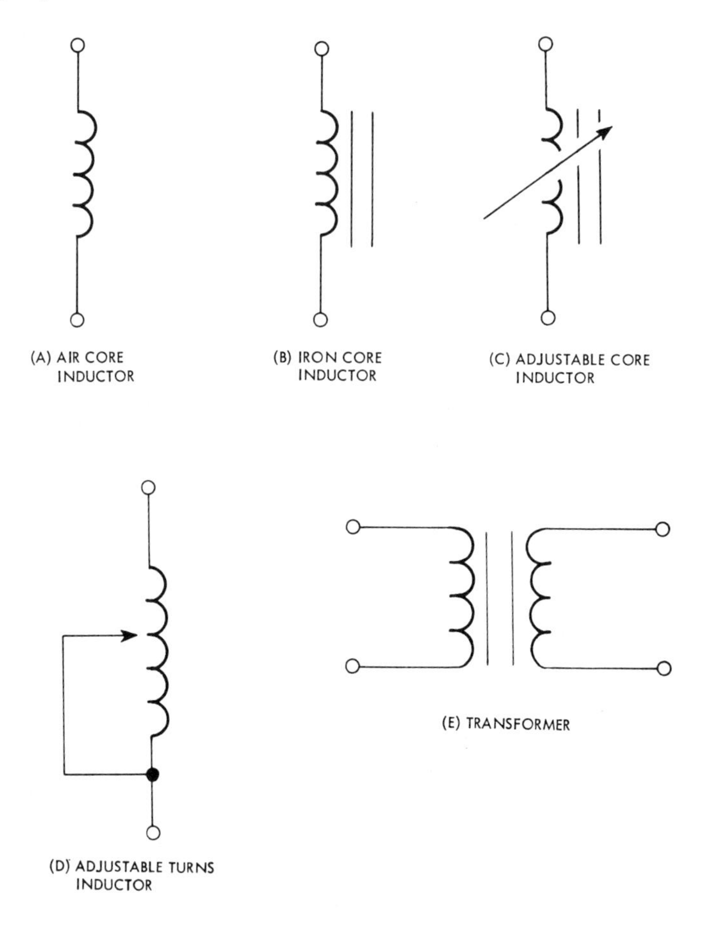

Figure 2-13 Inductor Symbols

The Transformer

The transformer (Figure 2-14) is a device that transfers power from one coil of wire to another. It does so by mutual inductance. Mutual inductance is the phenomenon that allows flux lines of an inductor to induce voltage in another inductor close to it when current changes.

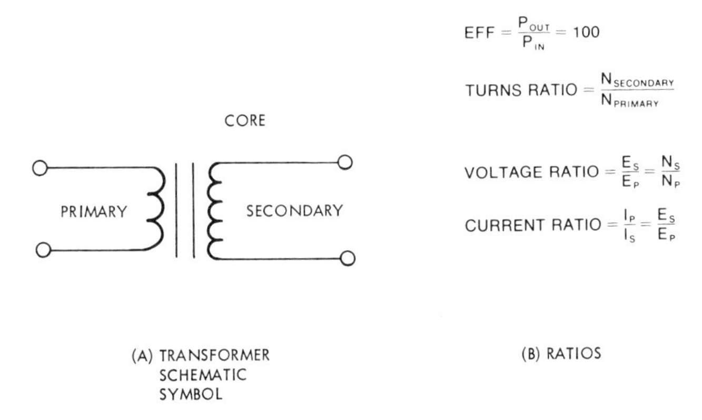

Figure 2-14 Transformer

A transformer is usually constructed to have a primary coil and one or several secondary coils. AC power is applied to the primary of the transformer and induced into the secondary, where it is used to produce current through a load. Power in the primary of a transformer is equal to power out of the secondary of the transformer with accepted losses.

One of these losses is eddy currents. *Eddy currents* are circular currents in an iron-core transformer that are induced in the transformer because its core is made of a conductive material (iron). The eddy currents represent a power loss that increases with frequency. *Hysteresis* is another power loss and is due to the additional power required to reverse the magnetic field as the alternations take place.

Other cores are powdered iron and laminated iron. These core types help to eliminate eddy currents. A *variac* is a transformer with variable inductance. It is wired so that the output can vary from 0 to a specified number of volts. Its primary and secondary coils are wound from a single wire. This variable type transformer is also called an *autotransformer*.

Transformer Calculations Efficiency in a transformer is calculated by the ratio of power out to power in. That is,

$$\text{Efficiency} = \frac{P_{\text{out}}}{P_{\text{in}}} \times 100 \qquad \text{(expressed as a percentage)}$$

A perfect transformer would have the same power out as in. This, of course, is never the case, so efficiency is under this value.

Other ratios that are peculiar to transformers are turns ratio, voltage ratio, and current ratio. Turns ratio has to do with the number of turns in the primary and secondary coils. That is,

$$\text{Turns ratio} = \frac{N_{\text{secondary}}}{N_{\text{primary}}}$$

A transformer whose secondary coil has a larger number of turns than the primary coil is called a *step-up transformer*. A transformer whose secondary coil has fewer number of turns than the primary coil is called a *step-down transformer*.

The voltage ratio is directly proportional to the turn ratio. That is,

$$\frac{E_{\text{secondary}}}{E_{\text{primary}}} = \frac{N_{\text{secondary}}}{N_{\text{primary}}}$$

or

$$\frac{E_S}{E_P} = \frac{N_S}{N_P}$$

Isolation Transformer The isolation transformer has the same basic schematic look as the transformer in Figure 2-14. The difference is that an isolation transformer has the same number of turns in the secondary as in the primary. Therefore, the voltage applied to the primary is the same as the voltage at the secondary. The purpose of the isolation transformer is to isolate the secondary and its load from the ac source line. This is a safety precaution. Engineers may be using a metal chassis for common.

Autotransformer The autotransformer is wound with a single wire for both primary and secondary. Figure 2-15 shows two types of autotransformers, the fixed and the variable. The reader will note that there is no isolation between the primary and secondary. The secondary is indeed connected to the ac power line. The fixed secondary autotransformer has one leg grounded and the other tied to the voltage level desired for the secondary. The variable secondary autotransformer has one leg grounded and the other variable picks off the desired secondary voltage. Care must be taken when working with autotransformers. Since the secondary is in direct connection with the ac power line, the engineer may mistake ground for a power leg.

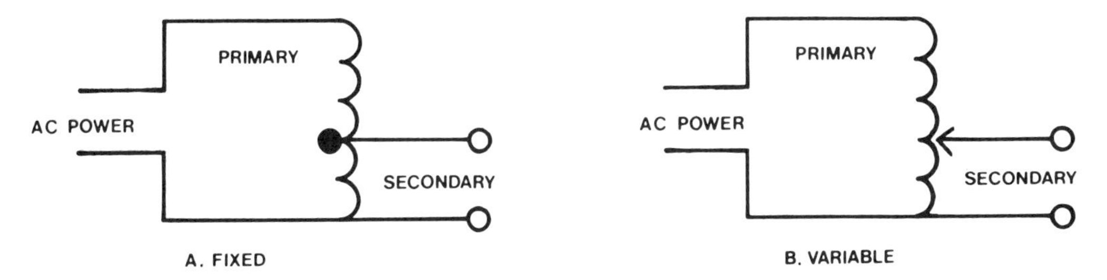

Figure 2-15 Autotransformer

Single-Phase Transformer Connections In Figure 2-16 a single-phase transformer is connected in series. When a pair of secondary windings are connected in series, their voltages are additive. For instance, if the secondary was rated at 115 V, 20 amperes (A) and was connected in series, the output would be $230 \times 20 = 4600$ volt-amperes (VA). If the same transformer was connected in parallel, the current would be additive. For instance, if the secondary was rated at 115 V, 20 A and was connected in parallel, the output would be $115 \times 40 = 4600$ VA.

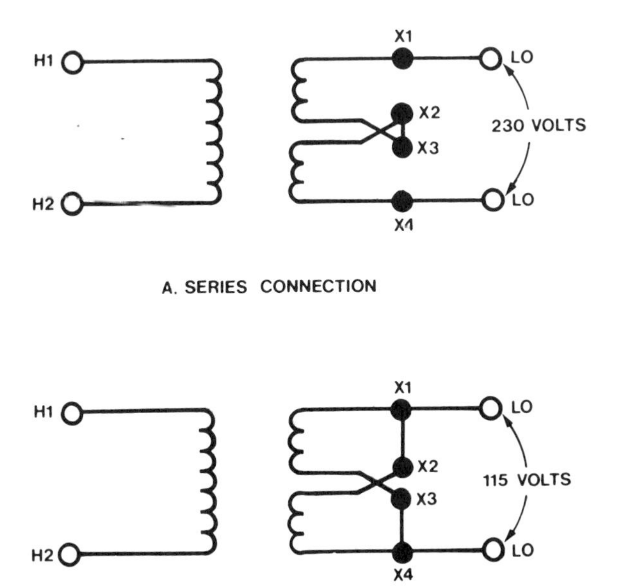

Figure 2-16 Single-Phase Transformer Connections

Three-Phase Transformer Connections As with single-phase transformer connections, the three-phase transformer may also be connected differently. Standardization throughout industry brought about the three-phase concept. The three-phase transformer, also called a *transformer bank*, has standard connections for primary and secondary windings called wye and delta. The primary may be connected by either wye or delta. The secondary may also have this choice.

Figure 2-17 is the wye connection. This is representative of a typical primary for power lines. Between each pair of input wires is 2200 Vac. Across each winding is 1270 Vac. To further define the schematic, the windings are 120° apart. There is a ratio of 1.73 between the input line voltage and each winding. If you multiply a winding voltage by 1.73, the result is the line voltage:

$$1270 \times 1.73 = 2197.1 \text{ V} \quad \text{(actual)}$$

Dividing the line voltage by 1.73 results in the winding voltage:

$$2200 \text{ V} \div 1.73 = 1271.6 \text{ V} \quad \text{(actual)}$$

Current in a wye winding is the same as the line current.

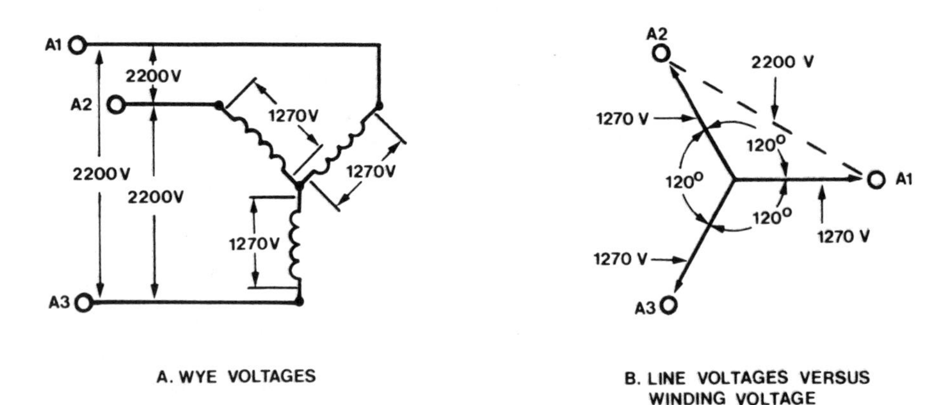

Figure 2-17 Wye Transformer Connection

Figure 2-18 is the delta connection. In the delta connection the line voltages and the winding voltages are the same. In the illustration, the line voltages are 2200 Vac and the winding voltages are 2200 Vac. Current from each line flows into two delta windings. The two currents are 120° apart. The line current is 1.73 times either one of the delta currents.

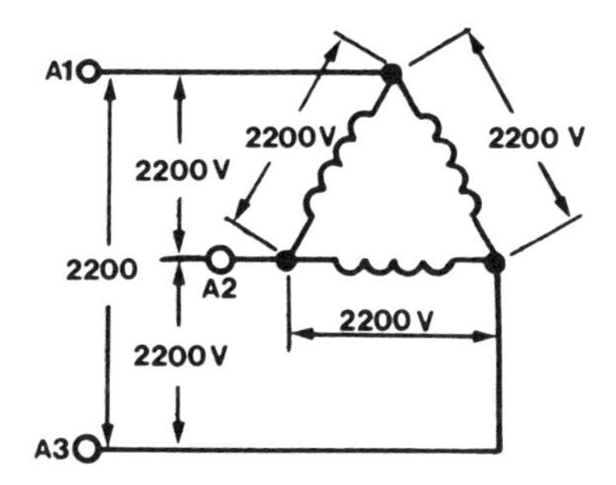

Figure 2-18 Delta Transformer Connection

As you may have determined, the primary windings and the secondary windings may be connected wye or delta to meet the requirements of an application. Since there are two ways to wind both, there are four combinations:

Primary		*Secondary*
Wye	to	Wye
Wye	to	Delta
Delta	to	Wye
Delta	to	Delta

Wye-to-Wye Transformer Connections Figure 2-19 is representative of the wye-to-wye transformer connection. In this configuration, supply-line voltage is 2200 Vac, which makes the primary winding voltage 1.73 × 2200 Vac = 1270 Vac. The turns ratio for this configuration is 10 : 1. The secondary winding voltages are therefore 127 Vac. The secondary line wires provide 1.73 × 127 = 220 Vac (approximately).

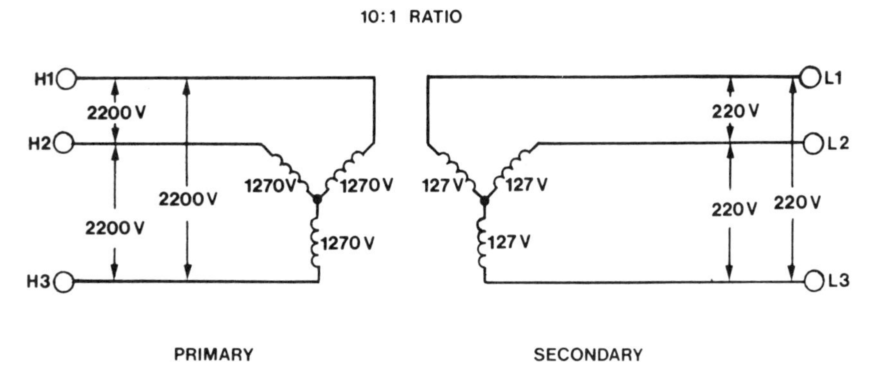

Figure 2-19 Wye-to-Wye Transformer Connection

The equation for secondary voltage calculation is as follows:

$$E_S = \frac{N_S}{N_P} \times E_P$$

where E_S = secondary line voltage across a single winding

E_P = primary line voltage across a single winding

$\frac{N_S}{N_P}$ = ratio of secondary to primary voltage (windings)

Wye-to-Delta Transformer Connections Figure 2-20 is representative of the wye-to-delta transformer connection. In this configuration, supply-line voltage is 2200 Vac, which makes the primary winding voltage 1.73 × 2200 Vac = 1270 Vac. The turns ratio for the configuration is 10 : 1. Therefore, the secondary winding voltages are 127 Vac. The secondary line wires provide 127 Vac (approximately).

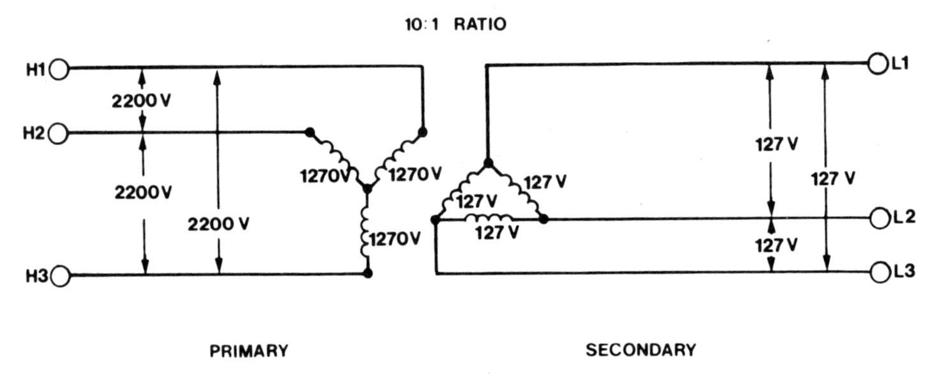

Figure 2-20 Wye-to-Delta Transformer Connection

The equation for secondary voltage calculation is as follows:

$$E_S = \frac{N_S}{N_P} \times E_P \div \sqrt{3}$$

where $\sqrt{3} = 1.732$

Delta-to-Wye Transformer Connections Figure 2-21 is representative of the delta-to-wye transformer connection. In this configuration, supply-line volt-

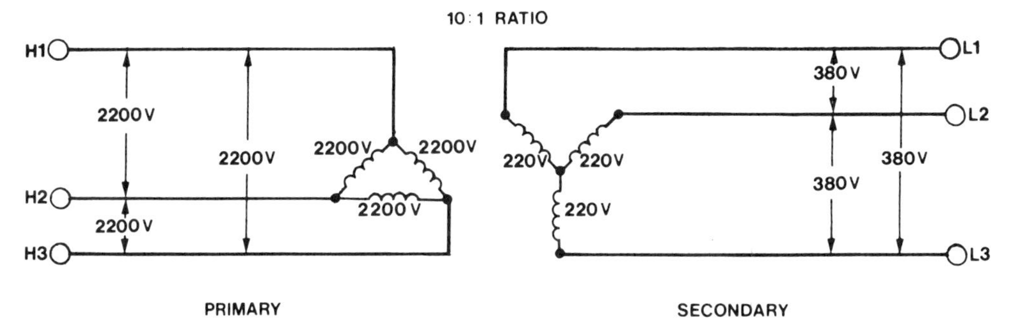

Figure 2-21 Delta-to-Wye Transformer Connection

age is 2200 Vac, which makes the primary winding voltage 2200 Vac. The turns ratio for this configuration is 10 : 1. Therefore, the secondary winding voltages are 2200 Vac. The secondary line wires provide 380 Vac (approximately).

The equation for secondary voltage calculation is as follows:

$$E_S = \frac{N_S}{N_P} \times E_P \times \sqrt{3}$$

Delta-to-Delta Transformer Connections Figure 2-22 is representative of the delta-to-delta transformer connection. In this configuration, supply-line voltage is 2200 Vac, which makes the primary windings 2200 Vac. The turns ratio of this configuration is 10 : 1. Therefore, the secondary winding voltages are 220 Vac. The secondary line wires provide 220 Vac.

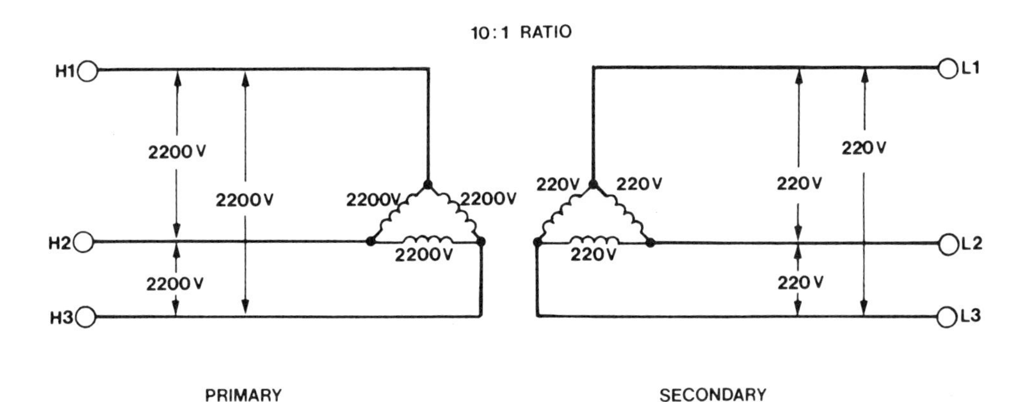

Figure 2-22 Delta-to-Delta Transformer Connection

The equation for secondary voltage calculation is as follows:

$$E_S = \frac{N_S}{N_P} \times E_P$$

Power Calculations for Three-Phase Circuits The total power for three phase is three times that of similar single phase. A general formula is as follows:

$$\text{Power (single phase)} = \text{voltage} \times \text{current}$$
$$P = E \times I$$
$$\text{Power (three phase)} = 3 \times \text{voltage} \times \text{current}$$
$$P = 3 \times E \times I$$

If the windings are wye connected, the line voltage should be divided by $\sqrt{3}$ or 1.732

$$\text{Power} = 3 \times \frac{E}{1.732} \times I$$

If the windings are delta connected, the line current should be divided by $\sqrt{3}$ or 1.732

$$\text{Power} = 3 \times E \times \frac{I}{1.732}$$

SWITCHES

There are probably more switch devices than any other electric device. They serve the purpose of turning things on and off. By doing this, control of industrial and other systems is accomplished.

Switch Functions

Switches are functionally used to open or close a circuit. When a switch is ON it is closed and in the *make* position. When a switch is OFF, it is open and in the *break* position. A switch position marked NC means that contact is normally closed and takes some action to open it. A switch position marked NO means that contact is normally open and requires some action to close it. There are four basic switch functions that a switch makes. These are (1) maintained contact, (2) momentary contact, (3) make before break, and (4) break before make.

Maintained Contact The basic switch type that most people are familiar with is the maintained contact switch. In Figure 2-23, the function of actuating these switches is shown. Pressure may be applied using several different mechanical means. The most common means are the press type, where pressure is applied on top of a button, and the toggle type, where pressure is applied to sides of a lever (see Figure 2-23a). In Figure 2-23b the switch has been pressed to the left, placing the switch contacts to an open contact. Once pressed, the contacts will latch and stay in the OFF position. Pressure applied on the lever to the right, as in Figure 2-23c, moves the contact from latched position to the ON position. The switch will latch and remain latched when pressure is removed. The switch contacts remain where they are until pressure is applied in the opposite direction.

A. PRESSURE APPLICATION

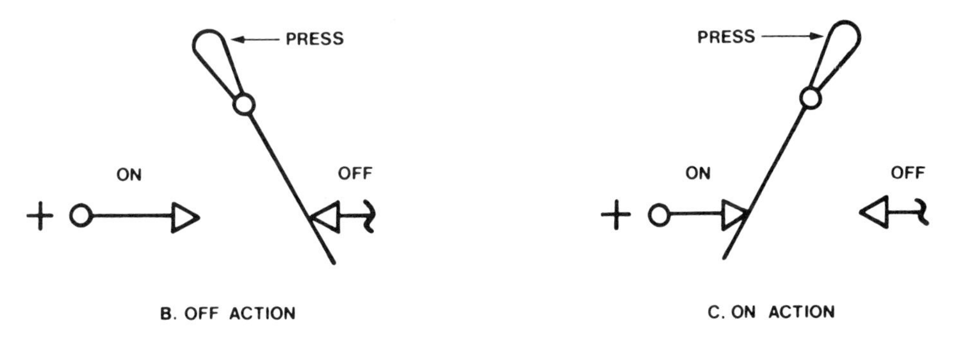

B. OFF ACTION

C. ON ACTION

Figure 2-23 Maintained Contact Switch

Momentary Contact The next major switch function is momentary contact. Two contacts are provided in this switch. One is NC (normally closed), which represents the break or OFF position. The lower contact is NO (normally open), which represents the OFF position. A movable contact attaches to a push button. Figure 2-24a illustrates the OFF position where the push button is upright and the movable contact is in the OFF (NC) position. Pressure is applied to the push button and the movable contact moves downward to the ON (NO) position and makes contact (see Figure 2-24b). The push button is released (see Figure 2-24c), and the movable contact returns to the OFF (NC) position. When released, spring action usually returns the push button to an upright position.

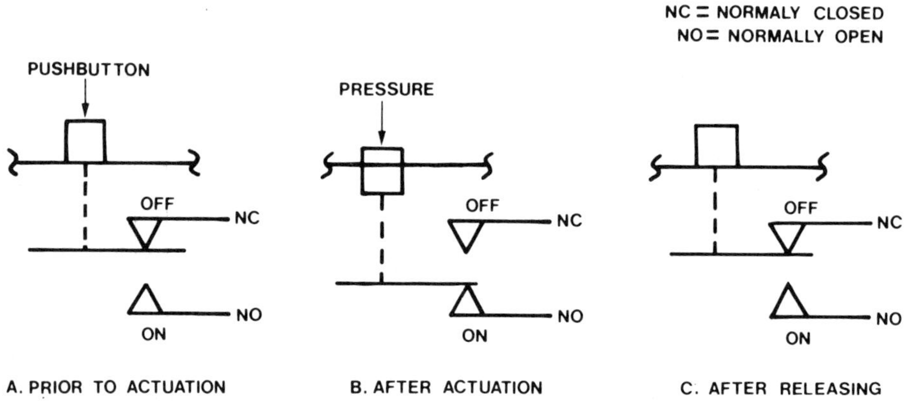

NC = NORMALY CLOSED
NO = NORMALLY OPEN

PUSHBUTTON
OFF
NC
NO
ON
PRESSURE
OFF
NC
NO
ON
OFF
NC
NO
ON

A. PRIOR TO ACTUATION
B. AFTER ACTUATION
C. AFTER RELEASING

Figure 2-24 Momentary Contact Switch

Make-before-break Function The make-before-break switch contact ensures that certain circuits are maintained until other circuits are energized. In Figure 2-25a, a make-before-break push-button switch sequence is shown prior to actuation. The blades of the switch are in the OFF (NC) position. Partial actuation of the switch moves one blade to the ON position contact while the second remains (see Figure 2-25b). Total actuation (Figure 2-25c) disconnects the blade from the OFF position. Note that the switch does not break OFF contact until the ON contact is ensured.

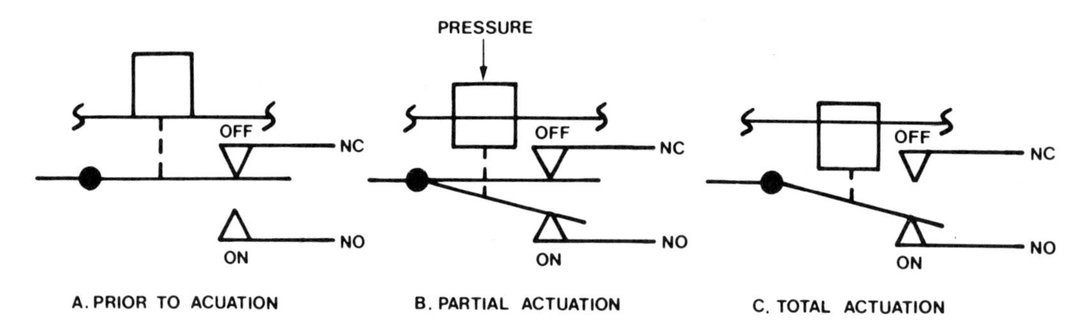

Figure 2-25 Make-before-Break Contact

Break-before-make Function Contrary to the make-before-break function, it may be required to break one circuit before making the second. Figure 2-26a represents a break-before-make contact prior to actuation. The blade of the switch is in the OFF (NC) position. Figure 2-26b shows the push button in partial actuation. Note that the switch blade has been separated from the OFF contact and lies between the OFF and ON contacts. Finally, in Figure 2-26c the push button has been totally actuated and contact is made with the ON (NC) position. Thus, the circuit was broken before making the second contact.

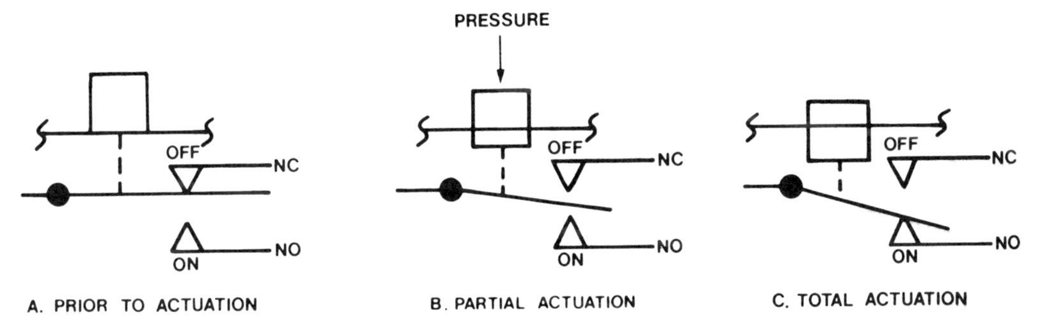

Figure 2-26 Break-before-Make Contact

Contact Types Basic contact forms have been standardized in industry. Some of them are illustrated in Figure 2-27. In the following list are the acronyms that are used to describe the contact function. For instance, SPST, the acronym for single pole, single throw, means that one end of the switch blade is fixed at a single pole and able to throw its other end in only one direction.

Acronym	*Description*
SPST	Single pole, single throw
SPDT	Single pole, double throw
DPST	Double pole, single throw
DPDT	Double pole, double throw

Other types and multiples of contacts are available. The author recommends that you survey manufacturers' catalogs for other contact types.

Spring Loading Toggle switches and push-button switches may be spring loaded to return the levers or buttons to the original position. This, of course, provides momentary action. In the Figure 2-28a, the spring-loading function of a push button is characterized. In the up position, the push-button switch is OFF and the spring is expanded upward. In the down (pressed) position, the push button sets the switch to ON and the spring contracts. When the push

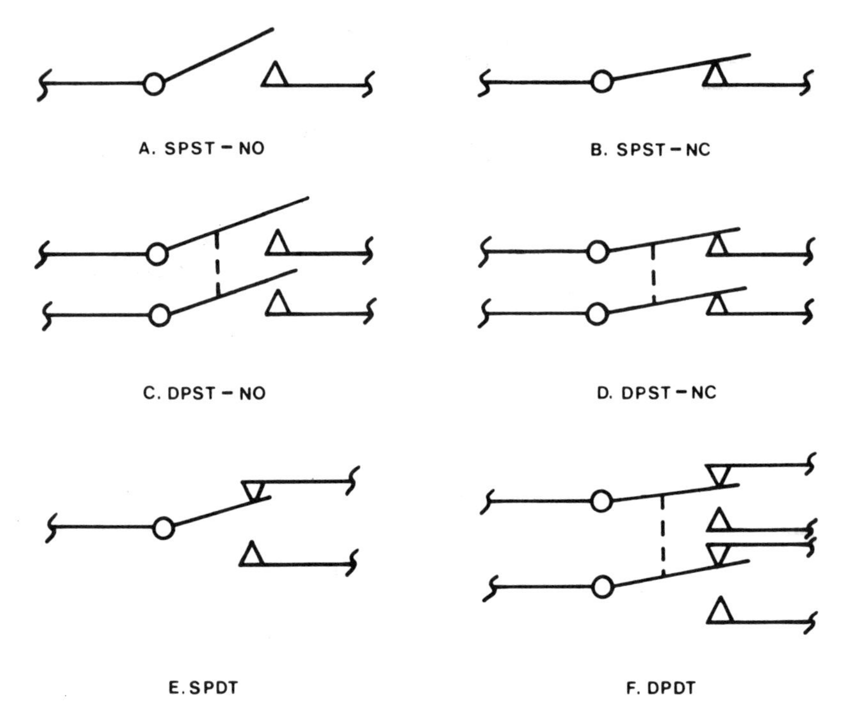

Figure 2-27 Switch Contact Forms

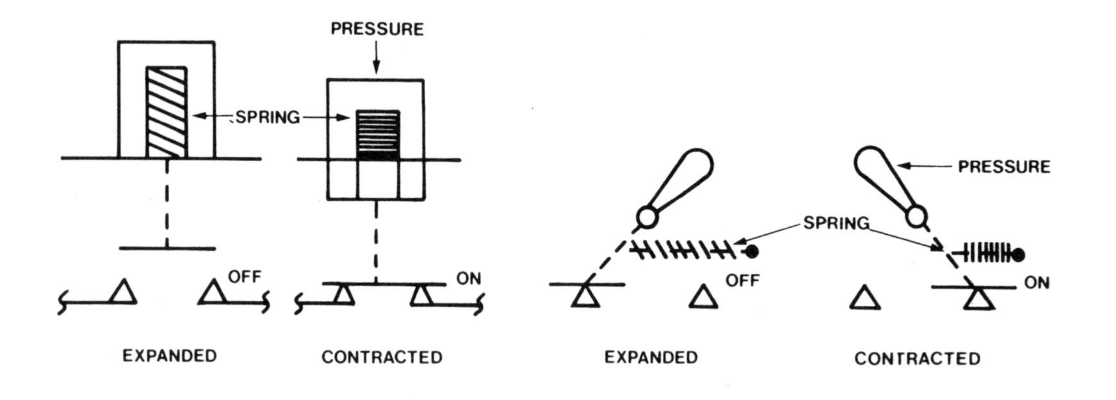

Figure 2-28 Spring Loading

button is released, the spring will expand upward, driving the switch to OFF position. In Figure 2-28b the toggle lever is to the right and the spring is expanded. The switch contacts are broken. Pressure to the left contracts the spring and makes the contacts. When pressure is released, the switch contacts break and the spring forces the switch back off.

Push-Button Switches

Push-button switches are those switches that require pressure on the top of a button to actuate. The push-button function is shown in Figures 2-25 and 2-26. A typical push-button switch is shown in Figure 2-29. The function shown in these figures could be momentary action or maintained action. An example of the momentary action push button is a doorbell. As long as you press the button, the doorbell will ring. An example of the maintained action push button is an oven light. Pressing the oven light switch turns the light on in the oven. It will remain on until you again press the push button and release the latch that maintains the switch on.

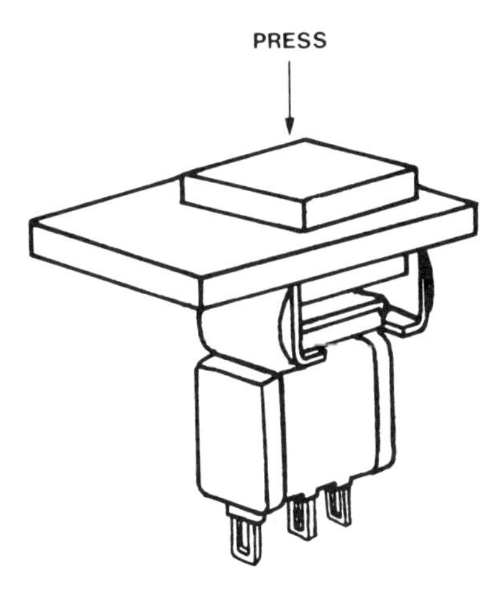

Figure 2-29 Typical Push-button Switch

Pull-Button Switches

Pull-button switches have much the same function as push button. The difference is that actuation takes place with a pulling operation. Typical of this type of switch is the automobile light switch. Pulling the switch on turns the lights on. Pushing the button back toward the dashboard releases the latch, opens the switch contacts, and turns the lights off.

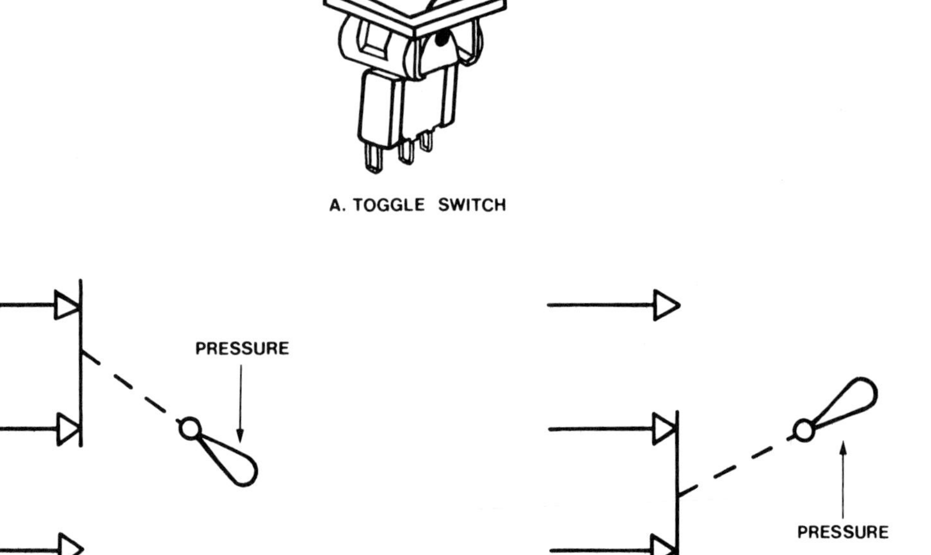

Figure 2-30 Typical Toggle Switch

Toggle Switches

A toggle switch, as shown in Figure 2-30a, is operated by pressing the lever on top to the right or left as the application demands. The toggle action was partially described in Figure 2-23. The toggle switch comes basically in two forms, two and three position. In Figure 2-30 the two-position toggle is illustrated. The toggle switch shown is the rocker type; it rolls on a shaft from left to right. In the OFF position, pressure is applied to the top of the switch and the top two contacts are made. In the ON position, pressure is applied to the lower side of the toggle lever and the lower two contacts are made.

The three-position toggle switch is illustrated in Figure 2-31. The center position is the OFF contact, while the end positions are ON contacts. The toggle switch shown is an SPDT switch. In Figure 2-31a, pressure applied downward on the toggle lever toggles the switch upward, making the top contact. Pressure in the opposite direction, as shown in Figure 2-31b, makes the lower contact. Centering the switch breaks the contacts (Figure 2-31c).

The switches shown in Figures 2-30 and 2-31 can be maintained or momentary contact. The switch in Figure 2-30 could be spring loaded to either side. The switch in Figure 2-31 could be spring loaded to the center.

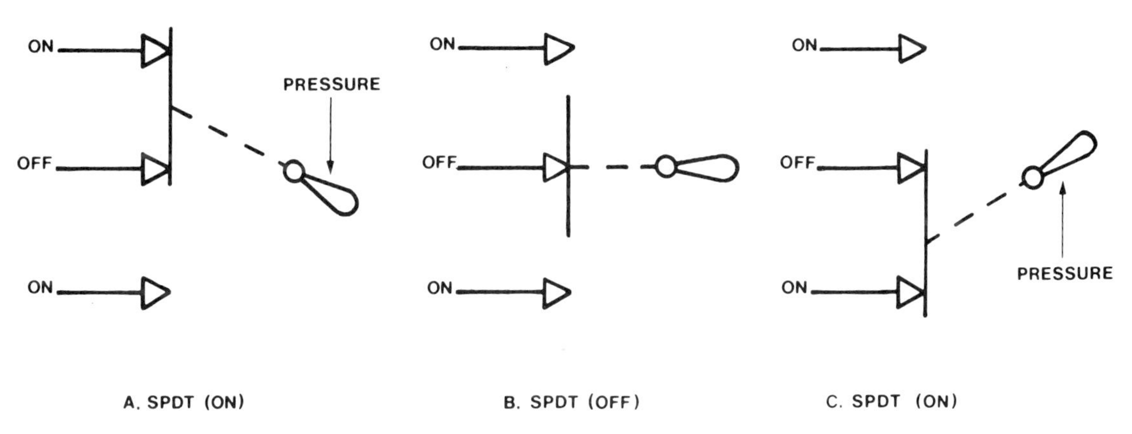

Figure 2-31 Center Off Position Toggle Switch

Multiposition Switch

The multiposition switch, also called a rotary switch, may be a master switch, a drum-type switch, or a selector. In any event, the multiposition switch is usually round and rotates from a center knob. There may be a few or a dozen or so switch positions, such as a frequency selector switch on a signal generator or a channel selector on a television set. The rotary switch may have several layers of terminals called *decks* connected to a common center shaft. In the Figure 2-32a, a single-layer eight-position rotary switch is illustrated. In Figure 2-32b, a three-layer (deck) rotary switch is shown. The multiposition switch allows many conditions to happen simultaneously in a system.

Slide Switches

Figure 2-33a shows a typical slide switch. Its function is extremely simple. Pressure to the left or right, as shown in Figure 2-33b, will turn the switch ON or OFF. The slide switch is used in small-current applications and works well on printed circuit boards.

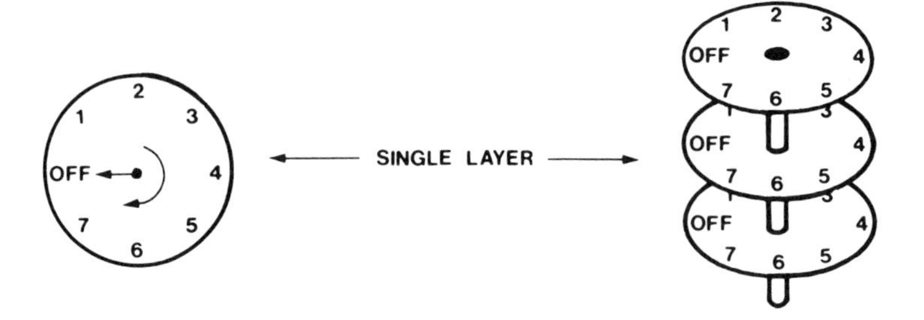

A. 8 POSITION ROTARY SWITCH

B. 3 LAYER (DECK) ROTARY SWITCH

Figure 2-32 Multiposition (Rotary) Switch

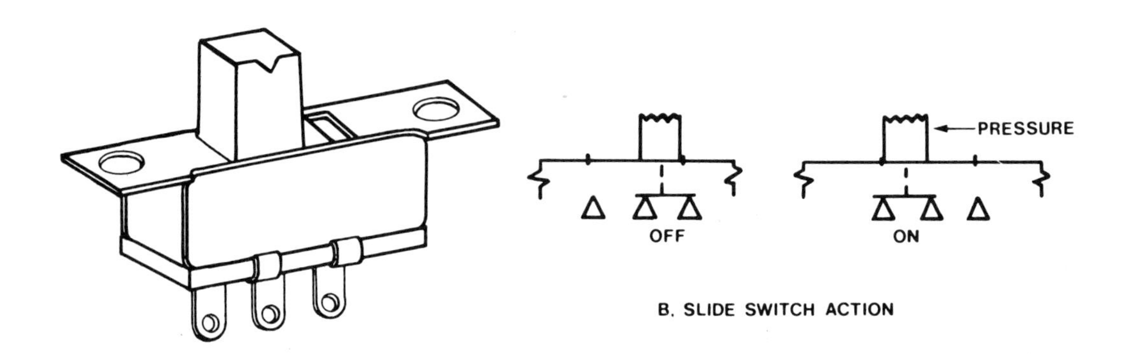

A. SLIDE SWITCH

B. SLIDE SWITCH ACTION

Figure 2-33 Slide Switch

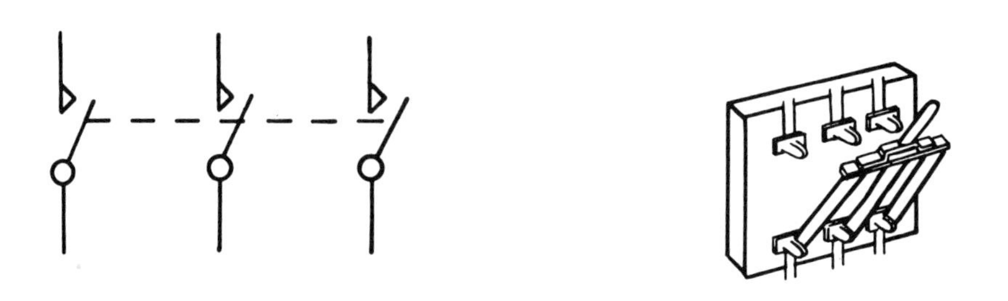

A. SINGLE THROW TRIPLE POLE

B. KNIFE ACTION

Figure 2-34 Knife Switch

Knife Switches

There are still some knife switches in use, as depicted in Figure 2-34. Notice that a single action allows the knife edge (three places) to meet the center of the contact clips. The knife switch has a disadvantage in that it is difficult to insulate the current-carrying parts of this switch. When closing the switch, it is possible to have a certain amount of arcing take place across the contacts. Large power switches of this type have the handle outside an enclosed case to protect the operator. This is an old-style device but should be recognized as being still in use.

Limit Switches

Limit switches are used to turn on a piece of equipment as the result of a motion of other equipment. For instance, when a machine cutting tool gets to a prescribed distance, the switch stops the motion of the tool or the table carrying the part. Furthermore, a limit switch may be placed on a machine to turn it off in the event of overtravel. The limit switch is used in motion mechanization of all types to limit motion and to protect machinery. Figure 2-35a shows limit switch contacts. When the roller is displaced left, the OFF contacts are made. When

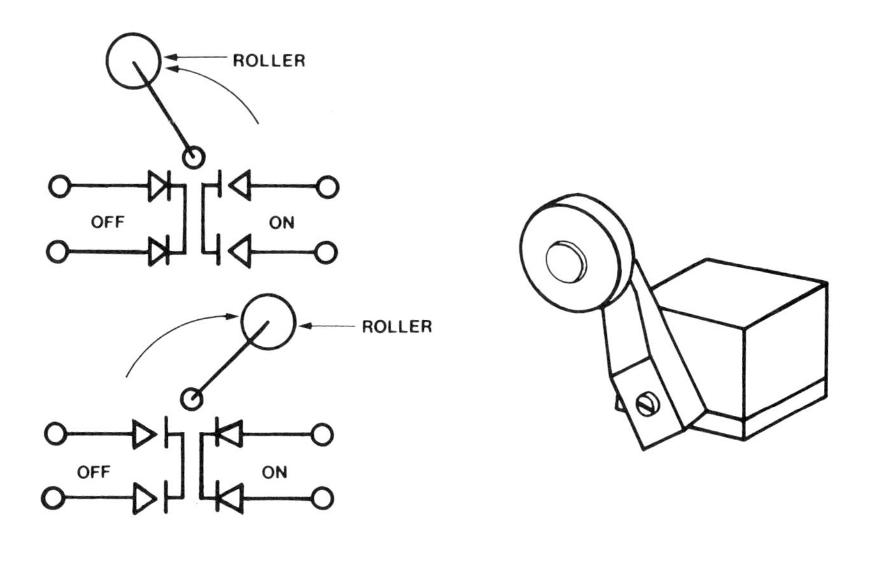

Figure 2-35 Limit Switch

the roller is displaced right, the ON contacts are made. Figure 2-35b is a typical limit switch with a roller function. Limit switches are not necessarily locked into a roller switching function. Push buttons, blades, precision poles called catwhiskers, and rotary actuators are used as actuators.

The circuitry shown in Figure 2-35 represents a single-action limit switch. In other words, it will only work in one direction. Another model can operate in both directions, with the center position being OFF.

Snap-Action Switches

The snap-action switch operates in much the same manner as the limit switch. The difference is that the snap-action switch is extremely sensitive to switching and operates in very small areas of several thousandths of an inch. Applications are precision doors, vending machines, and motor controls. As with the limit switch, the snap-action switch must be actuated by some force, though it be minimal. Contacts of these switches are gold-coated prisms. They are lightweight, require light force, and are subminiature. Figure 2-36 illustrates the blade type and the roller type.

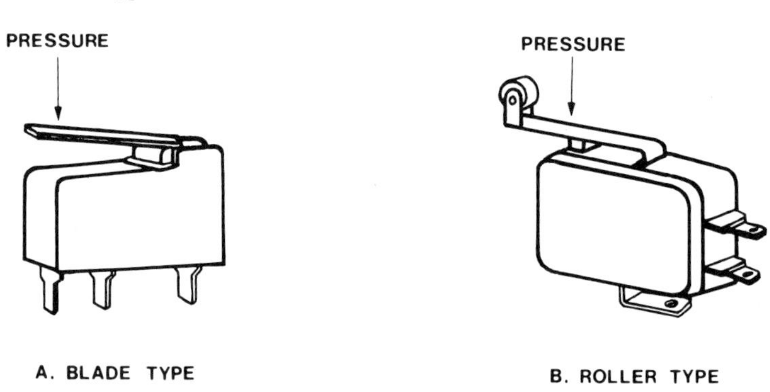

Figure 2-36 Snap-Action Switches

Special Switching Functions

The Figure 2-37 is a temperature-controlled switch. This device, called a thermal switch, operates with temperature. Contacts of the switch in the figure are closed. As temperature increases, the bimetallic strip bends and separates the contacts. This could be mechanized so that temperature would instead close the contacts.

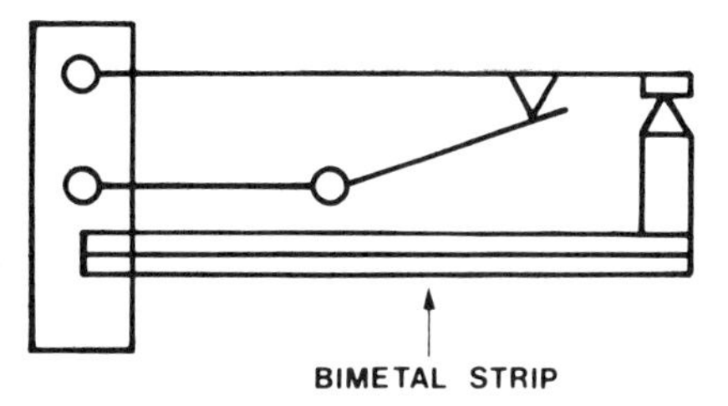

Figure 2-37 Temperature Function

Figure 2-38 is a magnetically controlled switch. The device has the switch contacts open in the illustration. Application of a magnetic field in the direction of the reeds will make the contacts. The reed switch is generally used as a relay with an electrical coil wrapped around the switch enclosure.

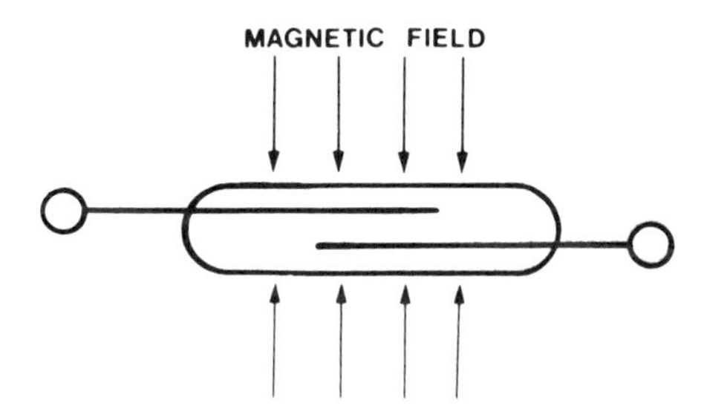

Figure 2-38 Magnetic Function

Mercury Switching Function

Mercury is a good conductor of electricity and, therefore, operates well for angle switching (see Figure 2-39). Mercury is placed in a glass envelope along with switch contacts. The envelope is placed at an angle on the machinery. When the machinery tilts at a specified level, the mercury flows and makes the contacts. The mercury in these switches is very pure. The glass envelope is evacuated and filled with hydrogen to prevent arcing and to keep the mercury from oxidizing. Contacts are made of nickel-iron, tungsten, and platinum. In Figure 2-39, a tilt angle makes and breaks the contacts.

Printed Circuit Board (PCB) Switch

The electronics industry is involved with miniaturization. The switch (Figure 2-40) is a little less than an inch long (0.9 in.). It is made to fit on a printed

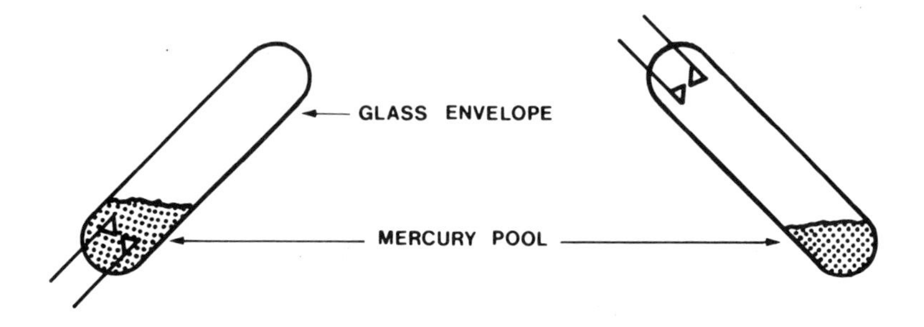

Figure 2-39 Mercury Switch

circuit board. Switches of the PCB type have either 14 or 16 pins, as in the drawing. This is a rotary type. A screwdriver slot in the numbered end allows selection of a switch position. Other selector types are as follows: an extended arm that can be attached to a knob, a lever that can be placed by a finger, and an adjusting plug that is pulled out, turned, and placed back in a selected position.

Another PCB switch is the dual in package (DIP). Switches are lever operated, usually from the top side of the package. As many as 10 or more switches may be placed in one package. They may be single or double pole and single or double throw.

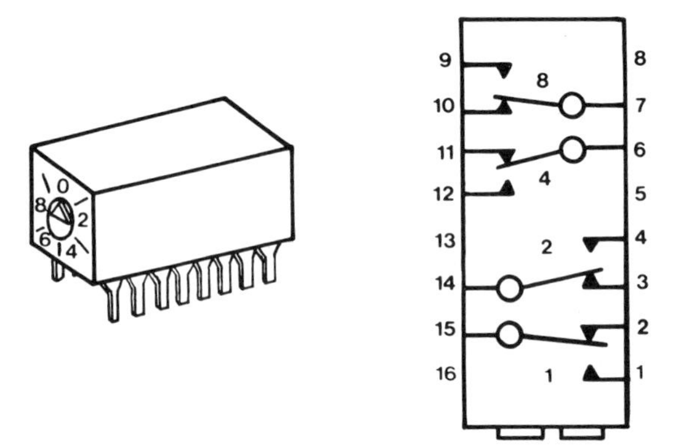

Figure 2-40 Printed Circuit Board Switch

Switches in General

The switches we have discussed are of the general variety. Switches are used to connect and disconnect power, to limit or curtail circuit action, to enable or disable circuitry, and to do many other things. Switches are built into other components such as transducers. Some switches are lighted. Some are numbered. The shape of the switch depends on its application and the space available. The cost of a switch is dependent on the amount of current it must carry and the richness of its package.

RELAYS

Relays are usually defined as electrically controlled switches. This is probably true in the real sense. Most relays are controlled by energizing an electrical coil, which, in turn, electromagnetically operates a movable slug and breaks or makes a set of contacts. It seems, in electronics, that the words "every" and "all" are not to be used. The moment we state "every," some condition arises that rebuts the situation. Inputs to relays may be voltage, current, temperature, light, magnetism, frequency, and so on. The operating principles may be based on similar subjects. There are some things that we can classify. Contacts may be classified along with some functions.

Electromagnetic Relay Function

Most relays operate by the electromagnetic function. Power is applied to a coil, such as in Figure 2-41c. The coil energizes, and an armature within the coil is attracted by the electromagnetic force. The armature moves downward and makes contact. Figure 2-41a is the functional drawing. This relay is called a clapper type. The plunger-type relay has the armature within the relay coil. As in the clapper type, power is applied to the coil energizing it. In this case, however, the coil is polarized with the armature in the make position. The armature is driven from the coil by electromagnetism and breaks contact. A DPST relay schematic is shown in Figure 2-41d. The plunger type is shown in the functional drawing as a single contact for simplification. See Figure 2-41b.

The electromagnetic relay is easy to troubleshoot and can have a great number of contacts. It is, however, slow to function, and the contacts may freeze up or burn off because of high current and arcing. Contacts also may chatter (open and close) before total closure takes place.

Reed Relay Function

The reed relay (Figure 2-42) is enclosed in a glass envelope. A pair of contacts (or more) called reeds are enclosed with an inert gas in the envelope. The contacts

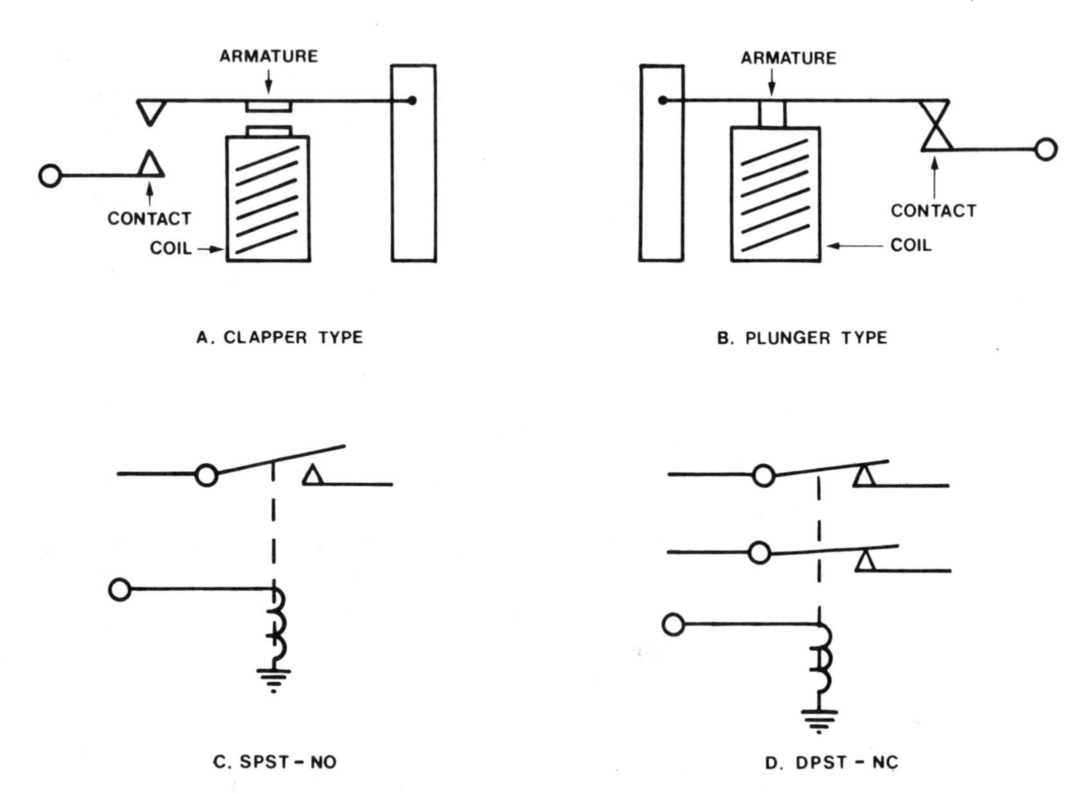

Figure 2-41 Electromagnetic Relays and Typical Schematics

may be open or closed (NO or NC). A permanent magnet is applied to the glass, and the magnetic field of the magnet causes the polarized reeds to make or break contact. One of the reeds is usually fixed. The reed relay may also operate with a coil of wire. In this case, power is applied to the coil, and the electromagnetic field causes the reeds to be drawn together.

Figure 2-42a shows a typical general-purpose electromagnetic reed relay. Figure 2-42b is a permanent-magnet reed relay. The reed relays in this figure are dry-reed relays.

Dry Blade Operation The blades of this relay are made from nickel–iron alloy, and the overlapping sections are plated with a precious metal such as gold. The gap between the blades is small (in the thousandths). The dry blade types have several types of bias. The simplest of the biases is to have two blades side by side with one fixed. A magnetic field will pull the movable blade to the fixed blade.

A second bias is mechanical in nature. A movable blade is attached between two fixed blades so that it presses against one of them (say the normally closed contact). A magnetic field applied will pull the movable blade from the NC to the NO contact.

A third bias is accomplished magnetically. A permanent magnet is placed next to a reed relay. The field of the magnet pulls a movable reed to one side away from a NO contact. A coil is energized, and its field overcomes the field of the permanent magnet and draws the movable blade to NC contacts.

Wet Blade Operation Wet blade reed switches have a pool of mercury within their glass envelope. The mercury is drawn by capillary action to the point of relay contact. Each time the coil is energized, the contacts are wet with mercury. In this way they have long life.

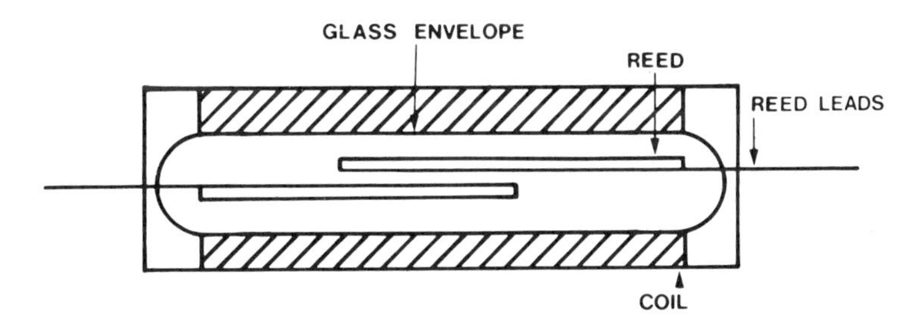

A ELECTROMAGNETIC REED RELAY

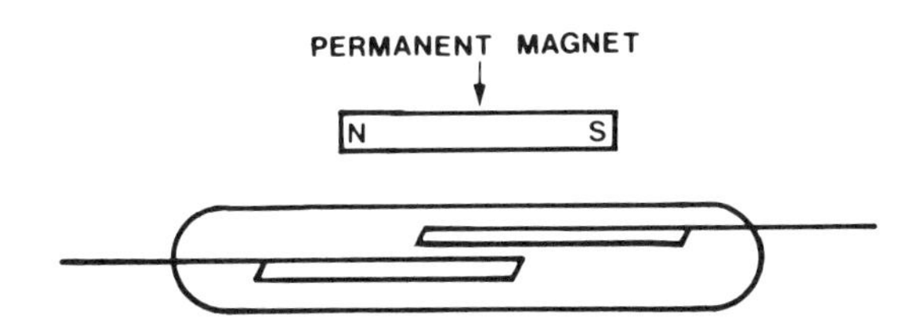

B. PERMANENT MAGNET REED RELAY

Figure 2-42 Reed Relays

Relay Contacts

Relay contacts are very similar to switch contacts. They are defined by their poles and their throws. A relay can either close or open contacts. Relay contacts are classified as switch contacts (refer to the section on switches in this chapter). Relay contacts are illustrated in Figure 2-43. Note that the relay contacts are normally open (NO) or normally closed (NC). This means that the relay, when energized, will either make or break the contacts.

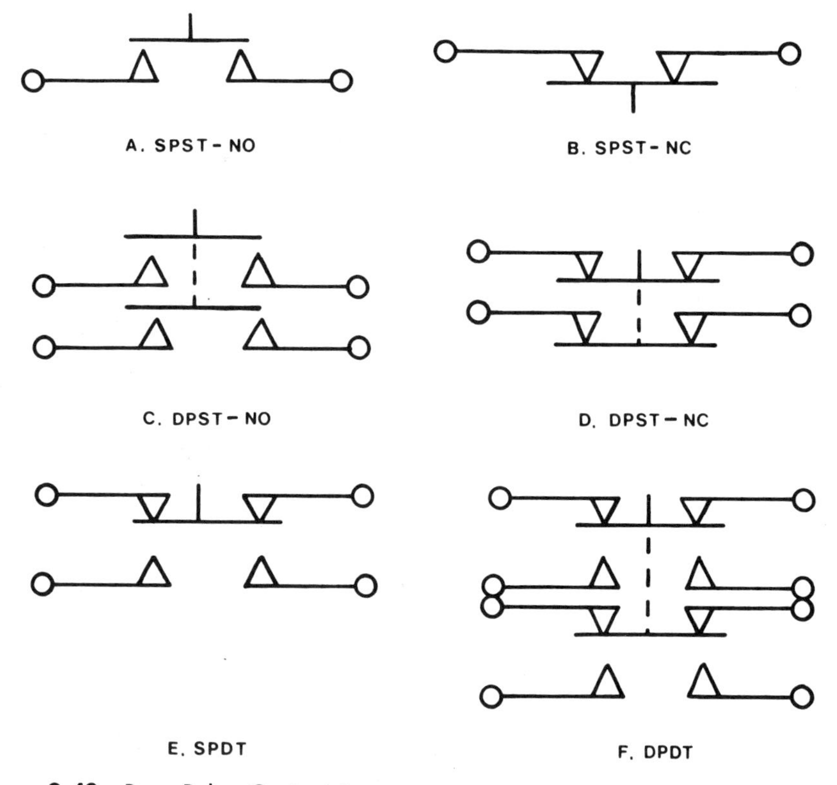

Figure 2-43 Base Relay Contact Forms

Relay contacts may also have make-before-break and break-before-make contacts. These two types are shown in Figure 2-44. With the break-before-make contacts in Figure 2-44a, a slug from the relay coil applies pressure to the contact blade, moving it away from a contact. A period of time exists between the time the blade reaches the second contact. With the make-before-break contacts, a slug from the relay coil makes a contact before it breaks away from the second contact (see Figure 2-44b).

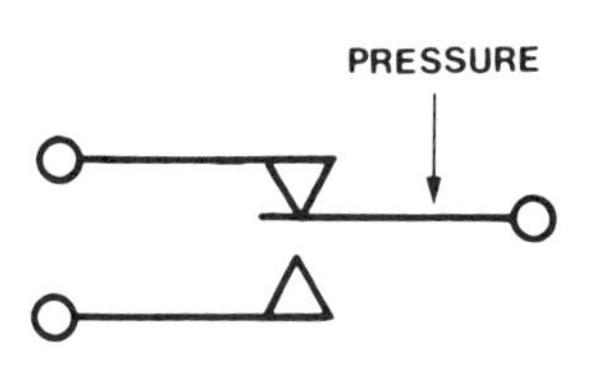

A. BREAK BEFORE MAKE

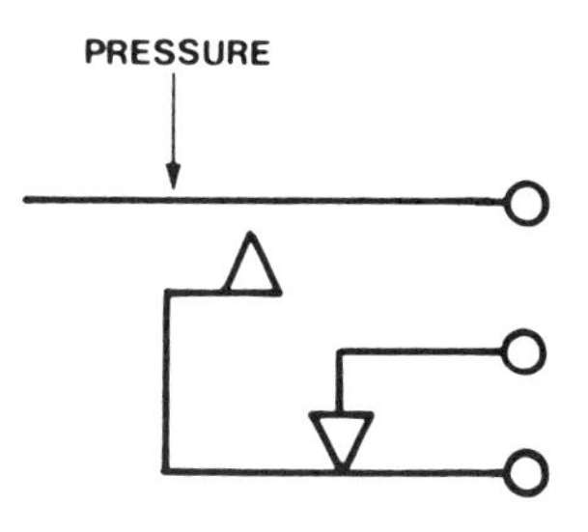

B. MAKE BEFORE BREAK

Figure 2-44 Break-and-Make Configurations

FUSES

It is necessary to protect the devices in a circuit in the event something occurs that will draw too much current. The job in most cases is accomplished with the fuse. The basic fuse is a small piece of metal that is calibrated to melt at specified current flows. This metal is encased in glass or in other insulating-type material. Figure 2-45 shows two fuse types.

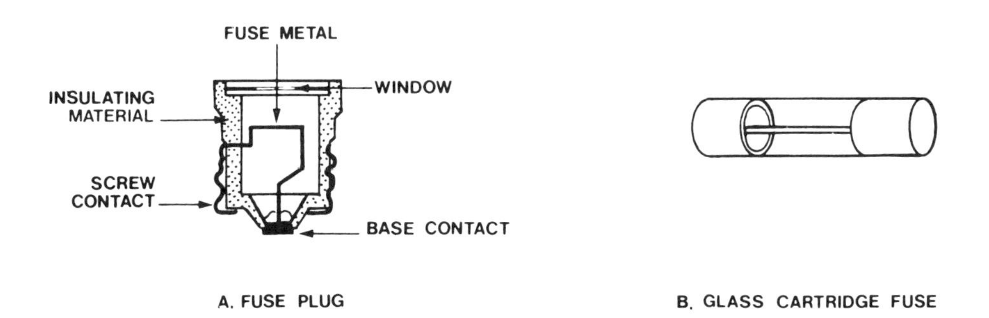

A. FUSE PLUG

B. GLASS CARTRIDGE FUSE

C. FUSE SYMBOL

D. FUSE HOLDERS

Figure 2-45 Fuse

Figure 2-45a shows the plug type. This type screws into a socket that is in series with the circuit. It becomes part of the circuit. The fuse metal is made of zinc alloy, which melts at a low temperature and is highly conductive. The fuse metal simply melts when the current in the circuit becomes greater than its rated value. This fuse type (plug) was a standard in most home applications (say 10 to 30 A) until the circuit breaker came into its own.

A second major fuse used in industrial and electronic applications is the glass cartridge type shown in Figure 2-45b. Next to the cartridge-type fuse are two typical fuse holders. Current-carrying capacity for the cartridge fuse may be as low as 1 or 2 mA and up to about 1 A.

Slow-blow fuses are used to protect the circuit in conditions where momentary current overload may be expected. A continuous current draw will "blow" the slow-blow fuse.

CIRCUIT BREAKER

The purpose of a circuit breaker is much the same as a fuse. The circuit breaker, however, will open the circuit when there is excessive current draw and then recover itself and the circuit. The fuse, when burnt out, must be replaced. As with the fuse, the circuit breaker is placed directly in series with the circuit. Figure 2-46a shows a typical circuit breaker.

Thermal Function

In Figure 2-46b the thermal function is represented by a set of contacts, a bimetal thermal element, and a spring. Current flows through the contacts and the

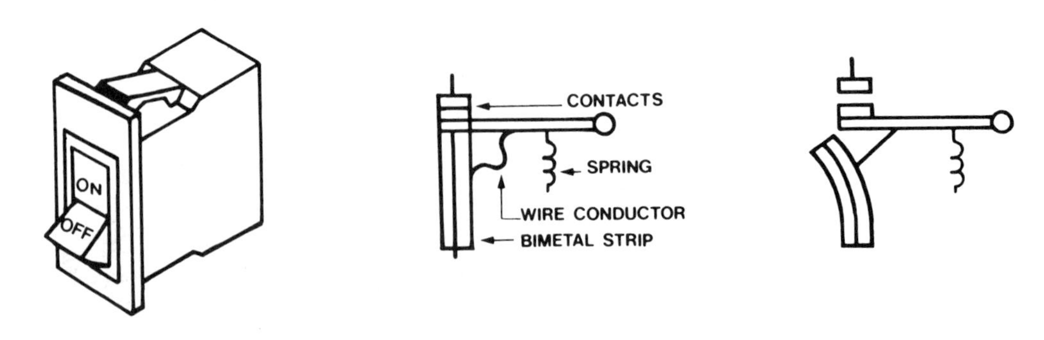

Figure 2-46 Circuit Breaker

bimetal strip. In the left diagram the circuit is intact. As current increases to over rated value, the bimetal strip bends because of the unequal expansion rate of the two metals.

When current increases enough, the metal bends and unlatches the contacts. A time delay is usually built in to compensate for momentary current surges (a few seconds for low current systems, more for large current systems). Reset may take place when the current is at normal.

Magnetic Function

A modified version of the thermal function is the magnetic function. This function has a magnet connected to the bimetallic strip with an adjacent, highly attractive member. When current increases and the bimetallic strip bends, it is aided by the magnet, which is attracted to the stationary member.

3

Industrial Electronic Components

Electronics components, by the simplest description, are all those electronic devices that are not electrical. We could also use the active–passive comparison; passive components would generally be electrical and active components would generate a voltage or amplify. Even then, there would be crossover components that would cause controversy. *Webster's Dictionary* says that the science of electronics is the behavior and control of electrons. By now, the reader has probably determined that we are moving in circles as far as a precise definition is concerned. However, we will discuss electronic components in this chapter, using the two general areas of tubes and solid-state devices as a guide. Then, in Chapter 8, we will cover sensing devices such as transducers, sensors, and detectors.

TUBE DEVICES

Around 1904, J. Ambrose Fleming, an Englishman, developed the two-element valve (the diode) while working for Marconi. Fleming had been a consultant to the Edison Electric Light Company of America, and remembered what was called the "Edison effect." He believed that this phenomenon could be of use to detect radio waves, and he was right.

Shortly after the two-element valve came into being, Lee De Forest created the greatest device known to electronics, the Audion. The Audion was simply a Fleming two-element tube with an inserted grid element. Yet this device was so revolutionary that it must be considered the epitome of all electronic-device inventions. Every amplifier device we have now has indeed followed the same concepts.

Legal problems plagued the Fleming and De Forest inventions. Marconi claimed that the Audion tube was a Fleming valve with a grid interposed; the patent was therefore Fleming's. So De Forest could not manufacture the tubes. De Forest owned the Audion patent; therefore, Marconi could not fabricate the Audion.

Around the same time (1907), the status of the tube as a detector was encroached upon by the crystal detector. G. W. Pickard created his first detector from silicon. Later detectors were made of galena, iron pyrites, and carborundum. The crystal detector was especially useful in detecting feeble signals and irregular pulses such as modulated continuous waves. The crystal detector was inexpensive and also increased the interest of the public in radio.

In 1912, De Forest and associates discovered the amplifying properties of the Audion by placing it in cascade (one stage after another). This work interested American Telephone & Telegraph (AT&T). Since De Forest had had problems with finances, he sold the rights to the Audion tube to AT&T for a sum that was embarrassing even in those days.

But there were problems with the Audion. The early Audion was a soft tube, and any plate voltage above 30 Vdc caused it to turn blue because of gas ionization. Gas modules interfered with the electron flow. In effect, the Audion's problems didn't help much in the competition with the solid crystal detector. However, the problems were overcome by Irving Langmuir, a physicist at General Electric laboratories who discovered a process for creating high vacuum in about 1912. Almost simultaneously, Western Electric (AT&T) did the same thing.

About the same time, the idea of feedback came into focus. De Forest and his associates had developed regenerative feedback to increase sensitivity in the Audion. However, Edwin Armstrong, a student at Columbia at the time, sought a patent for feedback and De Forest was back in court once more. The courts ruled for De Forest, but the patent didn't become his until 1934.

As a result of the hard vacuum tube and AT&T's work, De Forest triodes were used in 1914 to send messages around the world. World War I, of course, spurred activity in the radio field, and after the war, hard tubes were in such demand that everyone started making them, patent or no patent.

It must be remembered that, with the advancement of the tube, other devices such as the transformer, inductors, condensers, wire, and the manufacturing processes also underwent changes for the better. It also must be stated that the methods for hooking up the circuitry, such as the Hartley and Colpitts oscillators, the Armstrong superheterodyne receiver, and the Lacault Ultradyne, contributed greatly to electronics.

From this point onward, devices were advanced so quickly that it is difficult to determine just who created what and in what order. The number of inventions increased with the population and the wealth of the world, especially in the United States.

Probably the first major tube available from a major supplier was the Westinghouse VT series (VT1, VT2, etc.). This was used in regenerative circuits and operated from 22.5 Vdc for plate voltage. It was the best of the detectors available in 1917.

The RCA company was formed in 1919 and bought out the Marconi Company of America. It began to make the Marconi (Fleming) tube and warned others against infringing on the patent. In 1921, RCA released the UV200 (a detector) and UV201 (an amplifier). Both were triodes with brass shells. Both had heating filaments. AT&T was manufacturing with De Forest's patent. AT&T made a deal in 1920 to exchange privileges with RCA to keep them both out of court. The UV201A (with a thoriated filament) was developed soon after. This new tube required only a small amperage for its filament; it was very expensive and not very durable, which led to the overhaul of tubes and tube reactivators.

The next advance in tube manufacture was in low-current devices, which used only 60 mA of current. The B eliminator became popular in 1926, along with the Raytheon helium filamentless (cold tube). These were both direct heater tubes.

In 1927, the indirect heater came into being. This increased the interest in ac as the heater voltage. Battery-operated tubes were being pushed out of the picture except in isolated areas where no ac was available.

In 1928, the screen grid tube was released by several companies. This had tremendous amplifying power and was stable and sensitive. In fact, it was such a good amplifier that it caused stations to interfere with each other. The defect, called cross-modulation, was a worry to engineers until 1931, when variable mu tubes and the pentode tube arrived on the scene.

In 1933, the first multipurpose tube, the pentode, was released. This was sensitive, and still could be driven by a detector with high-power output. Also in 1933, the 6.3-Vac heater voltage was developed, along with the acorn tube, as a result of ultrahigh-frequency studies.

Metal tubes were created in 1935; they made higher amplification possible. The electron multiplier and secondary emission tubes followed soon after. Then came the cathode ray, the iconoscope, the beam-power tube, and the cold cathode. Further advancements were phototubes and thyratrons.

In most low-power applications, the tube has been replaced by solid-state devices. High-power applications are still used in the industrial market and will be at least through the 1980s. Let's look at some of these tubes and their operation.

Some Basic Tube Fundamentals

The entire principle of electron tube operation is based on the fact that electrons move freely in a vacuum (see Figure 3-1). The *Edison effect* states that free electrons in a vacuum will be attracted by a positive electrode and will be repelled by a negative electrode. Free electrons in a tube are called a negative space charge. The reason for the vacuum is that the electrodes would burn up easily in a space that has oxygen. In a vacuum, the elements will last for some time. In some cases, tubes are evacuated and then filled with some type of inert gas to increase the efficiency. In this gaseous atmosphere, free electrons combine with gas molecules to form positive ions and neutralize the space charge (negative).

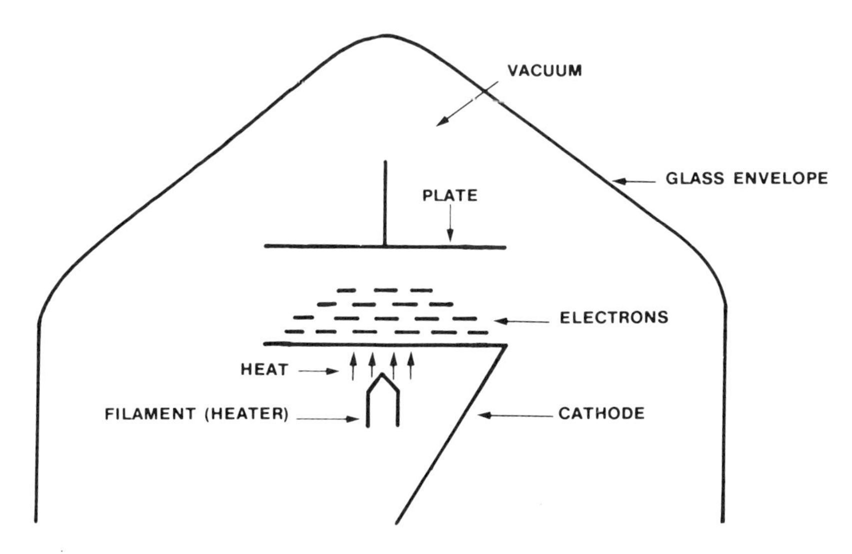

Figure 3-1 Function of an Electron Tube

Heaters are used to obtain a space charge. The heater heats the electrode called a *cathode*. The cathode is made of material such as tungsten, coated with an oxide. The cathode freely gives up electrons when heated. The action of the electrons is called *thermionic emission*. The process is often called *boiling*. A large potential is placed on a second electrode, called a *plate*, made of material such as nickel. The plate draws electrons from the space charge supplied by the cathode. Current in the form of electrons flows until the cathode heat or the plate potential is removed.

Diode Tube

The diode tube is the basic electron tube. It has two elements, the cathode and the plate. It also has a heater for the cathode. The heater may be part of the cathode (direct heating as shown in Figure 3-2a) or separate (indirect heating as shown in Figure 3-2b). The plate (also called an *anode*) provides a large

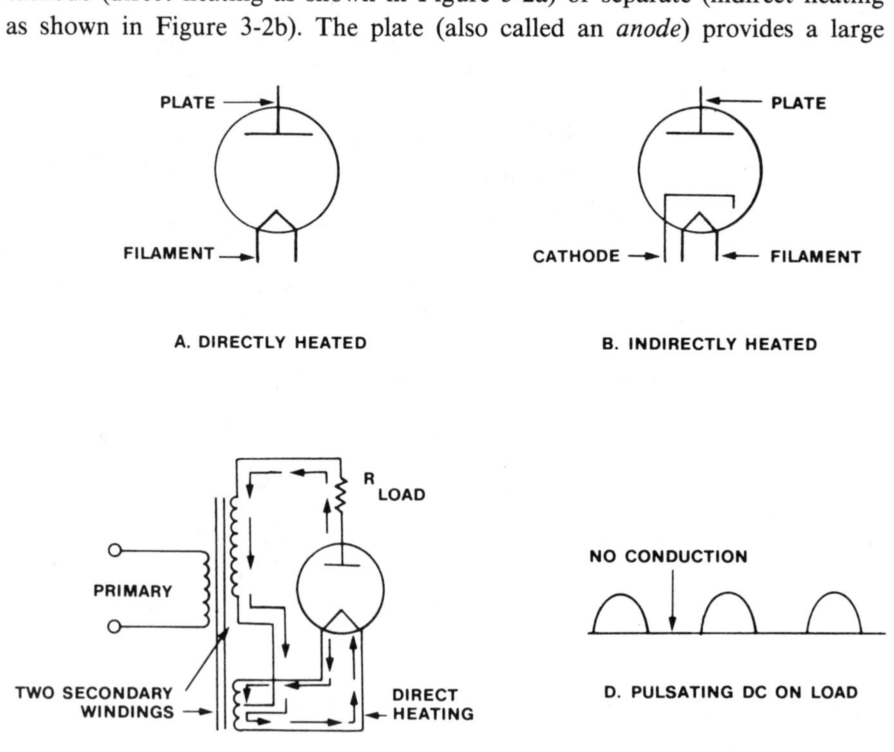

Figure 3-2 Diode Tube

positive potential. The heated cathode expels electrons into a cloud between the plate and cathode. A large positive potential on the plate will draw these electrons through the load. In Figure 3-2c, the tube and a transformer are illustrated together in a rectifier circuit. The circuit is a single-wave rectifier because it only conducts when the top side of the transformer is positive (every other ac alternation). When the plate is negative, it will not draw electrons from the cloud emitted by the cathode. The heater is on at all times in this circuit. The output voltage felt across the load resistor shown in Figure 3-2d is called *pulsating* dc because conduction does not take place during the negative half-cycle.

The full-wave rectifier circuit illustrated in Figure 3-3 uses a transformer and a tube called a *duodiode*. The tube has a single cathode and heater and plates P^1 and P^2. The tube is connected to the transformer in such a manner so that, when plate P^1 is positive, plate P^2 is negative. The cathode is always heated and the cloud of electrons is always present. When the top side of the transformer is positive, plate P^2 is positive and the rectifier conducts. When the lower side of the transformer is positive, plate P^1 is positive and the rectifier continues to conduct. During conduction, current flows through the load resistor. The voltage drop felt across the load resistor is represented in Figure 3-3b.

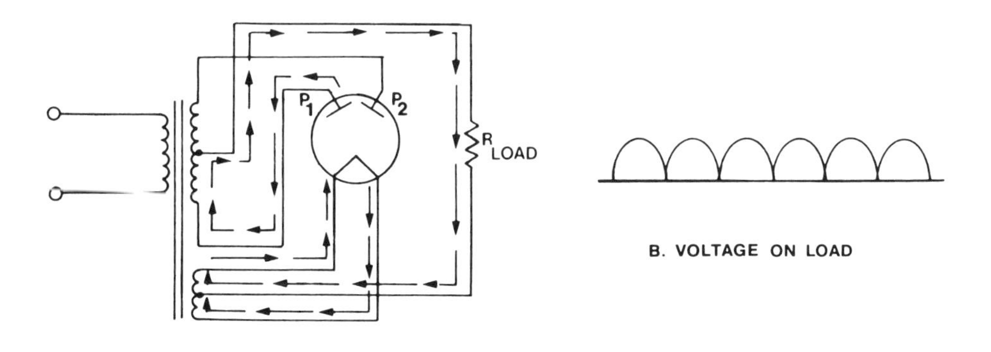

Figure 3-3 Full-Wave Rectifier Circuit

The space shown in Figures 3-2d and 3-3b may be flattened out (filled in) to approach a more desirable constant dc, with filtering. Filtering is accomplished by placing a capacitor in parallel with the load and/or an inductor in series with the load.

Tube rectifiers have been replaced by more efficient solid-state rectifiers.

Triode Tube

The triode tube (Figure 3-4) is a three-element tube. It is constructed much the same as the diode tube in an evacuated glass tube. The triode has a cathode for emitting electrons and a plate for collecting the electrons. Its ability to amplify is a result of a third added element, called a *grid*. The grid is a spiral-formed wire that physically fits between the plate and cathode (see Figure 3-4a for a triode symbol). The grid has a negative voltage applied. The negative voltage repels electrons emitted from the cathode, therefore preventing the electrons from reaching the plate. Thus, by varying the voltage applied to the grid, a varying current may be achieved to the plate. The more positive the voltage applied to the grid, the more current flows to the plate. A point of *saturation* will be reached if enough positive voltage is applied. The more negative the voltage applied to grid, the less current flows to the plate. A point of *cutoff* will be reached if enough negative voltage is applied. A varying voltage applied to the grid such as a sine wave will cause a varying current (sine wave) arriving at the plate. In Figure 3-4b, a signal is applied to a grid, which varies current flow to the plate and load. The signal output is a replica of the input signal, amplified and 180° out of phase.

The tube operates with direct current. A dc voltage is applied to the plate, cathode, and grid. The plate voltage is a high positive. The cathode is negative.

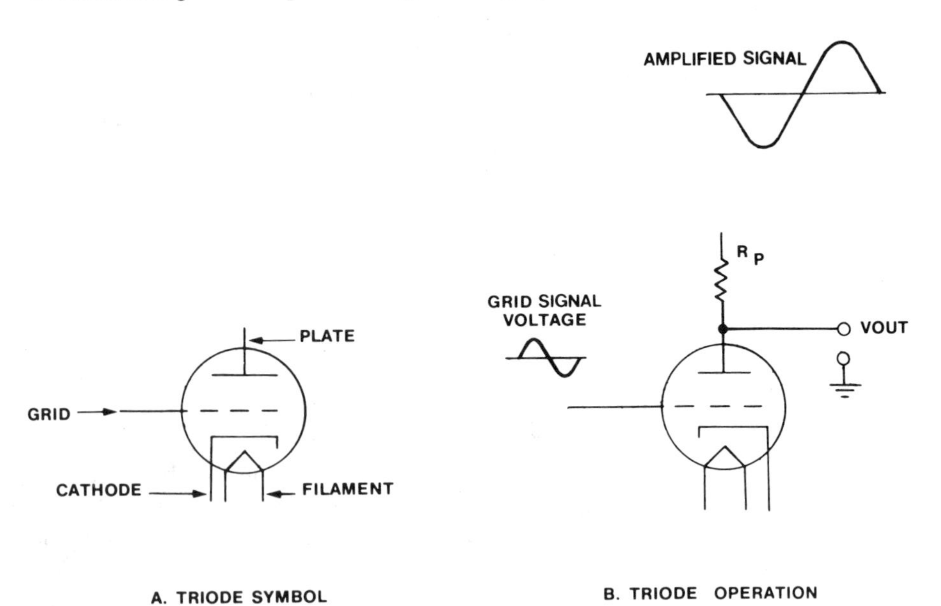

Figure 3-4 Triode Tube

The grid is negative in respect to the cathode. The dc voltage applied to the grid is called *bias*. Signal voltage is varying dc and ac sine wave.

Tetrode Tube

The tetrode tube is, in basic form, a triode tube and is used for amplification. Its symbol is shown in Figure 3-5a. The control grid is labeled 1, while a second grid has been added called a *screen grid*. The major purpose of the screen grid is to reduce interelectrode capacitance. This phenomena is shown in Figure 3-5b. Because the plate and control grid are close to each other and both are metal conductors, they create a capacitance with an air dielectric. At high frequencies the signal applied to the control grid may be coupled directly to the plate without being amplified. To eliminate this problem, a second grid, called the screen grid, is added. The screen grid is positive potential, although less positive than the plate. This potential adds to the plate potential, accelerating the electrons to the plate from the cathode. The screen grid by this action reduces interelectrode capacitance.

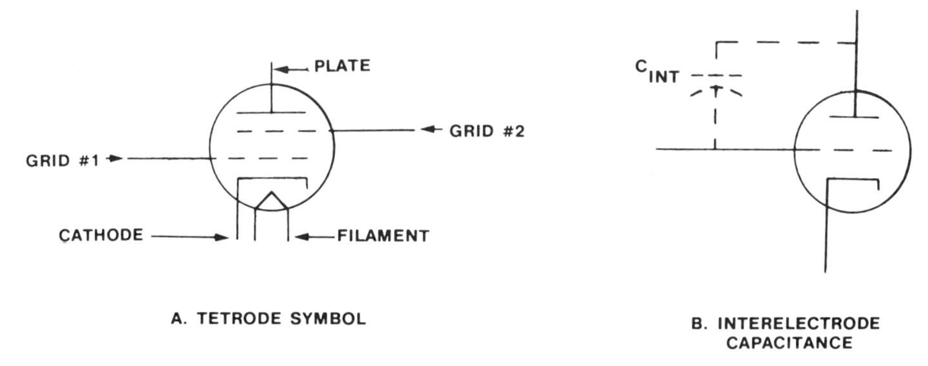

Figure 3-5 Tetrode Tube

Pentode Tube

The addition of the screen grid eliminated one problem (interelectrode capacitance), but in so doing added another. The screen grid accelerates the electrons from the cathode to the plate with such velocity that they knock other electrons from the plate into the electron cloud. These extra electrons are called *secondary emission*. They are, of course, negative and are attracted by the positive screen grid. This causes added screen current and decreases plate current. The effect is undesirable.

In Figure 3-6, a schematic symbol of the pentode tube is illustrated. An added element called a *suppressor grid* is added. The suppressor grid is connected to the cathode and has a negative potential. This negative potential repels the electrons driven from the plate due to secondary emission.

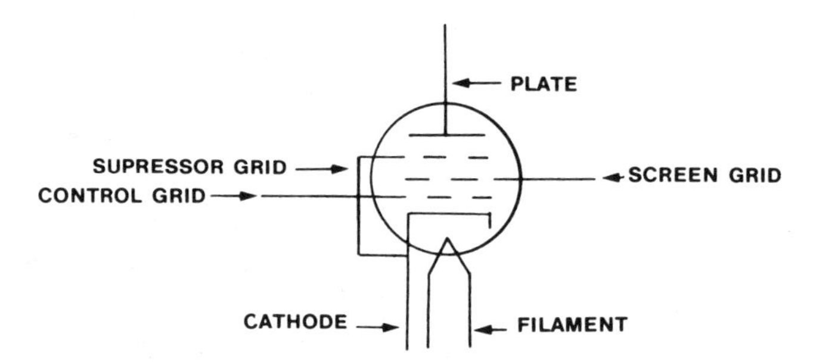

Figure 3-6 Pentode Tube

Some Tube Characteristics

As with all electronic devices, the tube has operating characteristics that are applicable to each tube model. The characteristics are either static or dynamic.

Static characteristics are those that relate to dc voltages without any signal applied. The major ones are plate characteristic curves and transfer characteristics. *Plate characteristic curves* are provided by manufacturers to tell the user what plate currents are expected with a fixed grid voltage and a changing plate voltage (see Figure 3-7). *Transfer characteristic curves* provide the user with details on what plate current to expect when plate voltage remains constant and grid voltage is varied (see Figure 3-8).

Dynamic characteristics are the amplification factor, dynamic plate resistance, and transconductance. *Amplification factor* is the ratio of a change in plate voltage to a change in grid voltage.

$$\mu = \frac{\Delta e_b}{\Delta e_g}$$

where

μ = amplification factor (mu)

Δe_b = change in plate voltage

Δe_g = change in grid voltage

The amplification factor is simply the tube's ability to amplify. It does not consider any circuit the tube may be applied to.

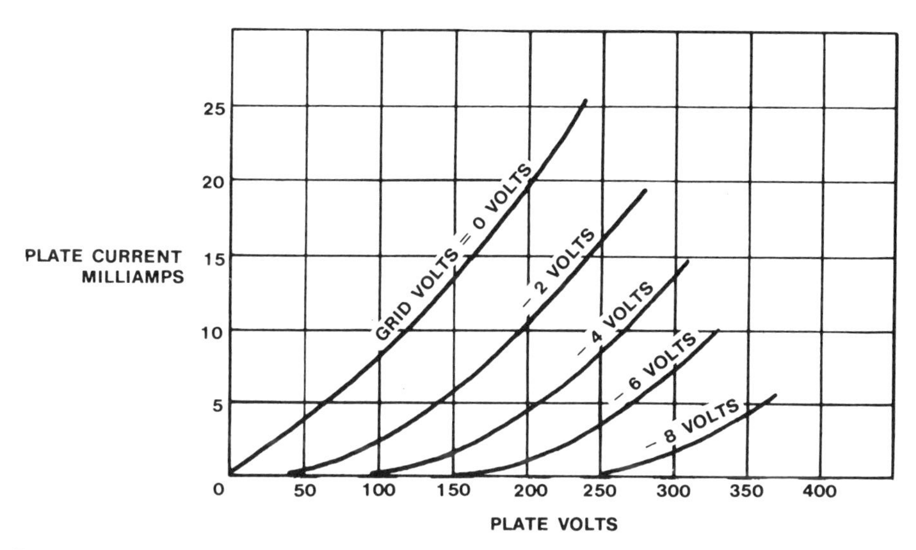

Figure 3-7 Plate Characteristic Curves

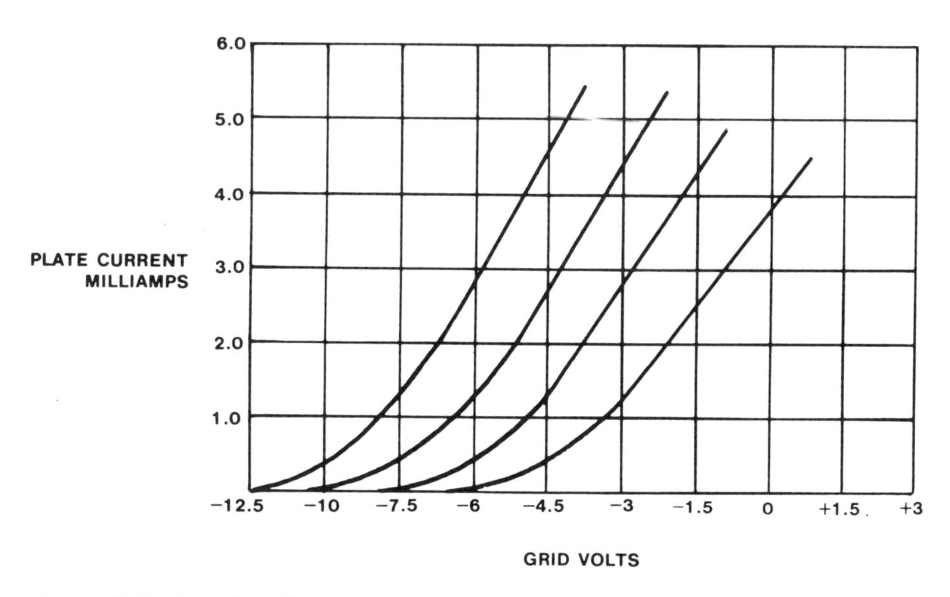

Figure 3-8 Transfer Characteristic Curves

Dynamic plate resistance is the ratio of a change in plate voltage to a change in plate current, with a constant grid voltage.

$$r_p = \frac{\Delta e_b}{\Delta i_b}$$

where r_p = dynamic plate resistance.

Dynamic plate resistance is a changing value because voltage signals and current change constantly.

Transconductance is the ratio of a change in plate current to a change in grid voltage with a constant plate voltage. There are two formulas for calculating transconductance (or *mutual conductance* as it is often called). They are as follows:

$$g_m = \frac{\Delta i_b}{\Delta e_g}$$

where g_m = transconductance

Δi_b = change in plate current

Δe_g = change in grid voltage

Also,

$$g_m = \frac{\mu}{r_p}$$

where μ = amplification factor

r_p = plate resistance

Mercury-Vapor Rectifier Tubes

Mercury-vapor tubes (Figure 3-9) are utilized where high current is required. Mercury-vapor tubes are constructed with a cathode and plate (anode) within a glass envelope. The envelope is evacuated, and filled with a small amount of mercury vapor. Often, some inert gas is added to the mercury vapor.

The mercury-vapor tube operates as a rectifier. In a vacuum, electrons are emitted from the cathode much faster than they can be gathered by the plate. Therefore, a cloud of electrons accumulates. The cloud is called a *space charge* and is negative. The space charge repels some electrons back to the cathode. In

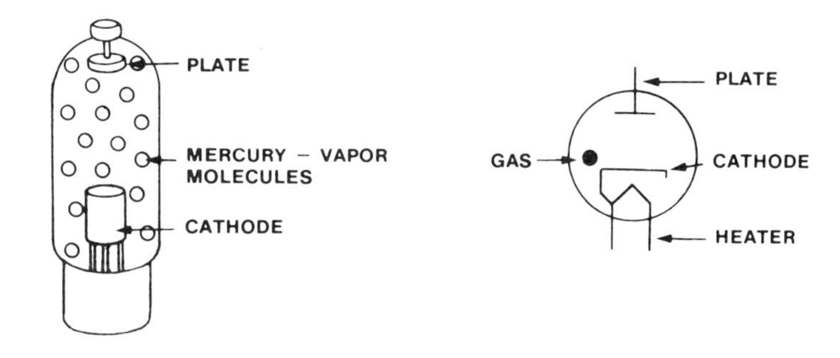

Figure 3-9 Mercury-Vapor Tube

a mercury vapor, these electrons collide with vapor molecules, thus separating some valence electrons. Therefore, some positive ions are created. These positive ions move toward the negative cathode, where they combine with new electrons. This action, called *deionization*, opposes the clouding of electrons. The free electrons left move toward the plate and become part of the plate current.

In the standard vacuum tube, plate voltage increases when plate current decreases, and vice versa. In the mercury-vapor tube, plate voltage remains constant with a change in plate current.

The mercury-vapor tube is used as a rectifier in circuits such as single- and full-wave rectifiers. It is often called a *phanotron*. The mercury vapor may be replaced by some gas. Argon or xenon may be used individually. These gases may also be used along with the mercury vapor. Operation of either of these is similar. Figure 3-9b is the schematic symbol for the mercury-vapor or gas-filled tube.

Thyratron Tube

The thyratron tube (Figure 3-10) is a gas- or vapor-filled rectifier that is grid controlled. The tube is used in high-current applications. The thyratron's primary use is in switching and control since it has no moving parts.

The thyratron is constructed with a plate, a cathode and heater, and a control grid. The glass envelope is evacuated and filled with gas or mercury vapor. The grid is negative and has a potential that allows the tube to be turned on (triggered) when the grid voltage is equal to the ionization potential of the gas or mercury vapor. The grid surrounds the cathode and serves as a passageway for electrons to the plate. Electrons must pass through a baffle in the center of

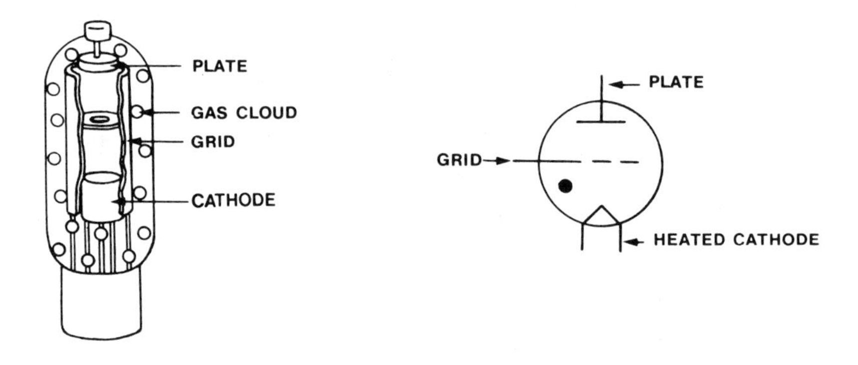

Figure 3-10 Thyratron Tube

the grid to reach the plate. The positive grid and its baffle can therefore control current flow until the desired moment.

A tetrode thyratron has a shield grid. The shield grid is used to minimize the small current that usually flows in the grid of a standard three-element thyratron. The tetrode thyratron has two baffles with a control grid mounted between.

The thyratron is essentially a triode. It operates as a controlled rectifier. Remember that the grid only controls turn-on time, whereas a triode tube amplifies. When the thyratron is triggered, the grid has no more control of current flow unless it returns to zero or the circuit is opened.

Ignitron Tube

The ignitron tube (Figure 3-11) is a mercury-vapor-filled rectifier tube used in power rectification high-voltage switching and welding control. The ignitron is constructed of plate (or anode), a cathode (mercury pool), and an ignitor. These are enclosed in an evacuated envelope. The envelope is a two-layer jacket for cooling water. The anode is made of graphite. The cathode is a pool of mercury. The ignitor is a pencil-shaped rod that is excited externally. The ignitor touches the mercury surface. In operation, a high positive voltage is applied to the plate. At this time, current does not flow because there are no electrons being emitted from the mercury pool. When the ignitor is positive, an arc is formed between the ignitor tip and the mercury pool. A high-voltage potential results between the ignitor and the mercury pool. The potential pulls electrons from the pool

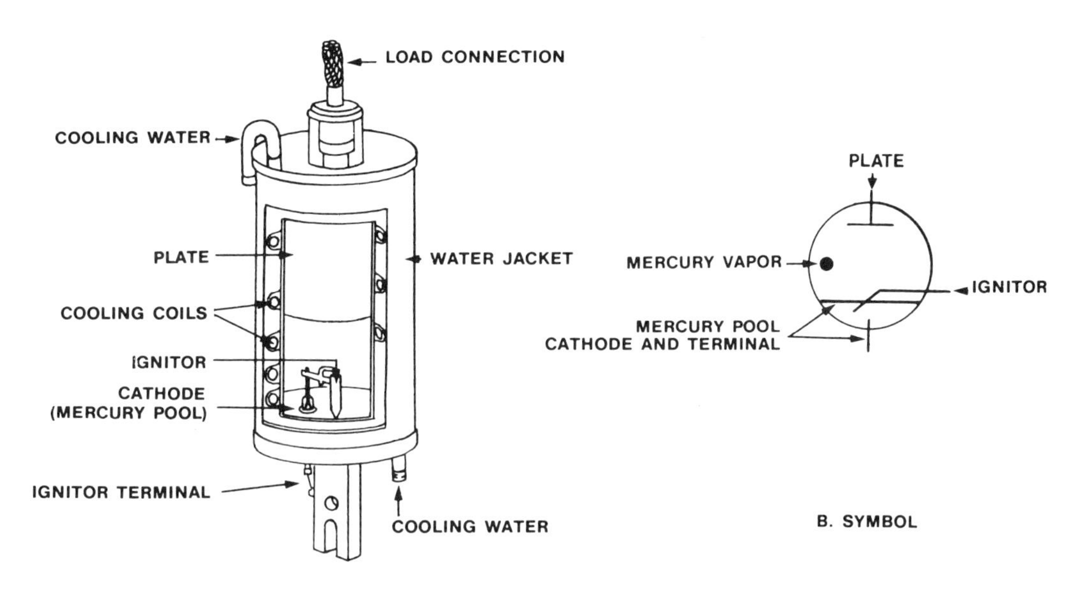

Figure 3-11 Ignitron Tube

and ionizes some of the mercury vapor. The positive ions accelerate to the mercury pool and free electrons by secondary emission. The secondary-emission electrons are drawn to the high positive of the plate, and a primary current flow (called an *arc*) is established between cathode and anode. The ignitor is not required until the next positive alternation.

Typical operating plate voltage and current of the ignitron is 700 A at 500 V and 600 A at 1200 V. Some ignitrons are available to produce much higher currents of up to 10,000 A.

SOLID-STATE DEVICES

For years, since the crystal detector and even farther back, scientists had been working on solutions to eliminate the frailties of the tube. In 1948, at Bell Laboratories, this problem was somewhat alleviated. The transistor was born, thanks to William Shockley, John Bardeen, and Walter Brattain, who received a Nobel prize for their efforts.

It must be assumed that solid-state devices have been around since the advent of the piezoelectrical devices, the crystal quartz, Rochelle salts, barium titanate, and lead titanate zirconate. This does not mean that the terms "solid-state device" has two meanings. Quite often the term is used synonymously with "semiconductor devices." Solid state is any electrical device that has no moving parts or gaps.

The world "semiconductor" implies something that is neither a great conductor nor an insulator. In electronics, the semiconductor is usually an atom that has four electrons in its outer valence band. These are primarily silicon and germanium. Quite often other atoms such as arsenic, phosphorus, and boron are also lumped into the general category of semiconductors. These are dopant materials used for the fabrication of semiconductor devices. These atoms have three or five electrons in their outer valence band, one less or one more than the silicon atom.

In any event, William Shockley of Bell Telephone Laboratories conceived the whole idea for the transistor around 1940. It wasn't until 1947 that his colleagues Brattain and Bardeen succeeded in creating the transistor. The device was a low-frequency amplifier that used a piece of silicon immersed in a salt solution. Then, in June 1948, the first transistor was created using germanium and three pointed wires. Two of the wires were connected to external power and the third to a battery providing bias, the same as does the grid in a vacuum tube.

In 1949, Shockley published the now-famous article "The Theory of pn Junctions in Semiconductors and pn Junction Transistors" in the Bell System Technical Journal. The devices used in experimentation were not very good. The idea of plunging wire into the material was problematical, because the crystal was hard and did not allow uniformity. A single crystal material was suggested by Shockley and the first of the commercial devices was made in 1951. Unwanted impurities plagued the manufacturers until 1954, when William G. Pfann, a metallurgist at Bell Laboratories, found a method of refining the silicon to purity. By 1955, excellent quality transistors existed. Then, later, in 1955, the diffusion technique was developed and the transistor was on its way. Leaders in fabrication techniques were Western Electric Company, Texas Instruments, Radio Corporation of America, General Electric, Westinghouse, and Raytheon.

Between 1955 and 1960, Leo Esaki, a Japanese physicist, in the course of his work for General Electric, created a gallium arsenide transistor that could handle frequencies of 4000 to 5000 megacycles. The types of transistors made since that time until the development of integrated circuits are almost uncountable. The transistor led the way to miniaturization and later microelectronics.

The transistor dominated the solid-state device world until 1958. It was in that year that the integrated circuit was conceived and constructed by Jack Kilby of Texas Instruments. The circuit was the first monolithic device.

Metal Oxide Rectifiers

Before the use of solid-state rectifiers as we know them today, metallic oxide rectifiers (Figure 3-12) were used. A small copper disk is protected on one side while the other side is exposed to air and extremely high heat. The unprotected side develops a coat of copper oxide. When placed in an electrical circuit, electrons flow readily from the copper through the oxide and the circuit, but not from the copper oxide to the copper. By placing this device in an ac circuit, current flows readily during one alternation and stops flow during the opposite alternation. This operation provides rectification.

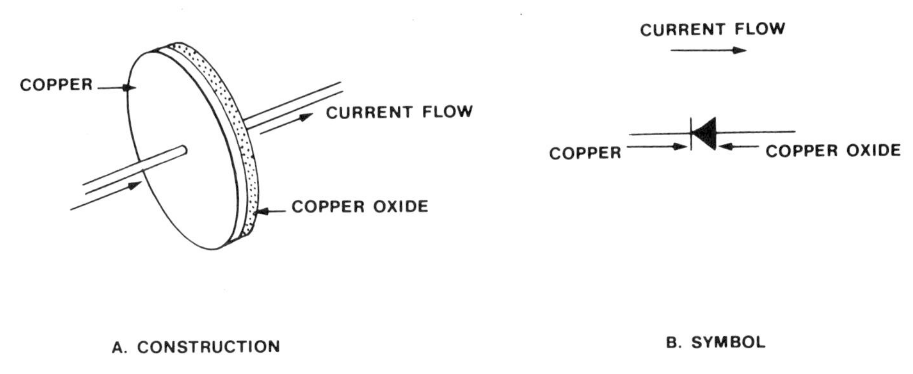

Figure 3-12 Metallic Oxide Rectifier

The selenium rectifier is a similar device. This device is manufactured by spraying selenium oxide on a disk of iron. Current flows readily from the iron to the oxide, but not in the reverse direction.

Some Semiconductor Physics

The atom is the smallest particle of a material that retains the identity of the material. For instance, a copper atom is the smallest part of a piece of copper wire. Let's review the two-dimensional model of a typical atom to reestablish some pertinent terms.

The atom consists of a nucleus with electrons in orbit, much the same as the planets orbit the sun. Within the nucleus are the proton and the neutron. The number of protons equals the number of orbiting electrons. The number of neutrons is not necessarily equal to the number of protons. The atomic weight of the atom is equal to the number of protons plus the number of neutrons. The weight may or may not be twice the number of electrons in orbit. For instance, the atom sodium has an atomic number (number of electrons) of 11, while its atomic weight is 22. In comparison, an atom of aluminum has 13 electrons, while its atomic weight is 27. The electron is very small. It is said to be $\frac{1}{1840}$ of the mass of a proton or a neutron.

Valence Band Model A valence band model (Figure 3-13) has been established by research. No matter how many electrons are involved, the number of electrons in any orbit follows the same pattern. For instance, the K shell has two electron positions possible, but no more. The L shell has possible two electron positions in its first ring and six electron positions in its second ring. The M, N, O, P, and Q shells follow this same pattern. For explanation purposes, let's consider the copper atom. The copper atom has 29 electrons, 29 protons, and 35 neutrons. Its atomic weight is approximately 64 (see Figure 3-14). Its K shell has 2 electrons. Its L shell has 8 electrons. Its M shell has 18 electrons and its N shell has but one electron in its valence. This adds to a total of 29. The outer ring is called a valence. In the two-dimensional illustration, it appears that electrons follow the same orbit. In actuality, the electrons all have their own orbit. The distance from the nucleus is the same for all electrons in the same shell. Other good conductors in the table of elements (aluminum, silver, and gold) have only one electron in the valence. A good conductor has this structure. It is possible to move the electron from the valence, thereby producing electron flow (electricity). Some force is used to remove the electrons, such as a magnetic field, friction, or heat. These electrons are considered to be free.

Just as some materials are good conductors, other materials are poor

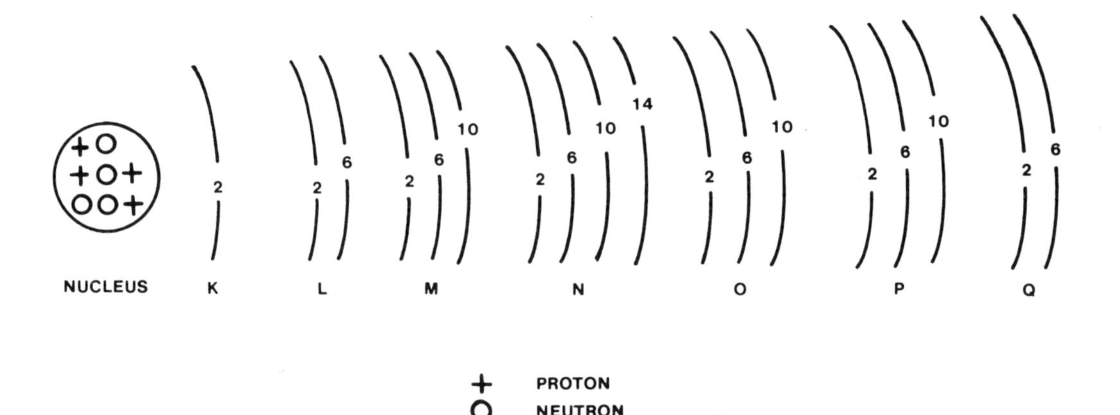

Figure 3-13 Valence Band Two-Dimensional Model

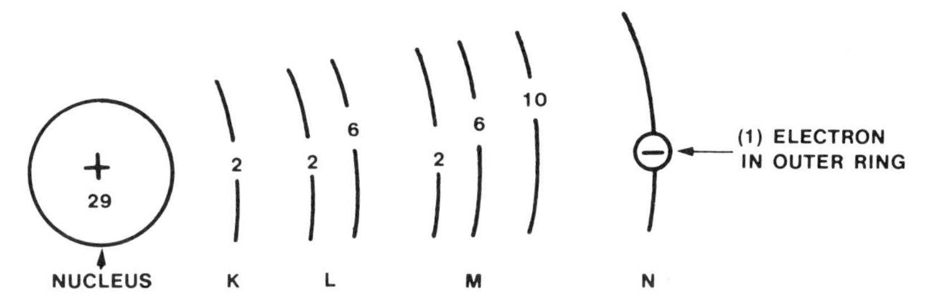

Figure 3-14 Valence Band Two-Dimensional Mode of Copper Atom

conductors. Poor conductors of electricity are called insulators. Insulators are usually compounds such as glass, mica, ceramics, and other substances. Since the atoms of these substances generally have a large number of electrons in the valence, the attraction of the electrons to the nucleus is strong.

Some special elements are neither good conductors or good insulators. These are called semiconductors. The two semiconductor elements in common use are silicon and germanium. The silicon atom has 14 electrons while the germanium atom has 32 electrons. Each of these atoms has 4 electrons in its outer shell, 2 in one ring, and 2 in their valence ring. Figure 3-15 is a two-dimensional model of the silicon atom. In Figure 3-15a, the atom of silicon is shown to have 14 electrons, 2 in its K shell, 8 in its L shell, and 2 apiece in the rings of the M shell. Semiconductors such as silicon or germanium have crystalline structures. Each silicon atom, as shown in Figure 3-15b, shares one of its electrons with each of its neighboring atoms. This is called diamond sharing, covalent bonding, or valence bonding. Pure silicon has this structure (called a *lattice*) throughout. Of course, pure silicon is very difficult to obtain (grow) and

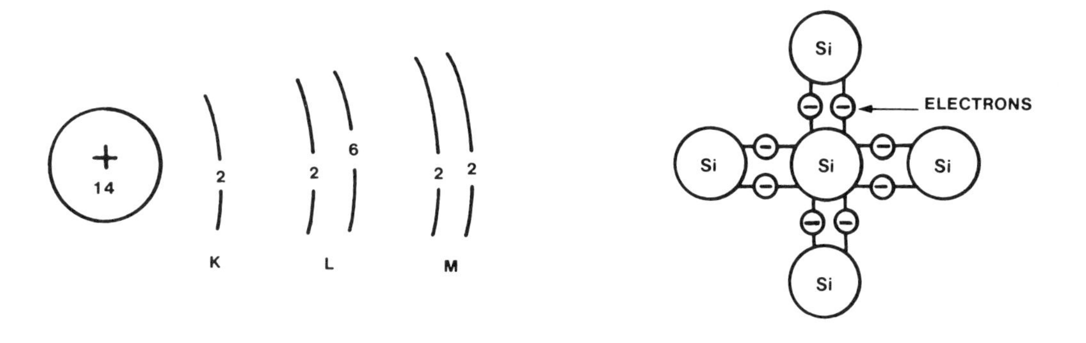

Figure 3-15 Two-Dimensional Model of Silicon Atom

in reality never exists. Silicon is called pure when it has less than 1 part per 10^{10} impurities. Electron movement is minimal. Pure silicon is called *intrinsic material.* The model in Figure 3-15b represents an intrinsic silicon crystal.

Doped Semiconductor Material Figure 3-16 compares two types of dopant, *N* and *P*. Doping a material is the addition of other atoms into the lattice structure of the material. This changes that structure. The *P*-type element (also called an acceptor atom) has three electrons in its valence. Figure 3-16a shows a model of silicon with an atom of *P*-type material (boron) in its crystal lattice structure. The reader will note that the acceptor atom has a space called a hole (absence of an electron). Typical donor elements (atoms) are boron, aluminum, and gallium. The holes are called majority carriers or positive charged carriers. Minority carriers in *P*-type material are electrons. The *N*-type element (also called a donor atom) has five electrons in its valence. Figure 3-16b shows a model of silicon with an atom of *N*-type material (phosphorus) in its crystal lattice structure. The reader will note that the donor atom has a loosely bound electron. Typical donor elements (atoms) are phosphorus, arsenic, and antimony. The free electrons are called majority carriers or conduction electrons. Minority carriers in *N*-type material are holes. *N*-type material is silicon that has been doped to be electrically conductive by means of mobile electrons and therefore is negative. Conversely, if the silicon has been doped for conduction by holes, it is said to be positive or *P*-type.

The type of dopant (*N* or *P*) is extremely important to determine bias

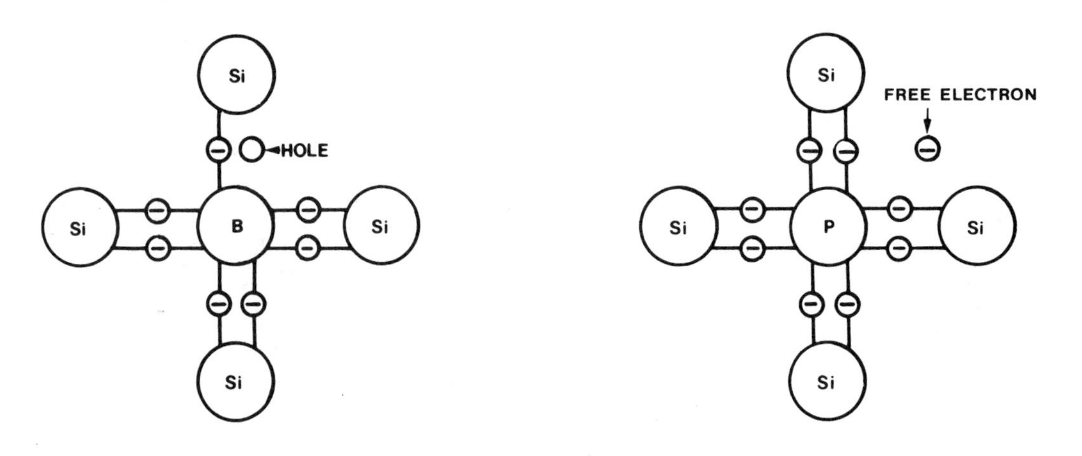

Figure 3-16 Dopants

polarity in the design of electrical circuits. Impurity dopant is added in the order of 1 or 2 parts per million. The exact doping level is extremely proprietary information. The second important characteristic of the dopant is resistivity. A semiconductor is, as its name implies, semi or partially conductive. The degree of conductivity of silicon is fixed by the amount of dopant that is added. The resistivity of the end product (semiconductor device) is determined by the type and amount of the dopant material added to the silicon. Typical conductive metals have a resistivity of 10^{-3} Ω/cm^2. Intrinsic (pure) silicon has a resistivity of 3×10^5 Ω/cm^2 at room temperature. *N*-type or *P*-type materials are semiconductors that have been doped to have an excess or a lack of electrons in the valences of their atoms. This semiconductor material is called *extrinsic*.

Energy Model The valence band model provided us with a space relationship. The energy model allows us to study the semiconductor in terms of energy. To interface the valence band model with the energy model, let's look at the comparison in Figure 3-17. This illustration compares the electron orbits with energy levels. The first orbit is the first energy level, and so forth.

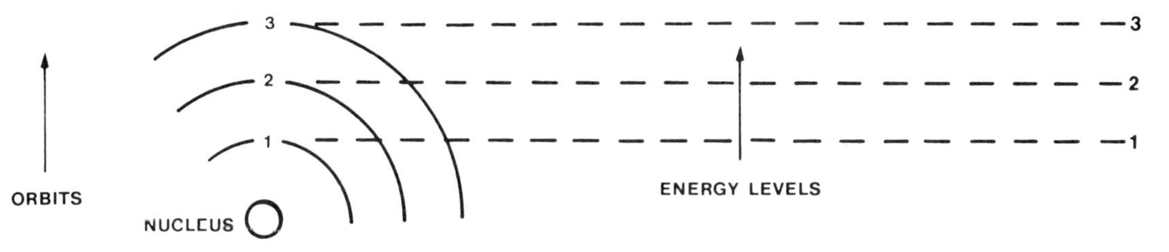

Figure 3-17 Comparison of Orbits and Energy Levels

It takes a certain amount of energy to move an electron from one orbit to another. These orbits are at distinct energy levels. Electrons cannot exist between these levels. Each energy level is a band of energy. Intrinsic silicon falls in Group IV of the periodic table of elements because it has four electrons in its valence. The significant energy bands in this intrinsic silicon material are the conduction band and the valence band. Valence electrons must be removed by some force and drawn across the forbidden gap. Electrons cannot exist in the gap.

There are several ways in which electrons can be drawn across the forbidden gap: the application of heat, the application of an electric field, and the application of light. To remove electrons from the valence band, it takes 1.1 electron volts (eV) of energy at room temperature. The forbidden gap decreases with an increase in temperature and increases as temperature decreases.

As you can see, when an electron leaves the valence band it leaves a positive charge called a hole. The hole may appear to move just as the electrons move to other atoms. Holes draw electrons from neighboring atoms, leaving another hole, and so on. It is imperative that a state of electrical equilibrium exist in an intrinsic crystal. Therefore, the number of electrons in the conduction band will equal the number of holes in the valence band. The Fermi level is a level in the forbidden (energy) gap where there is an equal opportunity for an electron to exist with a change in energy above the Fermi level or for a hole to exist with a change in energy below it. In short, the Fermi level is centered between the conduction and valence bands in intrinsic silicon. With impurities added, the Fermi level will move toward one or the other band.

Basic Solid-State Junctions Thee are three basic solid-state junctions (Figure 3-18): The *homojunction*, the *heterojunction*, and the *Schottky barrier*. All other junctions utilize some modification of these basic junctions.

The most used of all junctions is the homojunction. The junction is all silicon, which accounts for the name homojunction. A large area called *P*-type material is doped with donor atoms such as phosphorus. The barrier region, which is extremely thin, is created between the *N* and the *P* types of material. Electrons from *N*-type material move to fill the holes in the *P*-type material.

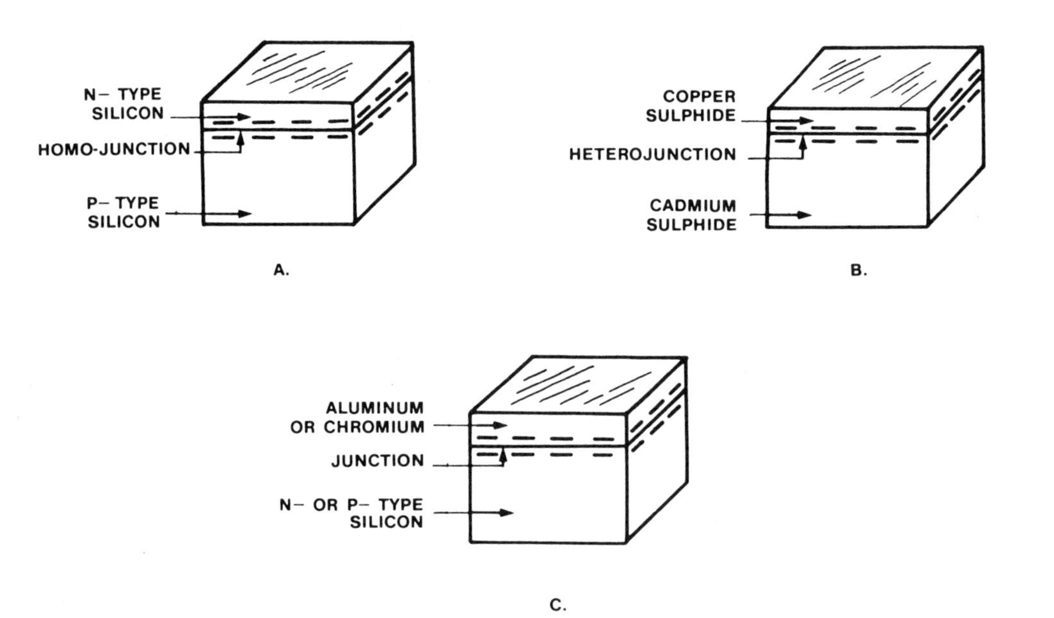

Figure 3-18 Basic Solid-State Junctions

The *N*-type material is left with a net positive charge. The extra electrons fill the holes in the *P*-type material with an excess of electrons. In this manner, a thin static electric charge is developed across the junction between the *N* and *P* types of materials. The barrier resists any movement of electrons through it. Therefore, only electrons of high energy can make it through. The barrier is called by at least two other names, the *depletion zone* and the *space charge*. Every junction must have a barrier.

When application of voltage is made to a junction, this barrier must be overcome. In silicon material the voltage required to overcome the junction barrier is 0.6 V approximately. Germanium junctions require approximately 0.2 V. The barrier voltage is called *hook voltage*.

Since junctions can be made of different materials than silicon, another cell type, the heterojunction, has found prominence. Junctions of this type are made from dissimilar metals. One such type is the cadmium sulfide–copper sulfide cell. As with semiconductors, one metal, such as copper sulfide, is deposited on the second metal, such as cadmium sulfide. The two metals, on contact, form a junction called the heterojunction. The heterojunction is widely used in the photodetection field. In the solar energy field, the heterojunction is a strong contender.

The Schottky barrier junction is made of metal and semiconductor material. The Schottky barrier junction is manufactured by depositing a thin metallic layer on silicon to create a junction. Figure 3-18 shows a layer of aluminum or chromium deposited on a substrate of *N*-type or *P*-type silicon. A junction is established. The Schottky photodiode is manufactured at low temperatures. Metal is deposited rather than diffused.

Manufacturing Processes

To create solid-state devices, certain technologies are required. Although there are many different technology skills and styles, some similar work is required by most. These are as follows:

1. Refining of the raw material
2. Growing of the silicon
3. Development of silicon sheets or wafers
4. Processing of the wafers into solid-state devices

The choice of who does what is, of course, left to the manufacturers. The industry is usually divided into three basic groups: those that provide the raw material, those that build the solid-state devices, and those that build the systems.

Silicon Preparation Silicon preparation is the purification process. Pure silicon is produced by several methods. The most prominent of these methods is the Siemen's process. The process essentially yields high-quality trichlorosilane from quartzite. It then decomposes the trichlorosilane. This leaves pure silicon. The process is expensive because it consumes heat energy. The result is a polycrystalline silicon (polysilicon).

A second method of purification is the Union Carbide process. This method prepares silane from metallurgical-grade silicon and produces near pure silicon by pyrolysis. The process provides a closed cycle, low energy consumption, and continuous operation.

Silicon Sheet Preparation The silicon must be in the form of a flat sheet to be fabricated into a solar cell. To derive a flat sheet, the silicon is melted down and grown into some form that will allow for the slicing of flat sheets. Other methods of growth are initially on flat sheets and therefore slicing is not required.

Crystal growing methods are reasonably fixed arts with only a few variations occurring in the industry from time to time. The Czochralski (CZ) method has been the industry standard for some time. This method grows a cylindrical ingot of several kilograms from a crucible of molten silicon. A seed is dipped into the molten silicon and withdrawn slowly as the seed and crucible are rotated.

The second most used method is float zone (FZ). A crucible is used similar to the CZ method. An RF heating coil is wound closely around a polysilicon rod. When current is applied to the RF coil, a molten float zone is established on the silicon rod. As the zone moves up (or down), the rod is purified and a crystal structure is formed.

The CZ and FZ methods both produce single-crystal ingots the shape of a large bologna that must be sliced.

Sheet Fabrication The CZ and FZ ingots require slicing. There are three basic types of slicers: the inside diameter (ID) saw slices one wafer at a time. The multiblade saw and multiwire saw slice 20 to 25 wafers at a time. The problem arising with slicing in any manner is loss of silicon due to sawdust (called kerf). Sheet fabrication also includes crystal alignment, grinding, polishing, and etching.

Solid-State Device Fabrication The development of solid-state devices has evolved into geometrics that are extremely small. The quality of the silicon substrates on which these devices are fabricated has inherently become more critical. Greater uniformity and decreased levels of impurities are demanded to increase device integrity and wafer yield. Fabrication of the solid-state device must start with the best possible single-crystal silicon wafer. It must be noted that there are as many procedures for the fabrication of solid-state devices as there are companies that produce them. The processes are all similar, however, with only a few of the more exacting details held as proprietary.

The fabrication of the silicon solid-state device begins with the growing of a single-crystal silicon ingot in an oven (crucible). The crystal is sliced, etched, and polished. The slices, now called *wafers*, are transported to device manufacturers who perform what is called the *planar process*. This process involves oxidation, selective dioxide removal, diffusion, and epitaxy. *Oxidation* is the formation of silicon dioxide layers on the silicon for photographic masking. Selective dioxide layers are removed with acids. *Diffusion* is the process of depositing dopant through dioxide windows. *Epitaxy* is the growing of one semiconductor upon another. The planar process places the pattern on the silicon wafer. The pattern is duplicated many times on the wafer (depending on the size of both wafer and the device). Individual devices are called *dies*. Each die is attached to a package to hold it in place. Wires are connected from the die to the outside of the package. The package is closed for protection. The package with die and wires is now a solid-state device.

Junction Diode

The most basic of all solid-state devices is the junction diode (Figure 3-19). The function of the diode has been explained under "Basic Solid-State Junctions" and illustrated in Figure 3-18. To be useful, the diode must be placed properly in a circuit. There are two ways the diode may be installed, reverse and forward. Installation is dependent on the polarity of the applied voltage. You may recall that acceptor silicon (*P*-type) and donor silicon (*N*-type) are selectively doped to provide an excess of holes and electrons, respectively (see Figure 3-16).

In the reverse bias condition (see Figure 3-19a), power is placed across a diode with its negative terminal connected to the *P*-type material and its positive terminal connected to the *N*-type material. In this reverse bias condition, holes in *P*-type material are attracted to the negative terminal. Electrons are attracted to the positive terminal. There is no current flow. A large depletion area exists across the junction. The diode is simply not conducting.

In the forward bias condition, power is placed across the diode with its positive terminal connected to the *P*-type material and its negative terminal connected to the *N*-type material. In this forward bias condition, electrons and holes are attracted across the junction. Other electrons leave the negative source and easily transit through the diode to cause current flow.

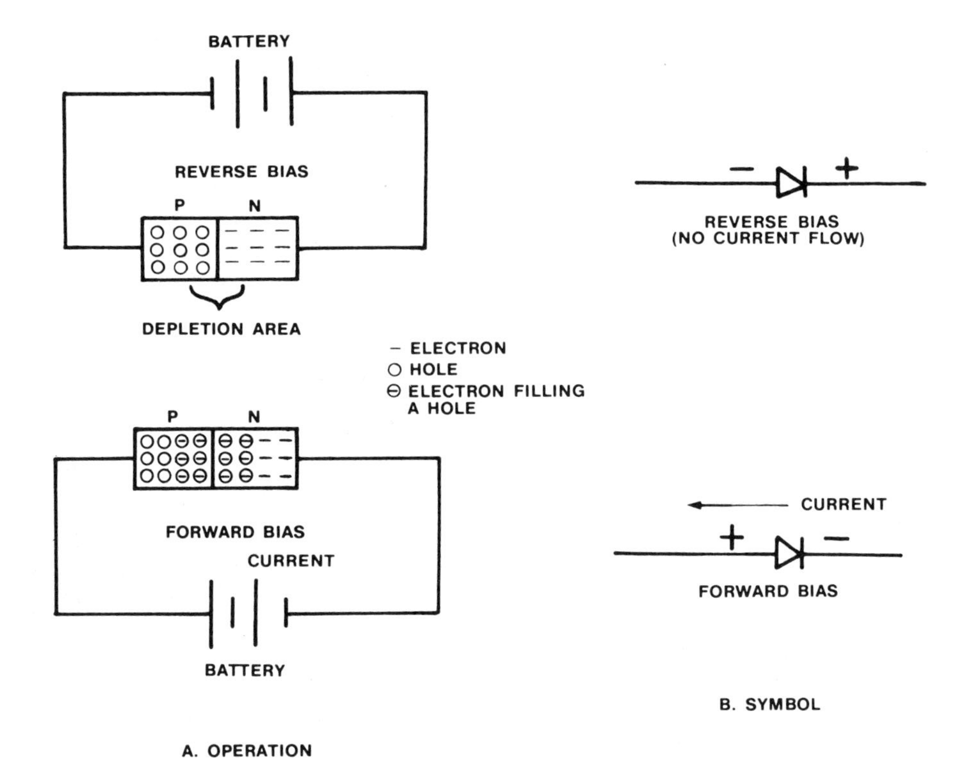

Figure 3-19 Junction Diode

The junction diode is made from silicon, germanium, or gallium arsenide in that order of importance. The junction diode is used for rectification and steering. Other applications are simply modifications of these.

Diode IE Curve The prominent characteristic curve for the diode is the forward and reverse voltage/current curve (Figure 3-20). With forward bias, (− on *N*-type and + on *P*-type material), the curve moves toward +. When the barrier potential (hook voltage) of the junction is overcome (0.6 V for silicon; 0.2 V for germanium), current will flow freely in the positive direction up to some specified destruction point. With reverse bias (+ on *N*-type and − on *P*-type material) the curve moves toward −, with very little current (leakage only) until the specified reverse voltage capability of the diode is reached. At that point (the knee), the diode will go into avalanche and current will increase in reverse until the diode is destroyed.

Typical Diodes Typical diode packages are illustrated in Figure 3-21. The diode package strength is usually equated to its current-carrying capability.

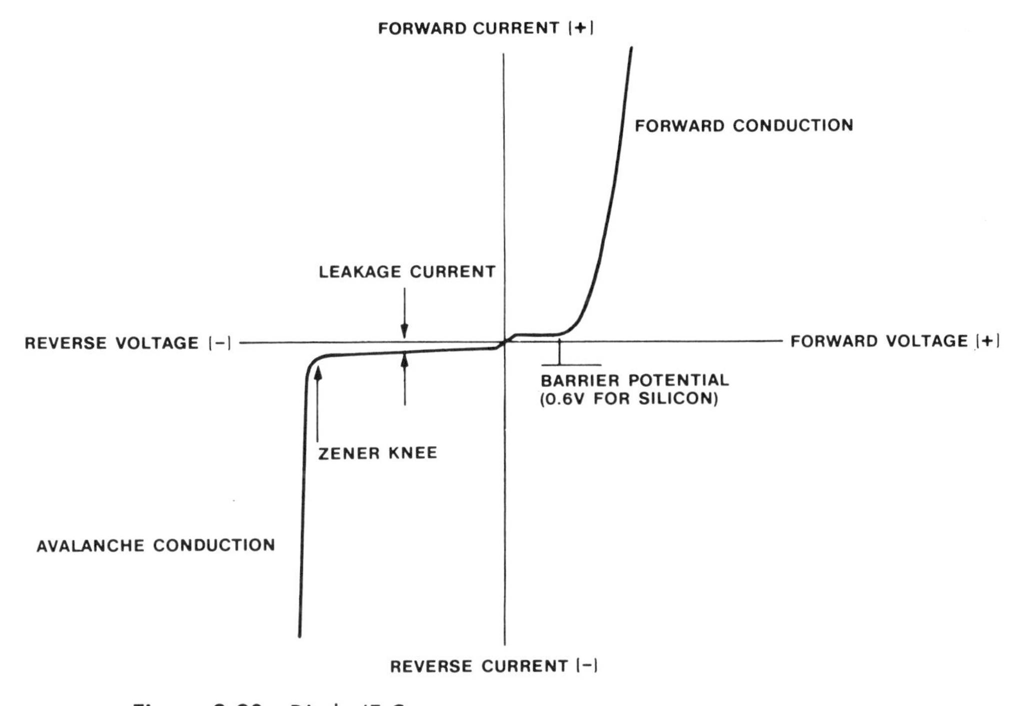

Figure 3-20 Diode *IE* Curve

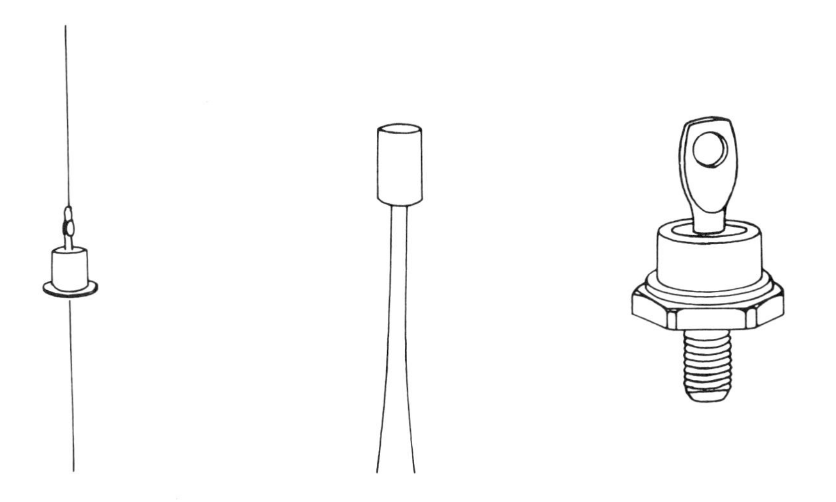

Figure 3-21 Typical Diode Packages

Zener Diode

The zener diode (Figure 3-22) is closely related to the junction diode. It is constructed with a single junction. When manufactured, the zener diode is specially doped so as to operate in reverse of a normal junction diode. On the diode curve shown in Figure 3-20, a zener knee is shown in the reverse direction. A standard diode will avalanche at that point and be destroyed. A zener diode when reaching the zener knee will conduct. Current will flow to a specified value. Voltage will remain at exactly the zener value. This last point portrays the usefulness of the zener as a voltage regulator. In Figure 3-22c the zener is placed in parallel with a load. A limiting resistor takes up the slack voltage between the supply and the zener. Remember that the zener voltage remains the same even if the source voltage varies. Therefore, the voltage on the load will remain constant. Note that the junction model is the same as that for the standard diode. Note also that the typical diode packages shown in Figure 3-21 are the same packages used for zeners.

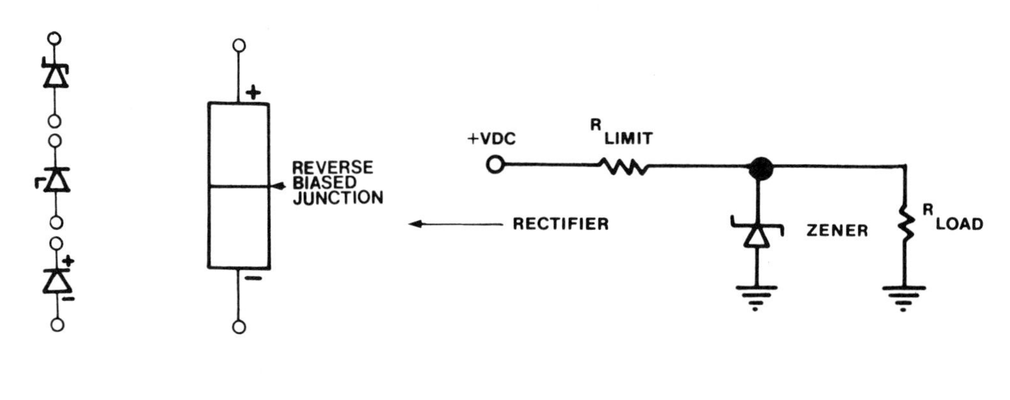

Figure 3-22 Zener Diode

The Diac

The diac (Figure 3-23) is a three-layer, two-junction diode. The diac model looks much the same as a transistor model (see Figure 3-23b). The diac is a two-terminal device and can be placed in a circuit in either direction. In other words, it is a bidirectional device. When power is first applied to the diac, it will not conduct. No matter which way the voltage potential is applied, one of the junctions will still be reverse biased. As voltage is increased, a conduction point will be reached called *breakover voltage.* Once reached, voltage is slightly reduced to normal operating voltage. A resistor is usually placed in series with the diac to limit current flow. Once conducting, the diac cannot turn off unless the circuit is broken or applied voltage is dropped to zero. The diac is one of the family of thyristors. The diac is used in ac control systems such as motor speed controls.

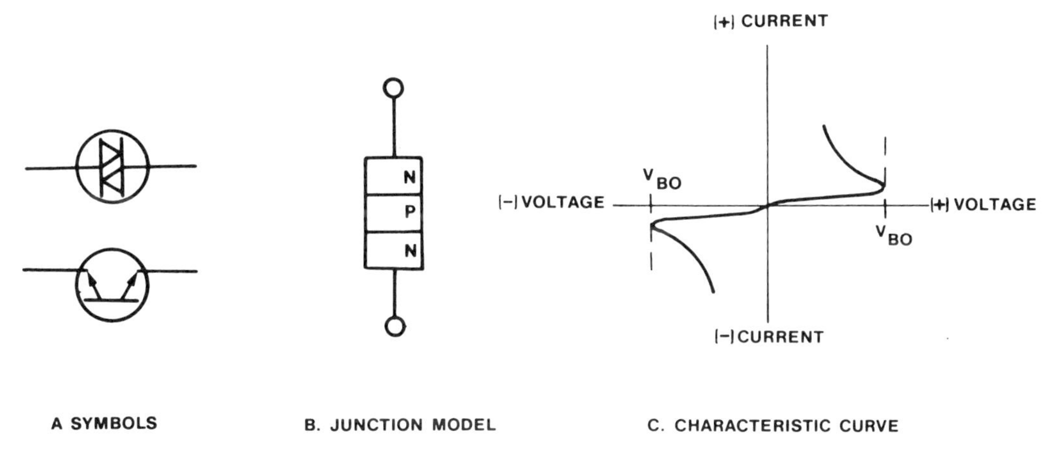

Figure 3-23 Diac

Silicon-Controlled Rectifier

The solid-state equivalent to the thyratron tube is the silicon-controlled rectifier (SCR) (Figure 3-24). The SCR is a four-layer, three-junction solid-state component. It is a rugged, high-power or low-power device. The device has cathode and anode terminals, along with a gate terminal. Positive potential is applied to the anode and negative to the cathode. Current does not flow through the SCR because the center junction is reverse biased. It will conduct when a positive voltage is applied to the gate terminal. This voltage may be in the form of a constant or a pulse voltage. It may be removed once an initial potential is applied. The voltage level applied to the gate is dependent on the values of anode-to-cathode voltages. The level of voltage that turns the SCR on is called *breakover voltage.* When this point has been reached, current breaks across the junctions and flow will continue from cathode to anode. This current is called *holding current.* Once turned on, the gate no longer has control of the SCR. Current will flow until the circuit is broken or the applied voltage is taken to zero. A current-limiting resistance is usually placed in series with the SCR anode for overload protection. The SCR is used for switching circuits and current control, such as motors or light circuits. The SCR package is similar to the diode with three terminals.

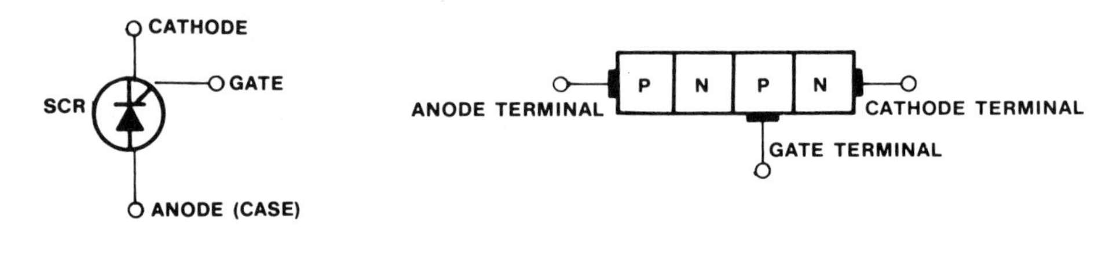

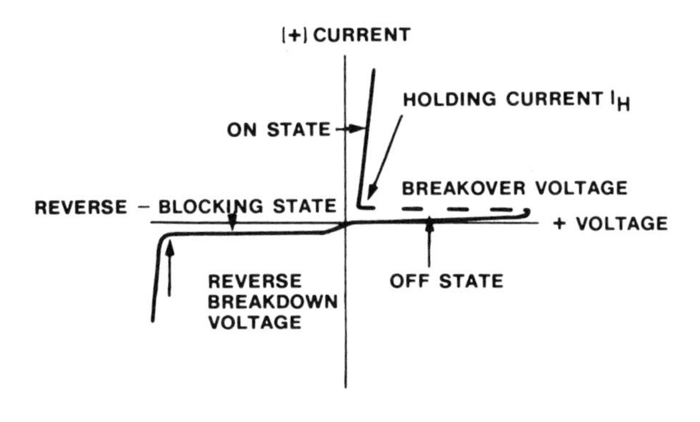

Figure 3-24 Silicon-Controlled Rectifier (SCR)

The Triac

The triac (Figure 3-25) is a back-to-back SCR. It may be connected in a circuit in either polarity. In other words, the triac is bidirectional the same as the diac. A predetermined gate voltage turns the triac on. An ac source of supply is used with the triac because it will operate for both half-cycles of the sine wave. The functional model is shown in Figure 3-25b. The characteristic curve is shown in Figure 3-25c. The triac is used for switching, control of current in motor speed control, and quite often in home light-dimmer circuits.

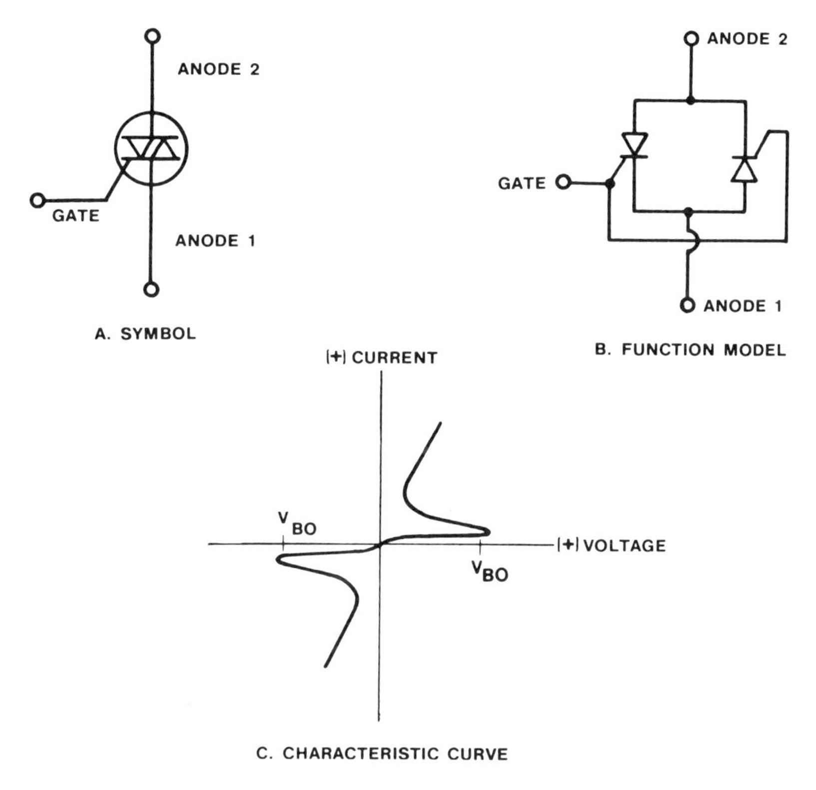

Figure 3-25 Triac

Silicon-Controlled Switch

The silicon-controlled switch (SCS) (Figure 3-26) is a thyristor device that utilizes the gate to control current flow. Note in the figure there are two gates, one to turn the switch on and one to turn it off. If the cathode gate G_C turns the gate on, then the anode gate G_A turns it off. A positive potential provides turn-on and turn-off gating. The SCS has lower power switching than an SCR. They are used in control switching.

Bipolar Transistor

The bipolar transistor is a two-junction, three-layer solid-state device used for switching and amplification. This device is described in detail in Chapter 5.

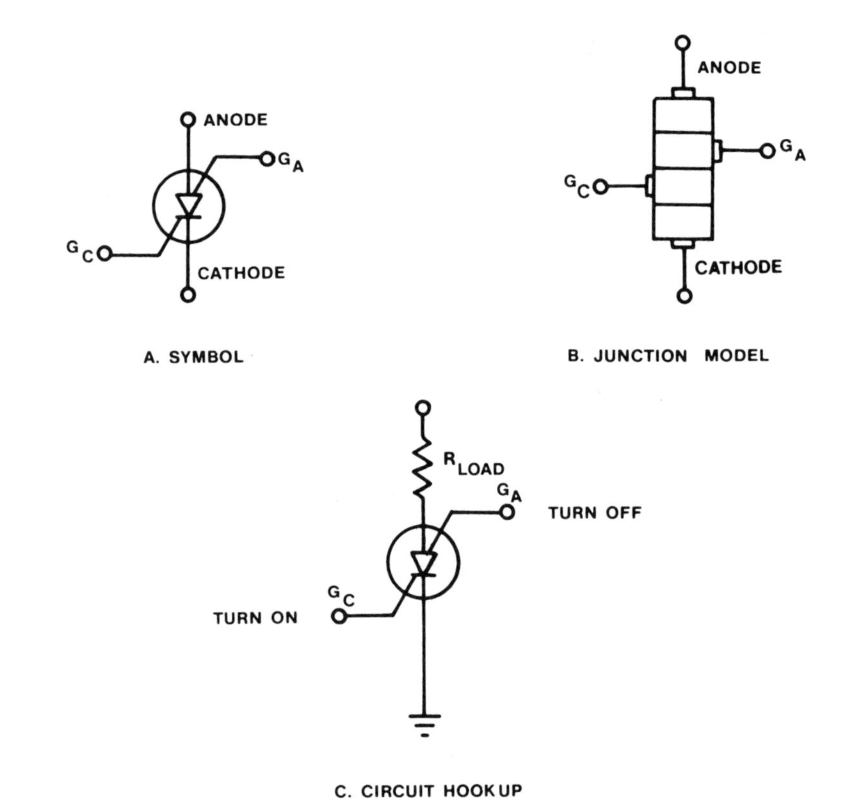

Figure 3-26 Silicon-Controlled Switch (SCS)

Field-Effect Transistors

There are two types of field-effect transistors (FET). These are the junction FET (J-FET) and the metal oxide silicon field-effect transistor (MOSFET). This device is described in detail in Chapter 5.

4

Direct-Current Power Supplies

A power supply is the source of power for operation of electrical and electronic equipment. It may be single, two, or three phase ac. It may be regulated or unregulated dc. The power may have to be converted from ac to ac, ac to dc, dc to dc, or dc to ac.

Ac is converted to another level of ac by means of a step-up or step-down transformer. We described this function in Chapter 2.

Ac is converted to dc by rectification, which we shall cover in this chapter.

Dc is converted to another level of dc by means of a voltage divider. The voltage divider is a series of resistors. Each resistance is chosen to provide a specific voltage drop and current-carrying capability. Dc may also be converted to a higher level of dc by means of a dc motor generator. The unit is driven by dc. The dc motor turns the dc generator, and the dc output may be changed a hundred fold.

Dc is converted to ac by means of an inverter. The inverter is a dc motor, which, in turn, drives an ac generator in the same housing.

In this chapter we shall dedicate the space to conversion of ac to dc. Since all electronic circuits, either digital or linear, require dc power for amplification and switching, this seems logical.

SOLID-STATE RECTIFIER CIRCUITS

Solid-state power supplies have replaced those using vacuum tubes. The reasons for this are obvious: there is no warm-up period or filament requirement for heating, power losses are low, and the newer power packs are smaller. Also, since the state of the art has expanded, prices for them are lower.

Semiconductor power supplies use the silicon rectifier, which is light, reasonably priced, and can be stacked in series to handle higher ac inputs. In fact, the solid-state rectifier can replace the electron tube diode in almost any modern application.

The circuitry for constructing solid-state power supplies is simple. Since no filament voltage is required, one basic power-supply trouble area is eliminated. While it is true that heat can be a problem with solid-state power rectifiers, proper heat sinks reduce this problem to a minimum. Transients during input or output switching or load changes may also cause solid-state unreliability, but the transient problem may be eliminated by proper circuit design.

Half-Wave Rectifier Circuit

The half-wave rectifier is the simplest of all the rectifiers. It consists of a transformer and one rectifier diode. The transformer is a step-down type because power for the circuit usually comes from a 115 or a 220 Vac source. Solid-state circuits are typically biased by low-voltage dc (example 15 Vdc). Therefore, the ac voltage must be transformed to a low level prior to rectification.

Transformers for power supplies are chosen to have a secondary average voltage output that is near the dc level required after rectification. Transformers are also chosen to be able to withstand the current demands of the load.

Diodes are chosen for their current capability and peak inverse voltage rating (PIV). Peak inverse voltage is defined as the maximum reverse-bias voltage that may be accepted by the rectifier. This voltage may vary somewhat with temperature. Actual construction of the power supply requires that the diode be installed on a heat sink to help dissipate some of the heat generated by current flow.

The half-wave rectifier is not very efficient. The reason is that the average dc derived by the circuit is low. In fact, the average voltage is only half that of the full-wave rectifier. Without filtering, the circuit is very inefficient to use under any application. Pulsating dc output ripple is determined by input frequency.

The circuit shown in Figure 4-1 is a typical half-wave rectifier. The transformer secondary has one lead tied to reference (ground). The second lead has a resistor R_S in series with the diode. The resistor is a current-limiting resistance. Its purpose is to limit the current flowing through the diode. If the secondary dc resistance in the transformer is great enough, the resistor is not needed.

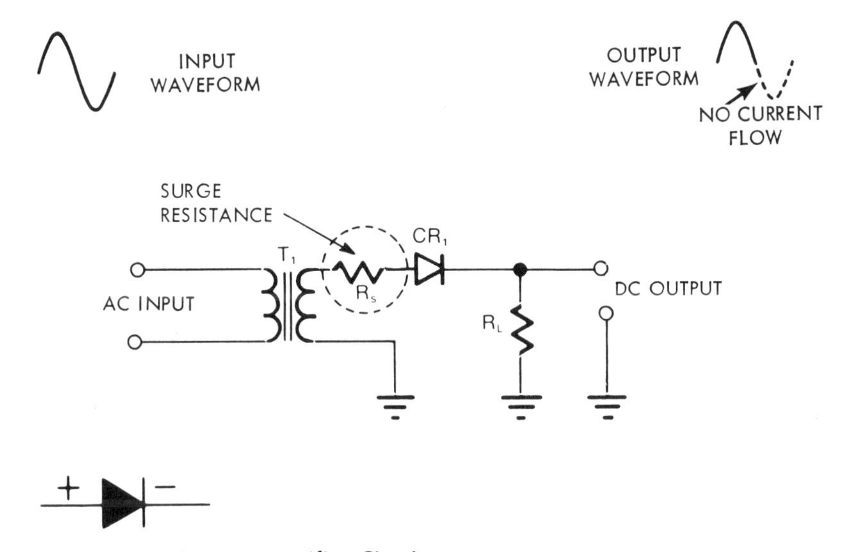

Figure 4-1 Half-Wave Rectifier Circuit

On the positive alternation of input voltage the diode is forward biased, and current flows through the diode and the load. On the negative alternation of the input voltage, the diode is reverse biased, and current stops flowing except for minor reverse current. This arrangement gives pulsating dc output in the positive direction. To obtain a negative dc pulse, the diode is placed in the circuit in the opposite direction.

The symbol on the bottom left of Figure 4-1 represents a forward-biased diode. To reverse bias the diode, polarity must be reversed.

Half-wave rectifiers have an output that is pulsating at the frequency of the primary ac input. The dc output pulse can be measured with an rms (effective) meter. Then the peak voltage output may be calculated as follows:

$$V_P = 1.414\ \mathrm{V_{EFF}}$$

where V_P = peak volts

V_{EFF} = effective volts

Full-Wave Rectifier Circuit

The full-wave rectifier is similar in operation to the half-wave rectifier. The transformer utilized has a center tap. Each leg of the center tap should be capable of an average ac voltage near the level expected of the dc output. The center tap is tied to reference (ground).

Two diodes are used. They are placed in the same direction in series with the transformer secondary winding and series current-limiting resistors. These limiting resistors R_{S1} and R_{S2} are used to limit current through diodes CR_1 and CR_2 if the secondary winding does not have enough resistance.

The full-wave rectifier is much more efficient than the single wave because it utilizes the secondary of the transformer for 360° of the input sine wave. Current flows through one diode for one half-cycle and the opposite diode for the second half-cycle. Transformer and diodes are chosen in the same manner as in half-wave rectifiers. Diodes in full-wave rectifiers must be mounted on heat sinks as in the single-wave configuration. Output of the full-wave rectifier is usually filtered, then regulated. Pulsating dc output ripple is twice the input frequency.

The circuit shown in Figure 4-2 is a typical full-wave rectifier. On the positive alternation of the input sine wave, the diode CR_1 is forward biased and

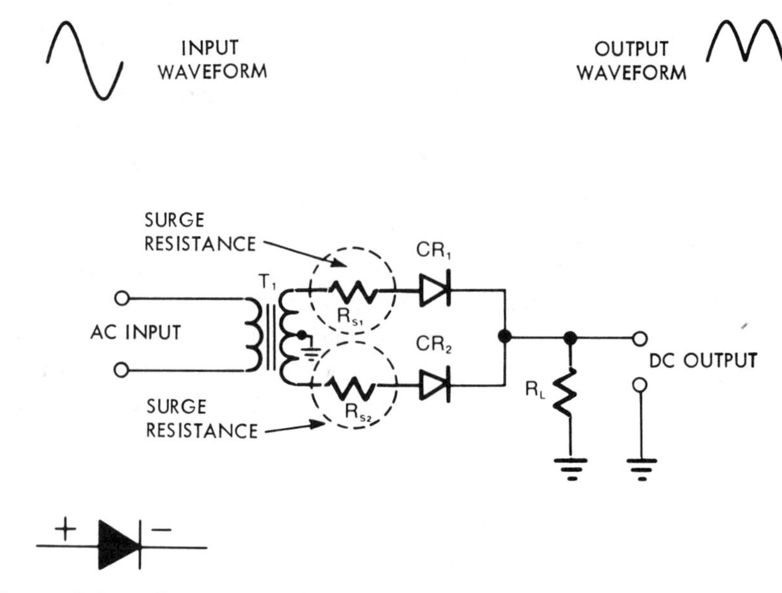

Figure 4-2 Full-Wave Rectifier Circuit

the diode CR_2 is reverse biased. Current flows through the load resistor, the diode CR_1, and the current-limiting resistor R_{S1} back to ground. On the negative alternation of the input sine wave, the diode CR_2 is forward biased and the diode CR_1 is reverse biased. Current flows through the load resistor, the diode CR_2, and the current limiting resistor R_{S2} back to ground.

The full-wave rectifier arrangement provides pulsating dc. The dc power is fed normally through a filter and a regulator. The full-wave rectifier is used in many high-voltage configurations where peak inverse rating is an important factor. In this event a pair of diodes may be placed in series with each winding to prevent rectifier breakdown.

The symbol on the bottom left of Figure 4-2 represents a forward-biased diode. To reverse bias the diode, polarity must be reversed.

Full-Wave Bridge Rectifier Circuit

The bridge rectifier is a full-wave rectifier. It relies on two rectifiers, one on each arm of the transformer, to conduct together on each alternation. The full-wave bridge rectifier does not require a center tap, as does the conventional full-wave rectifier. It conducts during the full secondary output of the transformer. There are four diodes used in the bridge. Each diode is connected in series with and operates at the same time as the diode on the opposite transformer winding.

In Figure 4-3, CR_1 and CR_3 are forward-biased when CR_2 and CR_4 are reverse biased. This arrangement allows the bridge rectifier to require only one-half the peak inverse rating of the conventional full-wave rectifier during conduction. The other advantage is that the bridge rectifier utilizes the entire secondary voltage at all times, making it more efficient.

The circuit shown in Figure 4-3 is a typical full-wave bridge rectifier. On the positive alternation, the diodes CR_2 and CR_4 are forward biased. The diodes

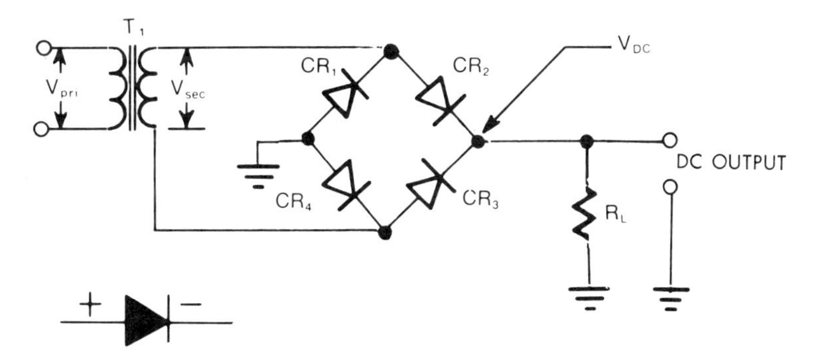

Figure 4-3 Full-Wave Bridge Rectifier Circuit

CR_1 and CR_3 are reverse biased. Current flows from ground through R_L and the diode CR_2 through the transformer secondary, the diode CR_4 and back to ground.

On the negative alternation, the diodes CR_1 and CR_3 are forward biased. The diodes CR_2 and CR_4 are reverse biased. Current flows from ground through R_L and the diode CR_3 through the transformer secondary, the diode CR_1, back to ground. This arrangement provides pulsating dc voltage. Input and output wave forms are identical to the wave forms for the standard full-wave rectifier shown in Figure 4-2.

The dc is normally fed through a filter and a regulator. The bridge rectifier can be designed with a center-tap transformer to provide two load voltages simultaneously. This circuit is seldom used, however, because of the current drain.

The symbol on the bottom left of Figure 4-3 represents a reverse-biased diode. To reverse bias the diode, polarity must be reversed.

Surge Resistance

In all power supplies there must be a method for limiting current through the rectifiers to a safe value. This is accomplished by either of two basic methods. The first method is by placing a surge (current-limiting) resistance in series with the semiconductor diode (rectifier). The surge resistance is selected after taking into account the peak current rating of the rectifier, the load, and other circuit specifics such as filtering or regulation. The second method for limiting current is by using a transformer with enough resistance in its secondary or a choke input filter in series with the load and the rectifier.

In Figure 4-1, a single-wave rectifier power supply is shown with a surge resistance in series with the rectifier. In Figure 4-2 a pair of surge resistors is used in series with the rectifiers in a full-wave rectifier power supply. A single surge resistance may be used in the ground lead at the transformer center tap in place of the other two resistors. In this manner the surge resistance would be common to both rectifier loops. Surge resistors are most commonly used in high-power supplies.

Switching Transients

Some of the failures of rectifiers in solid-state power supplies are due to switching transients caused by induction in the transformer. When currents in the primary of the power supply transformer are suddenly changed, large transient secondary voltages are developed, which result in inverse voltages harmful to the rectifier. In selenium rectifiers the magnetic energy of the transformer can be dissipated in the rectifier, but in silicon rectifiers the reverse current that causes rectifier breakdown is very low.

DC POWER SUPPLY FILTERING

Rectifier Output Ripple Voltage

Ripple voltage in a dc power supply is that peak area of the ac input to the power supply that was not removed by rectification or filtering. This voltage is usually very small (less than 10% of the dc), but in many cases it is necessary to remove it or at least be knowledgeable of how much ripple is present in the supply.

Ripple voltage may be measured with an ac voltmeter tied in parallel to the power-supply output. Ripple voltage is better measured with the use of an oscilloscope. With this instrument an actual picture of the ripple is seen, and fluctuations or distortions in the ripple pattern can be analyzed. A low-capacitance probe should be used in measurements.

In Figure 4-4 the dc output of a single-wave and a full-wave power supply are shown. The reader will note that the full-wave power supply has two times as many ripples as the half-wave power supply. If the input ac frequency for a half-wave rectifier is 60 hertz (Hz), the ripple frequency will have 60 peaks for each second. If the input ac frequency for a full-wave rectifier is 60 Hz, the ripple frequency would have 120 peaks for each second.

Ripple voltage is not measured in peaks, however, but in percentage. Ripple

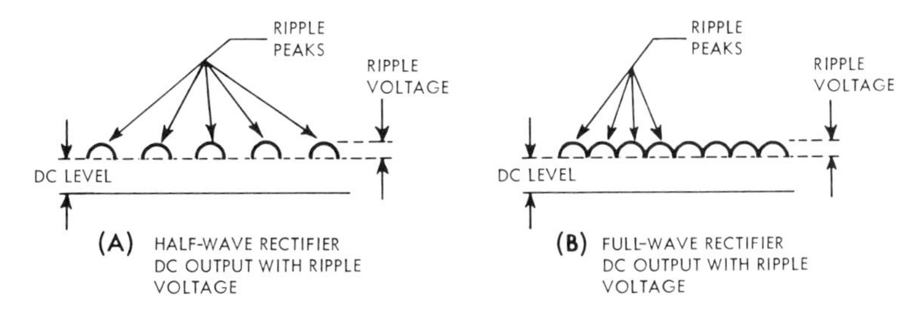

Figure 4-4 Rectifier Output Ripple Voltage

voltage percentage is calculated as a ratio between ac and dc voltage measured in parallel with the power supply; that is,

$$\text{Percent ripple} = \frac{\text{ac volts}}{\text{dc volts}} \times 100$$

Filtering

Filtering is a must for the solid-state dc power supply. The filter comes in many forms and arrangements and is used to smooth out the ripple voltage caused by rectification. Use of the filter is dependent on space, dc ripple requirements, and cost.

If a capacitor is used for filtering, it is placed in parallel with the load. The capacitor opposes a change in voltage, so a constant voltage is seen on the load. The larger the capacitor is, the better the filtering capability. There is a point of diminishing returns, however, where placing a larger capacitor in the circuit does not improve filtering.

The choke (inductor) is used in series with the load. The choke (inductor) opposes a change in current. The choke is seldom used in modern power supplies because it is heavy. Its original purpose was for current regulation. This task has been taken over by the power transistor. Figure 4-5a shows a capacitor used by itself to filter. Figure 4-5b is a pi-type filter. These are the most common filters used in today's power supplies.

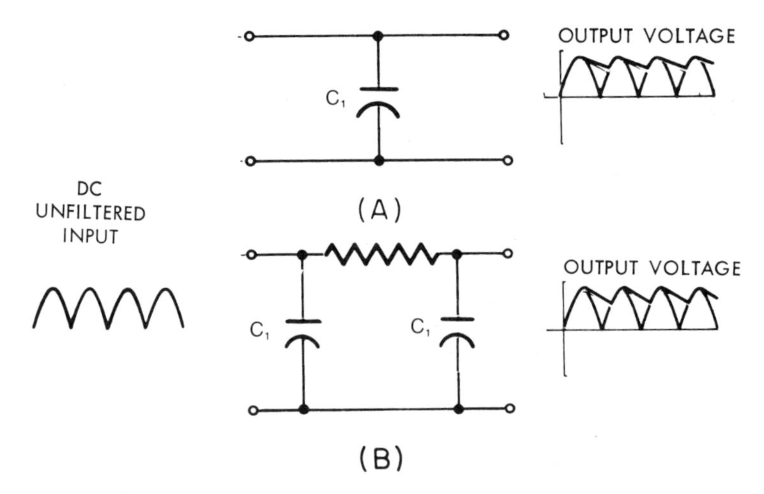

Figure 4-5 Filtering

Other arrangements are the L-filter, using one choke in series and one capacitor in parallel; the T-filter, using two chokes in series and a capacitor between them in parallel; and the pi filter, using two capacitors in parallel with a choke in series between them. Each does a respectable job of opposing current or voltage change. The choke filter has a disadvantage in that output voltage tends to be lower than with capacitor filters.

Ripple-free dc is demanded in many solid-state circuits. Ripples in power may cause switching transistors in computers to fire accidentally. Ripple dc can cause noise in audio and radio circuits. In some circuits where exact dc levels are required, ripple voltage may even cause some solid-state devices to be destroyed.

DC POWER-SUPPLY REGULATION

Regulation Percentage

The percentage is a specification provided with the supply. It describes the capacity of a power supply to provide a constant output with changes in load. This specification is calculated as follows:

$$\text{Regulation } \% = \frac{V_{NL} - V_L}{V_L} \times 100\%$$

whcre V_{NL} — no-load voltage
V_L = load voltage

A high percentage of regulation is usually undesirable, while a low percentage of regulation is the ideal. Load current determines voltage level at the power-supply output. If the load current is great, the problem of voltage regulation is also great. With a regulator, this loading can be compensated for.

Zener Diode as a Voltage Regulator

The most basic form of regulator is the zener diode regulator, mainly because it is simple to understand and use. Since its conception, it has been utilized alone and combined with other more complex regulator circuits. It has become the workhorse of the solid-state power-supply field of design.

The zener diode is closely related to the diode rectifier. Its purpose, however, is to act as a voltage regulator, either by itself or with a transistor-regulated power supply. The reason for its regulatory ability is its unique construction. The zener diode is constructed to operate in reverse of the normal conditions present in rectifier diodes. This reverse condition is made possible by carefully planned doping conditions when the diode is manufactured. The zener diode is basically a breakdown diode. It is installed so its bias is the reverse of the diode rectifier.

In Figure 4-6 the current–voltage characteristic curve of a zener diode is shown. This particular curve shows normal diode operation with the zener diode forward biased. In the reverse condition, the point at which the diode breaks down is called the zener *breakdown voltage* or the *avalanche breakdown*. As the voltage against the diode barrier approaches this zener point, current-carrying electrons tunnel through the barrier, causing the zener effect. At the same time, other electrons accelerate, knocking others loose from their valence rings in a process called *carrier multiplication*, causing the effect called *avalanche*. At that time the diode is said to be at *breakdown*. At this voltage point the current may increase greatly while the voltage remains the same. It is this reason that makes the breakdown diode a perfect voltage regulator.

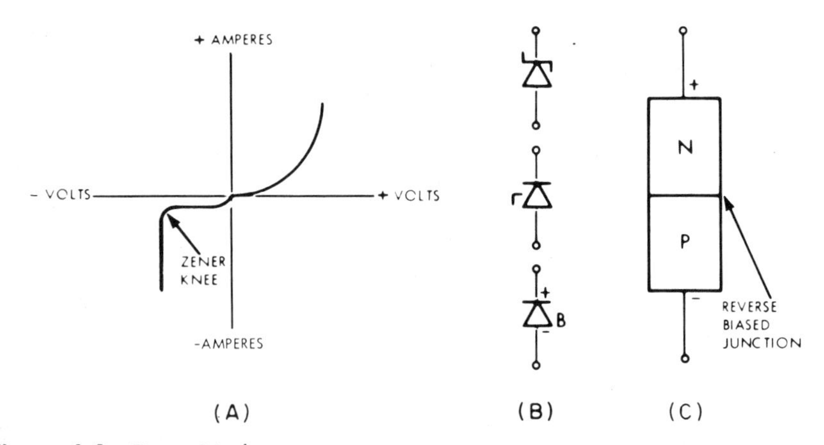

Figure 4-6 Zener Diode

In actuality, the zener point and the avalanche point occur at different levels. The avalanche occurs slightly after the "knee" of the zener. For all practical purposes, the two effects can be said to happen together. Figure 4-6b illustrates typical zener and breakdown diode schematic symbols. Figure 4-6c is a functional diagram of the zener diode. A zener diode is built into a package similar to normal rectifiers.

The zener diode is sometimes used alone for voltage regulation. In such cases, a limiting resistor R_S is always placed in series with the load and the shunted zeners. In this way the unregulated output is protected from overload current (see Figure 4-7).

The power rating of the limiting resistor R_S must be very high because it carries the load current and the current through the zeners. This makes the efficiency of the shunted zener regulator very low. Its simplicity makes up for its inefficiency. The zener diode is chosen for power-carrying capacity and its voltage setting. Large-current zeners require heat sinks.

Figure 4-7 shows the limiting resistance R_S in series with two zener diodes CR_1 and CR_2 and two loads. The zeners are regulating two load voltages at set levels. If the unregulated dc supply is large enough, it is possible to place several zeners in series with each other and provide multiple pickoffs similar to those illustrated.

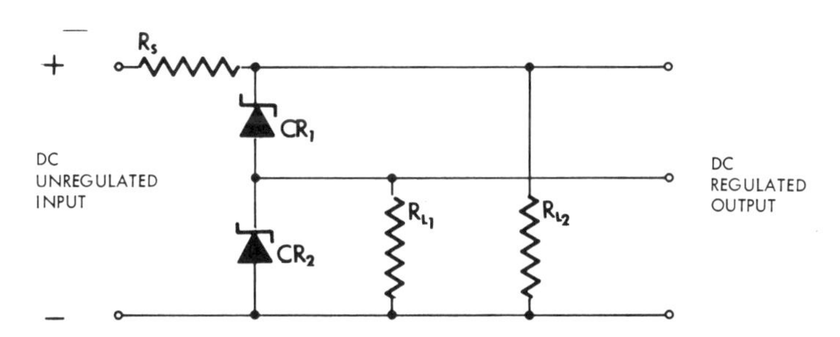

Figure 4-7 Dual Output Zener Diode Voltage Regulator

Transistor Shunt Regulator

The shunt regulator (Figure 4-8) consists of a current-limiting resistor R_1, a zener (voltage regulator) diode CR_1 and a power transistor Q_1. The transistor used in the illustration is a *PNP* power transistor. It is chosen for its current-carrying capabilities. The power transistor here must be placed on a heat sink as in the series regulator circuit. The zener diode is installed in the base circuit to provide a reference for bias. Bias for the base–emitter junction is provided by the load voltage and current-limiting resistor R_1. Note that the transistor is placed in the circuit in shunt (parallel) with the load. The moment a change is felt, regulation takes place.

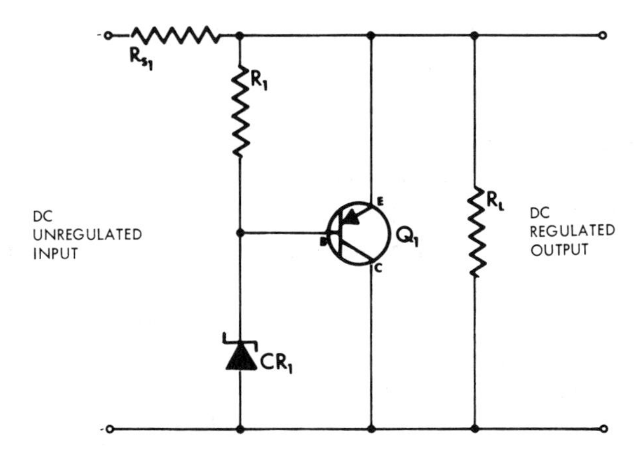

Figure 4-8 Transistor Shunt Regulator

Transistor Series Regulator

The series regulator (Figure 4-9) consists basically of a power transistor Q_1, a current-limiting resistor R_1, and a zener (voltage regulator) diode CR_1. The transistor used in the illustration is a *PNP* power transistor. It is chosen for its current-carrying capabilities. The power transistor must be placed on a heat sink when installed in the power regulator circuit. The heat sink is made of soft metal that absorbs heat. The zener diode on the base of the regulator, along with the load voltage, provides the base–emitter bias to control the transistor's operation. Note that the transistor is placed in the circuit in series with the load.

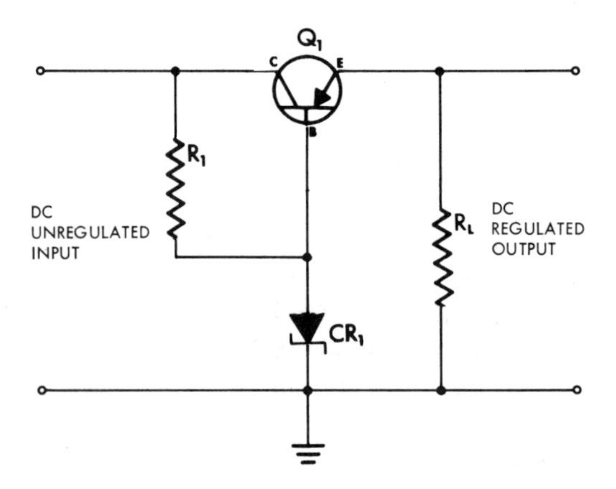

Figure 4-9 Transistor Series Regulator

Bias for the transistor Q_1 is provided by the load voltage and the zener diode CR_1. The moment a change is felt, regulation takes place. If you were to put a meter across the load, it would be difficult to impossible to see regulation happening.

A COMPLETE DC POWER SUPPLY

A full-wave rectifier power supply is illustrated in Figure 4-10. Other types can be used. For instance, a bridge rectifier could have been inserted left of V_{DC}.

The full-wave rectifier is used with a center-tap transformer (*T*1). It is, along with the bridge rectifier, the most used of the power supply types. The full-wave rectifier uses both alternations of the ac sine-wave input and provides efficient dc output.

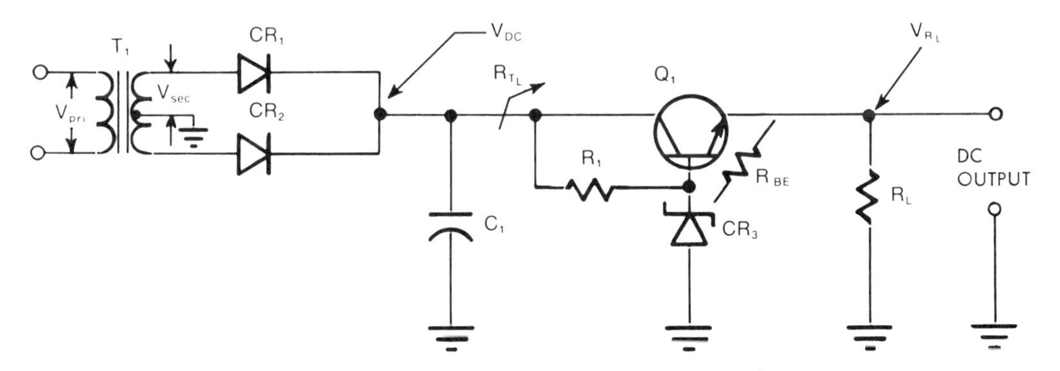

Figure 4-10 A Complete dc Power Supply

The full-wave rectifier power supply consists of a center-tap transformer (*T*1), a pair of rectifier diodes (*CR*1 and *CR*2), a filter capacitor (*C*1), a voltage-divider resistor (*R*1), a power transistor (*Q*1), and a zener diode regulator (*CR*3). Figure 4-10 is a circuit diagram for this type of power supply.

The diodes *CR*1 and *CR*2 are rectifiers and change the alternating current to pulsating direct current. The capacitor *C*1 is a filter. This device opposes a change in voltage and tends to flatten the ripple peaks of the pulsating direct current. The voltage-divider resistor divides the voltage from Vdc to the base of the transistor. The power transistor regulates current. The zener diode keeps a constant voltage for bias for the transistor.

There are, of course, much more complex power supplies. For instance, the shunt voltage regulator shown in Figure 4-8 could have been included. Or a more complex filter could have been included.

Choosing Power-Supply Components

The choice of power-supply components must be left up to the designer. Many components are available and the choices are not really as much a science as an art. One must choose what is available, and choices are usually concerned with cost. Whatever your personal requirements, the basic guidelines are set forth in the next several paragraphs.

Choosing a Transformer Power transformers are cataloged by manufacturers' specifications as to their volt-ampere (VA) ratings. Since power in the primary of the transformer is equal to power in the secondary of the transformer, determinations of primary and secondary voltages are selective. The only concern in the primary voltage is that the transformer primary match the available source ac voltage. The secondary ac voltage in power-supply transformers varies under standard values from 6 to 30 Vac (and more in some cases).

The dc voltage that can be derived from a power transformer is from 1.25 to 1.4 times the secondary rms voltage. This value will vary depending on the type of power supply and the method of filtering. The power rating of the secondary should be 1.25 to 1.4 times the dc voltage out of the rectifiers. In some cases it may be desirable to have a larger power (VA) rating.

Choosing Rectifier Diodes Diode specifications are written for the forward-biased diode and reverse-bias conditions. Forward characteristics that are specified are forward voltage drop and forward current. Reverse characteristics are maximum reverse voltage (peak inverse voltage) and reverse current.

Forward voltage drop is the amount of voltage that is dropped under a normal forward-bias condition. This voltage drop will vary up to around 0.6 V.

Forward current is the maximum current that can flow in the diode without damaging it. Normal diode forward current-carrying capability would be about twice the output dc current.

Maximum reverse voltage (peak inverse voltage) rating is the amount of reverse-bias (voltage) the diode can withstand before it breaks down. In design, a diode (rectifier) should be chosen that withstands twice the output dc voltage of the power supply.

Reverse current (leakage current) is a small amount of current that flows in reverse when the diode is reverse biased. This current should be as small as possible and, for the purpose of design, less than 1 mA.

Choosing a Capacitor Filter The capacitive input filter is the basic filter used in solid-state power supplies. In most cases it is the only type of filter required. It is always used in circuits where good regulation is not a requirement and in high-voltage applications. Capacitive inputs are recommended over choke input filtering because of simplicity. Values of capacitance are dependent on load current, ripple frequency, and ripple percentage required.

Load Current. Load current is the direct current (dc) that flows through the load. This current should not be misconstrued to be effective ac voltage (measured with an ac meter). Furthermore, it should not be related to peak current through power-supply rectifiers. The load current is the average value of dc current flowing through the load. Load current and load voltage are necessary to calculate load resistance (R_L), which, in turn, is used to calculate filter capacitance values. The load resistance formula is as follows:

$$R_{\text{LOAD}} = \frac{V_{\text{DC LOAD}}}{I_{\text{DC LOAD}}}$$

Ripple Frequency. The ripple frequency of a dc power supply for half-wave rectification is the same as the cycles per second (hertz). That is, a half-wave 60-Hz power supply would have a 60-Hz ripple frequency. Ripple frequency for a full-wave rectifier power supply would have a ripple frequency equal to the total negative and positive alternations. That is, a full-wave 60-Hz power supply would have a 120-Hz ripple frequency. Ripple frequency is represented by the symbol f_r.

Ripple Percentage. Output ripple is measured or calculated in percent of ripple. This output ripple may vary greatly from near 100% in half-wave rectifiers to as low as 2% or 3% in three-phase power-supply connections. Ripple percentage (r_p) is the ratio of the rms ripple voltage at the rectifier output to the dc voltage across the load. Ripple percent is calculated using the following formula:

$$\text{Ripple percent } (r_p) = \frac{V_{\text{rms}}}{V_{\text{DC}}} \times 100$$

The following table is a list of capacitance values for a 15-Vdc, full-wave, 1-A power supply with various ripple percentages.

Ripple (%)	*Capacitance (μF)*
1	12,500
2	6,250
3	4,166
4	3,125
5	2,500
10	1,250

Note: These are approximate values. Availability of standard sizes will prevail.

Choosing a Choke Input Filter The choke input filter is very effective in reducing ripple in power supplies designed for high-current applications. The penalty is weight, which is usually objectionable. The choke input filter is selected on the basis of the amount of current to be passed and the inductance (henrys) required to filter the power-supply ripple to an acceptable level. Typically, for most filter applications, 50 to 100 millihenries (mH) of inductance will provide satisfactory filtering. The limitation on the current passed is the point where the choke core becomes saturated and the ac impedance is reduced to the dc resistance of the winding. Additionally, winding wire size should be considered for proper rating. To ensure that the filter is operated within limits, the manufacturer's specifications should be consulted.

Choosing a Zener Diode Regulator The basic demand of a zener diode regulator is to choose a zener value equal to the regulated voltage need of the load. Other requirements are power and current capacities. Zener voltages range from very small, around 1.0 V, to very large, around 28 V. The zener voltage is the avalanche voltage at which the diode breaks down in a reverse-bias situation.

Depending on the severity of load changes, the zener diode will regulate voltage to the load at the rated value ±10% to 20%. If the voltage in reverse bias stays below the breakdown point, the zener diode will not go into avalanche; therefore, no regulation will be present. Furthermore, if the reverse-bias voltage lowers to a point 25% or so below the zener point, regulation will not be maintained. (This point varies with the zener diode and design application.)

Zener diodes normally have low current ratings, under 250 mA, although higher current ratings are available. Proper size heat sinks are required with high-current rated zeners.

Power capacity is sometimes given as maximum power dissipation. The maximum power rating is derived by multiplying the zener voltage by the maximum current rating.

Choosing a Power Transistor for a Regulator A transistor for a power-supply regulator is chosen in a similar manner to any circuit that has specific current and/or power requirements. The transistor must be able to withstand at least twice the peak power load that may be encountered. Most power supplies that have regulators are large enough so that they also must have heat sinks on which to mount the transistor.

If the transistor used is made of silicon, the voltage drop of the base–emitter junction is around 0.6 Vdc. This, of course, is simply a rule of thumb. Each manufacturer provides curves that depict this base–emitter voltage drop at different temperatures. However, in the case of power supplies the curves do not normally have such exact requirements.

The current gain (beta) of the transistor used should be from 50 to 100. This is also very flexible, and most power transistors have the ability to do the job.

Choosing a Heat Sink Heat sinks are rather complex components to deal with. Basically, the heat sink must be able to dissipate enough heat to prevent the transistor or the diodes from burning up. Power transistors and diodes are usually installed with recommended heat sinks. There are nomographs for choosing the heat sinks. These are also given in transistor/diode specifications. Since this is a complex determination, it is recommended that heat sinks be used as directed by these specifications. This practice will ensure a tested and tried design.

If expense is a problem, choose a finned heat sink as large as can be used within reason. The area of the heat sink, along with the material the sink is made from, are the basic factors to be considered. Again, on-shelf, tested and tried heat sinks are recommended.

Heat Sinks

Probably the foremost factor contributing to rectifier failure is heat. Any time current flows through a semiconductor, heat is generated. Since rectifiers in power supplies must handle large currents, it is expected that large amounts of heat will be developed. Therefore, there must be some method of removing the heat from the semiconductor. It is obvious that heat cannot be eliminated, so the next best thing to do is to dissipate it into the air. This is accomplished by the use of heat sinks, which are essentially miniature radiators made of thermally conductive metal. Figure 4-11 illustrates typical heat sinks.

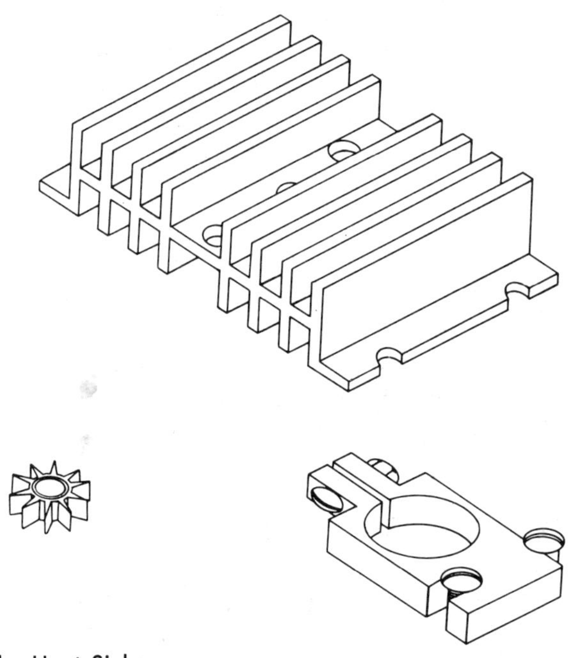

Figure 4-11 Heat Sinks

Each diode (rectifier) is rated as to its maximum *PN*-junction temperature rating. Design of the heat sink should provide enough cooling so the junction temperature never reaches this maximum. In actuality, it is recommended that temperatures never reach more than half the maximum.

The other factor in selecting a heat sink is its thermal resistance. Thermal resistance is calculated in terms of degrees celsius divided by watts (or °C/W). Heat is transferred from the junction to the rectifier case and is designated θja or θjc by the manufacturer's specifications. From the case, heat moves into the heat sink. This heat transfer or path from case to heat sink is designated θcs. Finally, the heat transfer or path from heat sink to ambient air is designated θsa. The total thermal resistance is the total of these factors or $R_{\text{THERMAL}} = \theta ja \pm \theta cs \pm \theta sa$. Manufacturers' specifications provide the thermal resistances of their heat sinks. These should be accepted rather than attempting to calculate the thermal resistance oneself.

Between the rectifier and the heat sink, an insulator made of mica or anodized aluminum should be installed. Heat sinks should be chosen by their cooling capability and the size or shape of the container that is required to contain them.

POWER SUPPLY LOADING CONTROL

Regulators are used at the supply output to give constant voltage for approximately 30% load variations. This is desirable to prevent output voltage fluctuations due to load changes or input voltage variations. Additionally, the regulators serve to reduce ripple to a negligible level. This is possible because the emitter of the regulation transistor follows the base, which is zener controlled. Therefore, voltage fluctuations at the collector are not seen at the emitter. If a power supply is designed without a regulator at the output, the ripple level and voltage output will vary in proportion to the load applied.

5

Discrete and Integrated Amplifiers

In a paper written by John R. Ragazzini entitled "Analysis of Problems in Dynamics," the words *operational amplifier* were coined. The article was printed by the IRE in May 1947 and is a classic in electronics. The paper defined the work of the National Defense Research Council (1943/1944) and described the properties of the op amp. It also covered the work of a technical aide, George A. Philbrick, during these times. Philbrick helped develop the first plug-in operational amplifier using vacuum tubes.

The first modular solid-state operational amplifiers were introduced by Burr-Brown Research Corporation and G. A. Philbrick Researches, Inc., in 1962. Since that time the op amp has been the workhorse of linear systems.

The first op amp that was provided to the public (and accepted by the public) was Fairchild Semiconductor's μA702. The μA702 was produced in 1963. The μA702 had ridiculous supply voltage requirements (+12 and −6 Vdc). The μA709 introduced the next major change in op-amp design. It was fabricated and issued in 1965 by Fairchild Semiconductor. It had higher gain, larger range, and a more acceptable power-supply requirement (±15 Vdc).

National Semiconductor created the LM 101 in 1967. It had gain to 160K, short-circuit protection, and simplified frequency compensation (external). National Semiconductor provided this later internally with the hybrid LH 101. In December 1968, the LM101A was devised. This device provided better input control over temperature and lower offset currents. National Semiconductor also introduced the LM107, which had the frequency compensation capacitor built into the silicon chip. The LM107 came out at the same time as the LM101A.

In 1968, Fairchild Semiconductor issued the μA748. The device had essentially the same performance characteristics as the μA741. The difference was external frequency compensation.

The first multiple op-amp device was Raytheon Semiconductor's RC4558 in 1974. Later in 1974, the LM324 quad op amp became public. This is a National Semiconductor device. The LM324 is especially useful for low-power consumption. The most useful attribute of the LM324, according to some engineers, is its single power-supply requirement.

The first FET input op amp was the RCA CA3130. With this addition to the op-amp family, extremely low input currents were achieved. Its power can be supplied by a 5- to 15-Vdc single-supply system.

In July of 1975, National Semiconductor announced the LF355. This was the first device created using ion implantation in an op amp.

Texas Instruments introduced the TL084 op amp in October 1976. It is a quad JFET input op amp; it also is an ion-implant JFET. Low bias current and high speed are two of its attributes.

Since 1976, the types of op amps have increased almost daily. Today we have a variety of op amps that will provide the user with essentially anything needed, such as high common-mode rejection, low-input current frequency compensation, and short-circuit protection. The designer simply expresses a need and is supplied with the correct type.

The reader may wish to look further into the history of transistors and integrated circuits. The author recommends that this information be requested from the various major contributors to solid-state devices. You may start with the Historical Division at Bell Laboratories.

In this chapter we shall look at the functions and operation of the amplifier, transistor, FET, and operational amplifier. You must remember that this will be a cursory look, for each of these subjects is vast.

BIPOLAR TRANSISTOR

The bipolar transistor (Figure 5-1) is a two-junction, three-layer, solid-state device constructed on a single crystal. It has two basic functions, switching and amplification. Switching functions are used in digital circuitry. Amplification is performed in linear and/or analog circuitry. The transistor uses both *N*- and *P*-type silicon. A thin layer of *P*-type material is sandwiched between thick layers

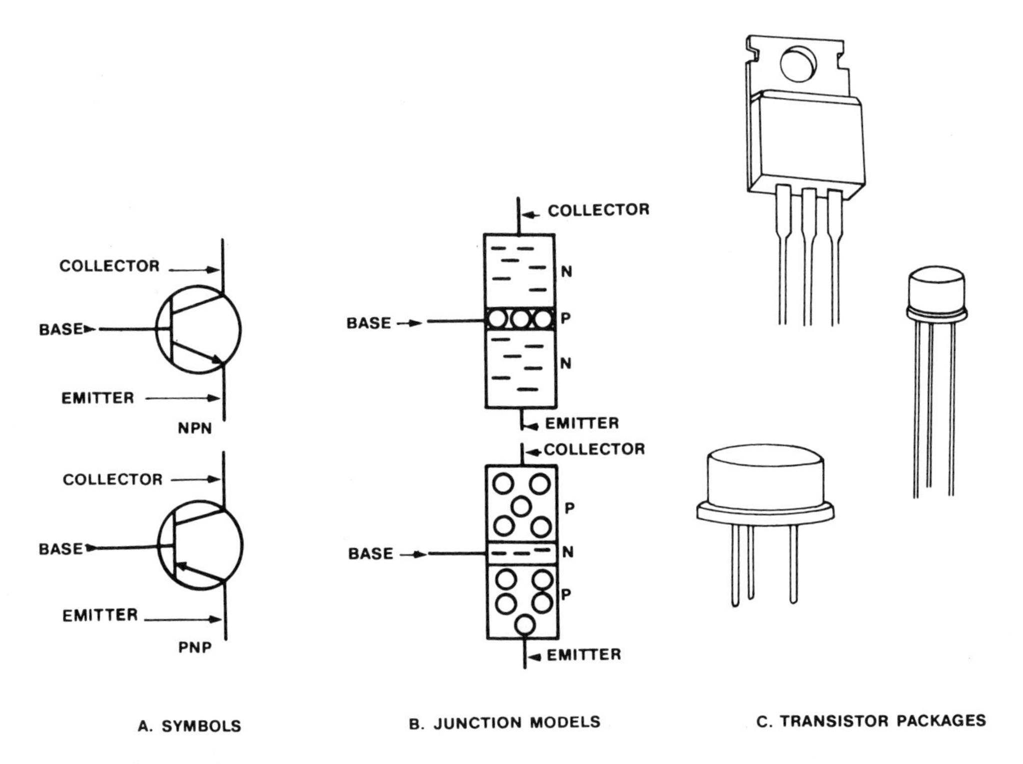

Figure 5-1 Transistor

of *N*-type material to form an *NPN* transistor. A thin layer of *N*-type material is sandwiched between thick layers of *P*-type material to form a *PNP* transistor.

The transistor has three terminals. The thick layers on the ends are *collector* and *emitter*. The thin layer in the center is the *base*. The base is extremely thin, around 1 micrometer (μm) or smaller. The shape of the transistor package is usually dependent on the amount of current flow it can handle, along with the application.

The transistor has for many years been the mainstay of the electronic field for amplifying. Since the development of the operational amplifier integrated circuit, the transistor amplifier has taken a far second place in amplification. The author believes, however, that the solid-state discrete amplifier will be around for several more years. Therefore, some basic transistor amplifier fundamentals should be explained.

Amplifier Configurations

There are three basic transistor amplifier configurations. These are the common emitter (CE), common base (CB) and common collector (CC). The three configurations are illustrated in the Figure 5-2. Transistor leads are labeled B for base, E for emitter, and C for collector. The configurations have been drawn with all the external parts intentionally omitted to improve the clarity of this explanation.

The first configuration is the common emitter (CE). The reason that it is called a common emitter is that the input signal is placed between the base with emitter reference (ground) and the output is taken from the collector with emitter reference (ground). The emitter is common to both the input and output signals; therefore the circuit is titled common emitter (CE). The signal output is out of phase with the input signal.

The second configuration is the common collector (CC). The reason that it is called a common collector is that the input signal is placed between the

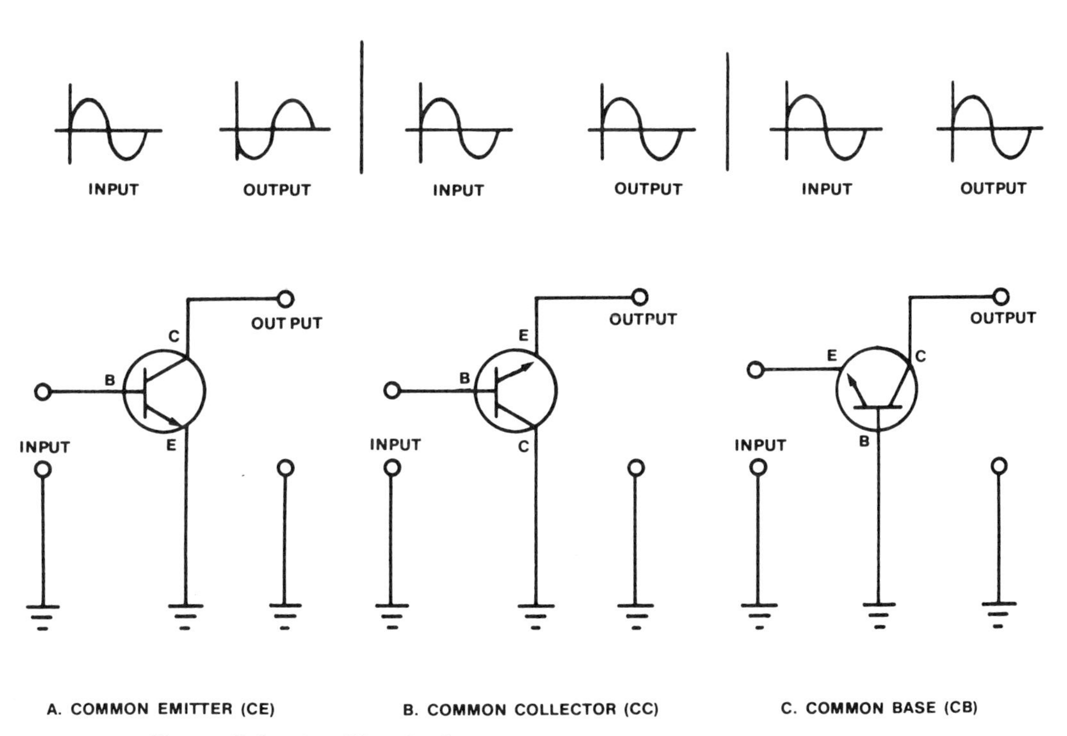

Figure 5-2 Amplifier Configurations

base with collector reference (ground). The collector is common to both the input and output signals; therefore, the circuit is titled common collector (CC). The signal output is in phase with the input signal. This configuration is often called an *emitter follower* because the signal output is felt in parallel with an emitter resistor.

The third configuration is the common base (CB). The reason it is called a common base is that the input signal is placed between the emitter with base reference (ground). The base is common to both the input and output signals; therefore the circuit is titled common base (CB). The signal output is in phase with the input signal.

Current Flow in Amplifier Configurations

Each amplifier configuration has three basic paths of direct current flow. The quantity of current flow is determined by the design of the configuration and the current gain (β) of the transistor chosen. However, there is a fundamental relationship of the currents. They are illustrated in the configurations shown in the Figure 5-3. Again the configurations have been drawn with all the external parts intentionally omitted to improve the clarity of the explanation.

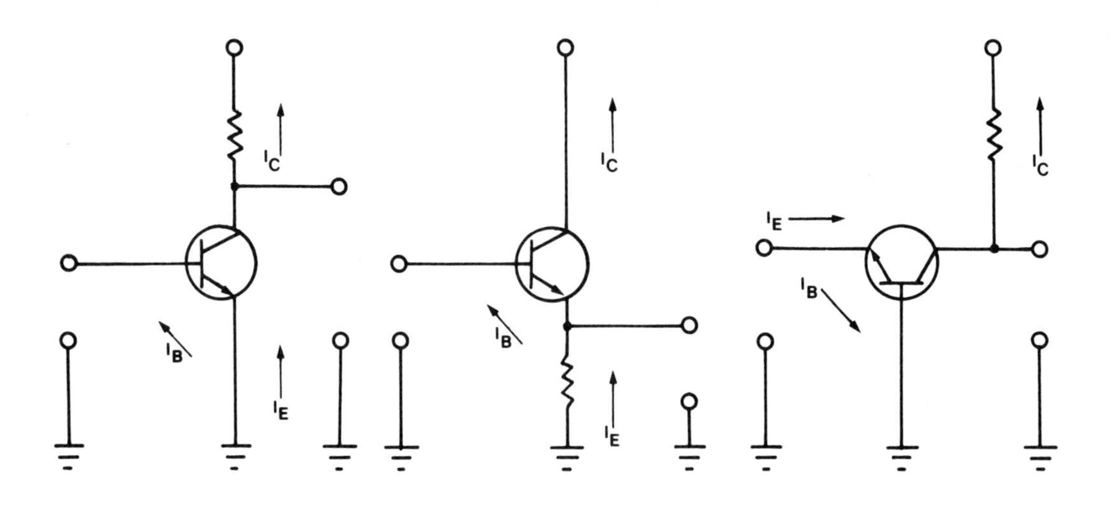

Figure 5-3 Current Flow in Amplifier Configurations

The configurations are, from left to right, the common emitter (CE), common collector (CC), and common base (CB). In each configuration there are three direct currents: emitter current (I_E), base current (I_B), and collector current (I_C).

Regardless of the configuration, the direct currents follow a predictable path and a simple equation:

$$I_E = I_B + I_C$$

Furthermore, the path (though simplified) for alternating current (signal) is the same:

$$i_E = i_B + i_C$$

The signal current is superimposed on the direct current during amplification. When the transistor is not amplifying, the direct currents flow at static levels because there are no signals present.

The base current, whether direct current or signal current, can be calculated as follows:

$$I_B = \frac{I_E}{\beta}$$

$$i_B = \frac{i_E}{\beta}$$

where β = beta = the current gain of the transistor.

Transistor Biasing Techniques

Each amplifier must abide by biasing rules. Biasing is the establishment of direct-current operating levels. The basic two rules for biasing are thus (1) the collector (output) junction must have reverse bias and (2) the base emitter junction must be forward biased. Let's consider several amplifier situations. In Figure 5-4 three configurations are shown; the CE, CC, and CB. Most of the external parts have been removed for clarity of explanation. *NPN* transistors are used in all three examples.

In the Figure 5-4, the common-emitter amplifier has +9 Vdc applied across a resistor to reverse bias the *P*-type material on its collector. The +5 Vdc applied to the base forward biases the base (*P*-type material) because the base must be positive (+) in relation to the emitter.

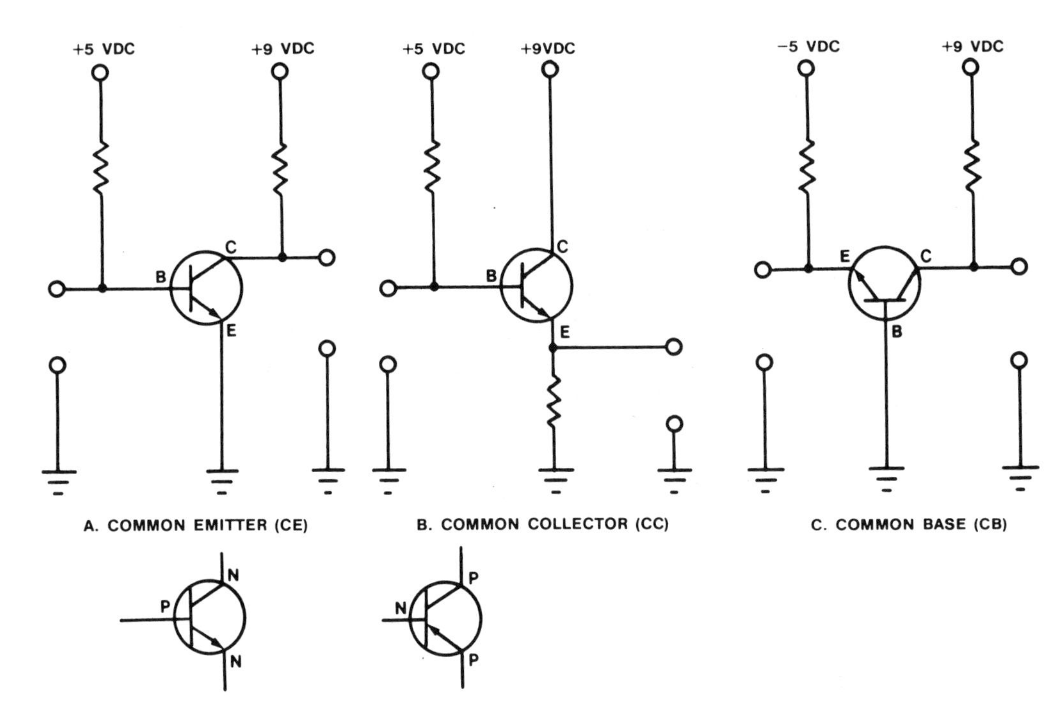

Figure 5-4 Transistor Amplifier Biasing

The second configuration is a common collector. There are +9 Vdc at the collector (*P*-type material), providing reverse bias. The base, as with the common emitter, has +5 Vdc applied, which makes it positive in relation to the emitter (*N*-type material). Thus the base emitter is forward biased.

The third configuration is a common base. As in the other two configurations, the collector is reverse biased with the +9 Vdc. The base–emitter junction is forward biased with −5 Vdc applied to the emitter.

Below the configurations, the *NPN* and *PNP* transistors are labeled to define the solid-state material in their construction. If a *PNP* transistor were used in any of the configurations, the polarities of applied voltage would have to be reversed.

Transistor Amplifier Thermal Runaway

Most problems that transistors have stem from an action called thermal runaway (see Figure 5-5). At the collector, power is dissipated into heat, which causes collector current to increase. An increase in collector current will cause forward junction bias to increase, further increasing collector current. This action is cumulative, thereby causing still more heat and finally thermal runaway and transistor destruction.

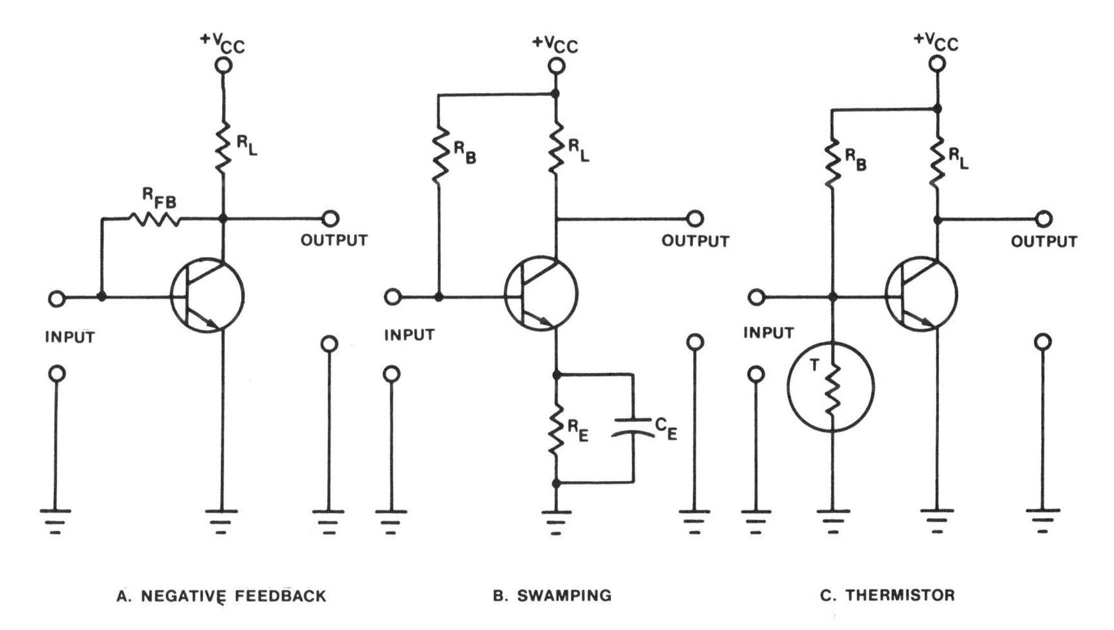

Figure 5-5 Transistor Amplifier Thermal Runaway Bias Stabilization

The problem is not impossible to handle. There are two methods for preventing thermal runaway. One method is to ensure that the transistor is operated well within the limits suggested by the manufacturer. The other method is by bias stability.

In Figure 5-5, three types of bias stability are illustrated. Figure 5-5a shows negative feedback. This method returns some of the output signal to the base, reducing gain and thereby reducing thermal runaway. In Figure 5-5b, a swamping resistor is placed into the emitter circuit. This resistance provides the larger part of the base–emitter input resistance, thereby reducing the problem of thermal runaway. Finally, the thermistor (another solid-state device) is placed in the base circuit of Figure 5-5c. The thermistor changes resistance with temperature change, thereby holding a constant bias on the base.

The best configuration for thermal stability is the common base, but this configuration may not be able to meet the requirements of the amplifier as to gain, input and output impedance, or phase inversions.

Common-Emitter (CE) Amplifier Configuration

The CE amplifier configuration is the most used of the amplifiers because it has high voltage and current gain (therefore, high power gain). The CE amplifier has a phase reversal between input signal and output signal. For this reason, two stages are often used. If phase is not a problem, the configuration is ideal.

Figure 5-6 is a typical common-emitter amplifier. The input signal is fed into the base in reference to the emitter. The output signal is taken off the collector in reference to the emitter.

Let's consider the dc circuit analysis first. Positive V_{CC} supplies reverse bias to the collector. The point Y in reference to ground is approximately one-

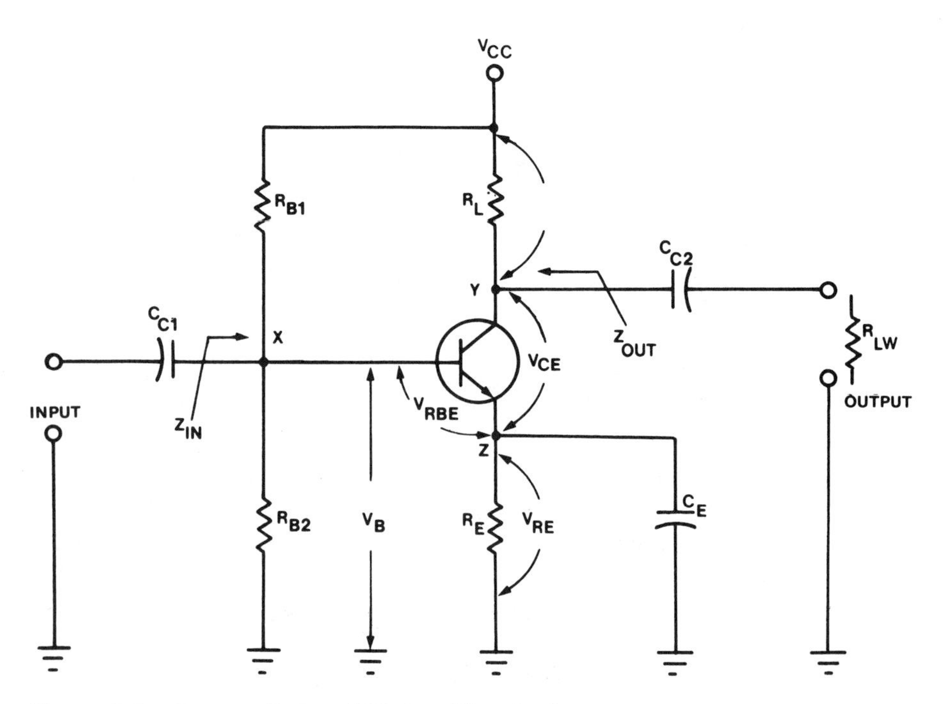

Figure 5-6 Common-Emitter (CE) Amplifier Configuration

half the dc voltage V_{CC}. The dc voltage drops on the output leg of the amplifier sum up to applied voltage:

$$V_{CC} = V_{RL} + V_{CE} + V_{RE}$$

R_{B1} and R_{B2} are a voltage divider for forward bias on the base–emitter junction. Bias voltage V_B is equal to the base–emitter junction voltage V_{RBE}, plus the emitter voltage V_{RE}. Bias voltage V_B can also be calculated using the voltage-divider formula in the following manner:

$$V_B = V_{CC}\frac{R_{B2}}{R_{B1} + R_{B2}}$$

The base–emitter junction voltage is approximately 0.6 Vdc if the transistor is made of silicon and 0.2 Vdc if the transistor is made of germanium. These values are approximates and are dependent on and vary with temperature.

The coupling capacitors C_{C1} and C_{C2} are charged to dc voltages at points X and Y, respectively. The signal is placed on the input X in relation to ac ground. Since the emitter resistor R_E is bypassed by the capacitor C_E, the point Z is essentially ac ground. This means that the signal will be felt only across the base–emitter junction. Junction dynamic resistance is calculated as an approximation using the following formula:

$$r_{be} = \frac{25}{I_E}$$

where 25 = Shockley constant

I_E = dc operating current in milliamperes

This formula was developed by William Shockley to present an estimate of dynamic base–emitter resistance for analysis purposes.

Once the dynamic resistance r_{be} of the base–emitter junction has been estimated, the input impedance can be calculated. Input impedance is an arrangement of parallel paths to ac ground looking into the amplifier at point X.

$$Z_{in} = R_{B1} \parallel R_{B2} \parallel \beta r_{be}$$

where β = current gain of the transistor.

Output impedance Z_{OUT} is the ac ground path looking back into the amplifier at point Y. Since there is but one path to ground for ac (signal), the output impedance is simply equal to R_L.

The voltage gain A_v of the amplifier is a ratio of output signal to input signal. This first formula is applicable for the voltage gain with a working load.

$$A_v = \frac{R_{l\ ac}}{r_{be}}$$

where $R_{l\ ac} = R_L$ in parallel with R_{LW}

$R_{LW} =$ working load resistance

This second formula is used for calculating voltage gain without a working load.

$$A_v = \frac{R_L}{r_{be}}$$

Common-Emitter (CE) Amplifier without a Bypass Capacitor Without a bypass capacitor, the CE amplifier is a low voltage gain amplifier (see Figure 5-6). The voltage gain of this amplifier is again the ratio of the input signal to the output signal. However, the input signal is felt across the base–emitter junction and the emitter resistor.

$$A_v = \frac{R_{l\ ac}}{r_{be} + R_E}$$

where $R_{l\ ac} = R_L$ in parallel with R_{LW}

$R_{LW} =$ Working load resistance

The input impedance also changes somewhat from the bypassed CE amplifier. Considering that the bypass capacitor has been removed, the ac path to ground is across the emitter resistor R_E.

$$Z_{\text{in}} = R_{B1} \parallel R_{B2} \parallel \beta(r_{be} = R_E)$$

Common-Collector (CC) Amplifier Configuration

The CC amplifier configuration, also called the emitter follower (Figure 5-7), is used for impedance matching and for voltage gains of less than unity (1). It has good current gain, similar to the common emitter. The CC amplifier has no phase reversal between input and output signal. This is, of course, a good point if phase is a problem.

Figure 5-7 is a typical common-collector amplifier. The input signal is fed into the base in reference to the emitter. The output is taken off the emitter resistor. This is why the configuration is often called an emitter follower.

Let's consider the dc circuit analysis first. Positive V_{CC} supplies reverse bias to the collector. The dc voltage drops on the output leg of the amplifier sum up to applied voltage.

$$V_{CC} = V_{CE} + V_{RE}$$

R_{B1} and R_{B2} are a voltage divider for forward bias on the base–emitter junction.

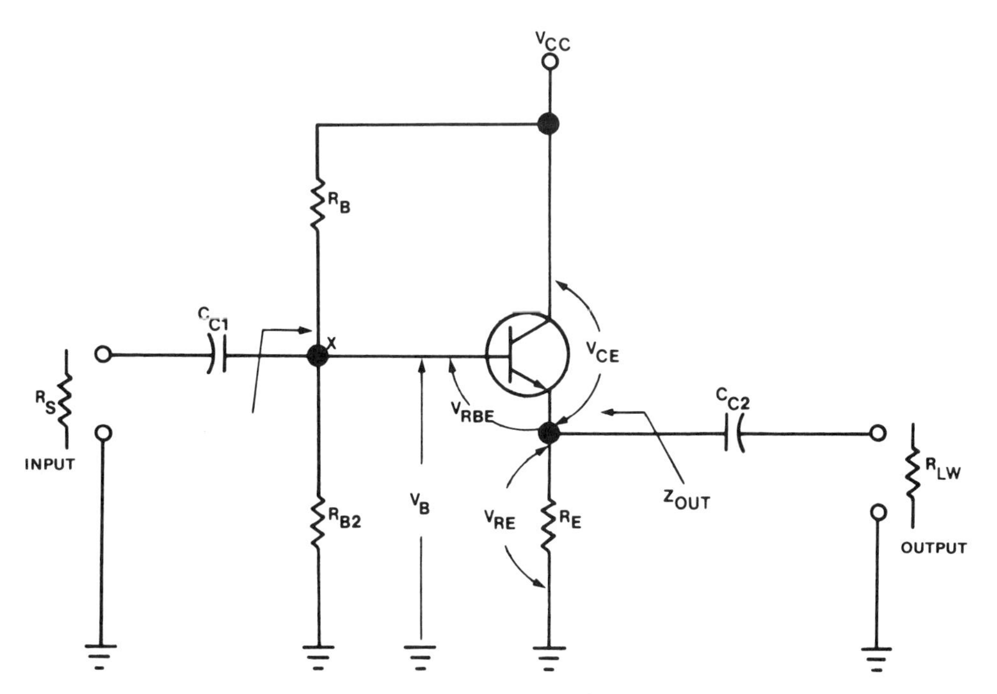

Figure 5-7 Common-Collector (CC) Amplifier Configuration

Bias voltage V_B is equal to the base–emitter junction voltage V_{RBE}, plus the emitter voltage V_{RE}. Bias voltage V_B can also be calculated using the voltage-divider formula in the following manner:

$$V_B = V_{CC}\frac{R_B}{R_{B1} + R_{B2}}$$

The base–emitter junction voltage is approximately 0.6 Vdc if the transistor is made of silicon and 0.2 Vdc if the transistor is made of germanium. These values are approximations and are dependent on and vary with temperature.

The coupling capacitors C_{C1} and C_{C2} are charged to dc voltages at points X and Y, respectively. The signal is placed on the input at point X in relation to ac ground. Since the output is in parallel with the emitter resistor, this signal is essentially the same as the input, less the base–emitter dynamic resistance r_{be}.

Junction dynamic resistance (r_{be}) is calculated as an approximation using the following formula:

$$r_{be} = \frac{25}{I_E}$$

where 25 = Shockley constant

I_E = dc operating current in milliamperes

Once the dynamic resistance r_{be} of the base–emitter junction has been estimated, the input impedance can be calculated. Input impedance is a parallel path to ac ground looking into the amplifier at point X.

$$Z_{in} = R_{B1} \parallel R_{B2} \parallel \beta(r_{be} + R_E \parallel R_{LW})$$

where β = current gain of the transistor.

Output impedance Z_{out} is the ac ground path looking back into the amplifier at point Y.

$$Z_{out} = R_E \parallel r_{be} + \frac{R_{B1} \parallel R_{B2} \parallel R_S}{\beta}$$

where R_S = input source resistance.

The voltage gain A_v of the amplifier is a ratio of the output signal to the input signal.

$$A_v < 1$$

The voltage gain is actually less than unity. All the signal input is felt across the emitter resistor R_E and the parallel working load resistor R_{LW}, except for a minor amount dropped across the base–emitter junction.

Common-Base (CB) Amplifier Configuration

The CB amplifier configuration (Figure 5-8) has a high voltage gain but a current gain of less than unity (1). Current gain in a common-base amplifier is called alpha (α). Alpha is simply a ratio of input current to output current. The signal is applied to the emitter and taken from the collector. Therefore

$$\text{alpha } (\alpha) = \frac{I_C}{I_E}$$

Since the emitter current includes the collector current, the current gain must be less than unity (1).

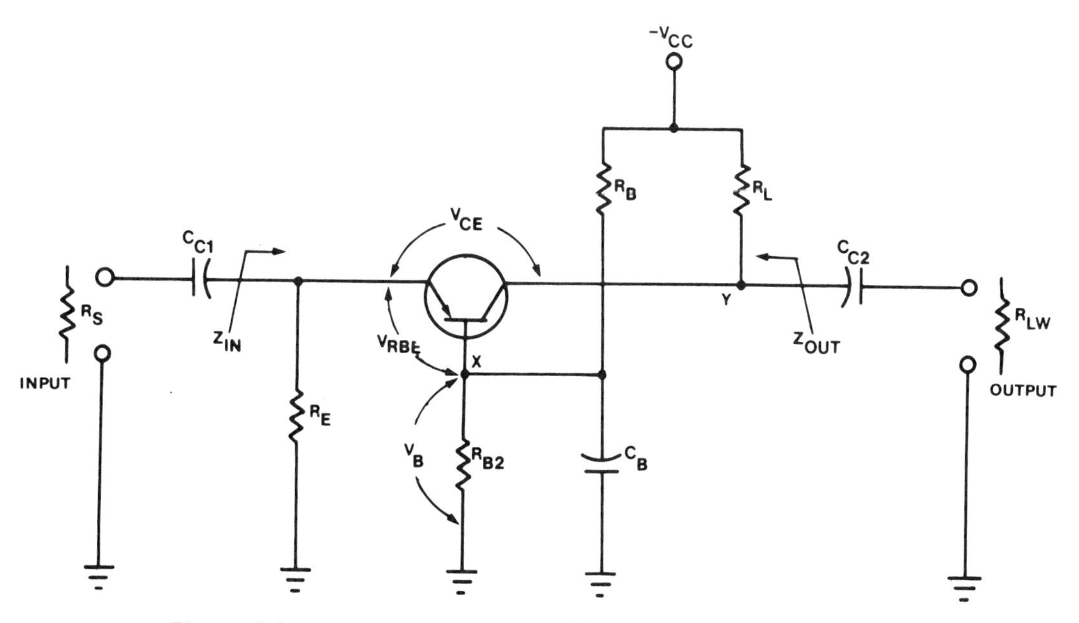

Figure 5-8 Common-Base (CB) Amplifier Configuration

Alpha (α) may be converted to the more useful beta (β) with the following equation:

$$\beta = \frac{\alpha}{1 - \alpha}$$

The CB amplifier has no phase reversal between input signal and output signal. Figure 5-8 is a typical common-base amplifier. We have used a *PNP*-type transistor in this configuration. Therefore, a negative power supply is required. The input signal is fed into the emitter in reference to the base. The base is coupled to ac ground with capacitor C_B. The output signal is taken off the collector in reference to the base.

Let's consider the dc circuit analysis first. Negative V_{CC} supplies reverse bias to the collector. The point Y in reference to ground is approximately one-half the dc voltage $-V_{CC}$. V_{CE} is the collector-emitter dc voltage operating range.

The resistors R_{B1} and R_{B2} are a voltage divider for forward bias on the base–emitter junction. Bias voltage V_B is equal to the base–emitter junction voltage plus the voltage drop across emitter resistor R_E. Bias voltage can also be calculated using the voltage-divider formula:

$$V_B = -V_{CC}\frac{R_{B2}}{R_{B1} + R_{B2}}$$

The base–emitter junction voltage is approximately 0.6 Vdc if the transistor is made of silicon and 0.2 Vdc if the transistor is made of germanium. These values are approximate and are dependent on and vary with temperature.

C_{C1} and C_{C2} are capacitors that couple the input signals into the amplifier and out of the amplifier. The signal is placed on the emitter resistor, which is in parallel with the base–emitter junction r_{be}. The base is at ac ground. The junction dynamic resistance r_{be} can be calculated using the following formula:

$$r_{be} = \frac{25}{I_E}$$

where 25 = Shockley constant

I_E = dc operating current in milliamperes

Once the dynamic resistance r_{be} of the base–emitter junction has been estimated, the input impedance can be calculated. Input impedance sees two parallel paths to ac ground.

$$Z_{\text{in}} = R_E \parallel r_{be}$$

and $$Z_{\text{in}} \simeq r_{be} \text{ (because } r_{be} \text{ is so small)}$$

Output impedance Z_{out} is the ac ground path looking back into the amplifier at point Y. Since there is but one path to ground for ac (signal), the output impedance is simply equal to R_L.

The voltage gain of the amplifier is a ratio of output signal to input signal.

The following formula is applicable for the voltage gain with a working load.

$$A_v = \frac{R_{l\ ac}}{r_{be}}$$

where $R_{l\ ac} = R_L$ in parallel with R_{LW}

$R_{LW} =$ working load resistance

A second formula is used for calculating voltage gain without a working load.

$$A_v = \frac{R_L}{r_{be}}$$

UNIJUNCTION TRANSISTOR (UJT)

The unijunction transistor is a two-layer, three-terminal device. It is used as a relaxation oscillator and as a timing device to trigger SCRs. The unijunction is not an amplifier but does have large current gain. This device could have been included with SCRs in Chapter 3.

The symbol for the UJT is illustrated in Figure 5-9a. The junction model is shown in Figure 5-9b. Note that the model has a comparatively large bar of N-type material and a small P-type section forming the junction. In operation, a positive potential is applied to base 2 and a negative to base 1. Very little current flows because of the resistance of the bar of N-type material. When a

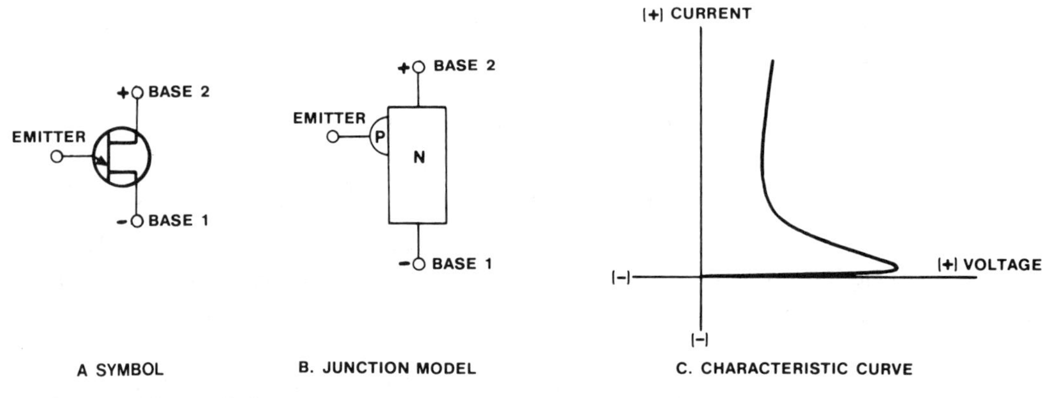

Figure 5-9 Unijunction Transistor

positive potential is applied to the emitter, it forward biases the junction of the emitter and base 1. This essentially decreases the resistance between the emitter and base 1 and, in turn, the resistance between base 1 and 2. A lowering of resistance in the *N*-bar increases the current flow from base 1 to base 2. More emitter voltage causes more current through the bases.

In Figure 5-9c, a characteristic curve of the operation is shown. When voltage is applied to the emitter in a large enough forward (positive) direction, it will overcome a reverse bias provided by the base 1 and 2 potentials. The resistance in the *N* material sets up a voltage divider, which reverse biases the junction. The resistance ratio is called the *intrinsic standoff ratio.* It is this action that allows the UJT to be triggered. The UJT is packaged the same as a bipolar transistor.

UJT Applications

UJT is used either as an oscillator or a trigger. In Figure 5-10a, the UJT is applied as a relaxation oscillator. Power is applied to the circuit. The capacitor C_1 begins to charge through the variable resistance R_V. The UJT is biased (+) on base 2 and (−) on base 1. The junction between the emitter and the bases is reverse biased. There is no conduction. As the capacitor reaches a predesignated charge, current flows from base 1 to base 2. A second current path develops when the capacitor discharges through R_{B1} and the emitter–base 1 junction. The output is a pulse across resistor R_{B1}. A similar wave form may also be taken from the terminal at base 2. This pulse would be 180° out of phase with the wave form taken from base 1. The capacitor discharges to some level below triggering, and the cycle begins again.

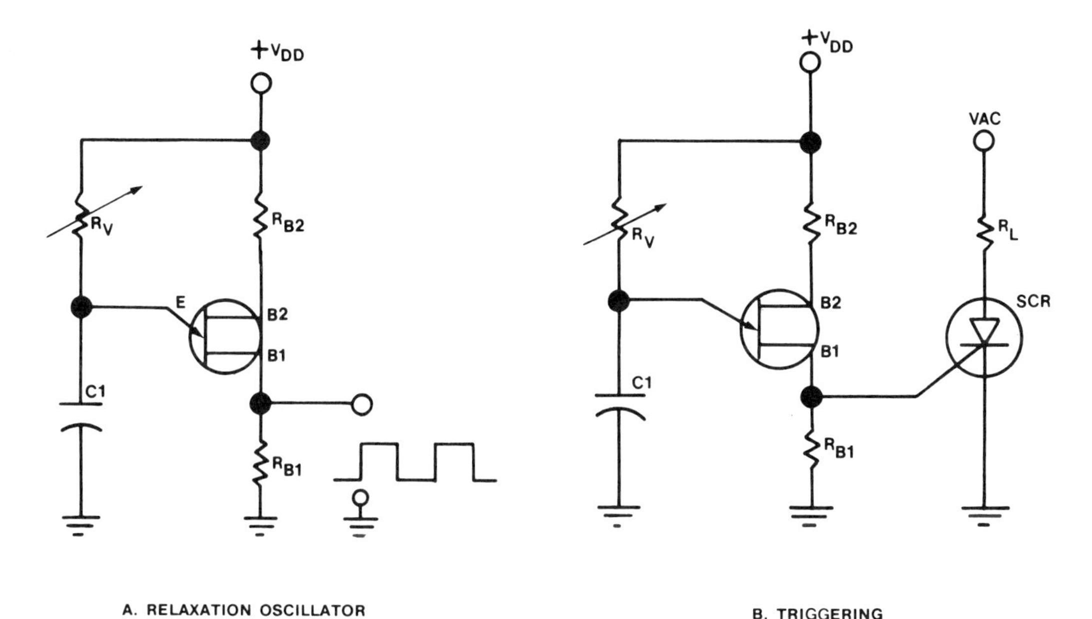

Figure 5-10 UJT In Application

The second application in Figure 5-10b is triggering. In this case, the UJT is used as a trigger for a silicon-controlled rectifier (SCR). In the circuit, the relaxation oscillator just explained is used for triggering. The frequency of the output of the oscillator is dependent on the values of R_V and C_1. The pulse turns on the SCR only when its anode is positive. Current is applied through the load. You may recall that the SCR, when turned on, will not turn off until power is removed, regardless of its trigger condition. Since the power source is alternating, the SCR will be off during negative alternations. It will be on when the pulse is positive and the UJT triggers its gate.

JUNCTION FIELD-EFFECT TRANSISTOR (JFET)

The junction field-effect transistor (JFET) is a simple amplifying device made of a sandwich of *N*- and *P*-type materials. The JFET symbol is shown in Figure 5-11a. Its junction model is illustrated in Figure 5-11b. The JFET is used for circuitry requiring very large input impedances. For instance, the integrated circuit operational amplifier often has a first stage made of a pair of JFET circuits. The JFET may also be used as a high-frequency amplifier and a switch. JFETs are packed in typical bipolar transistor cases.

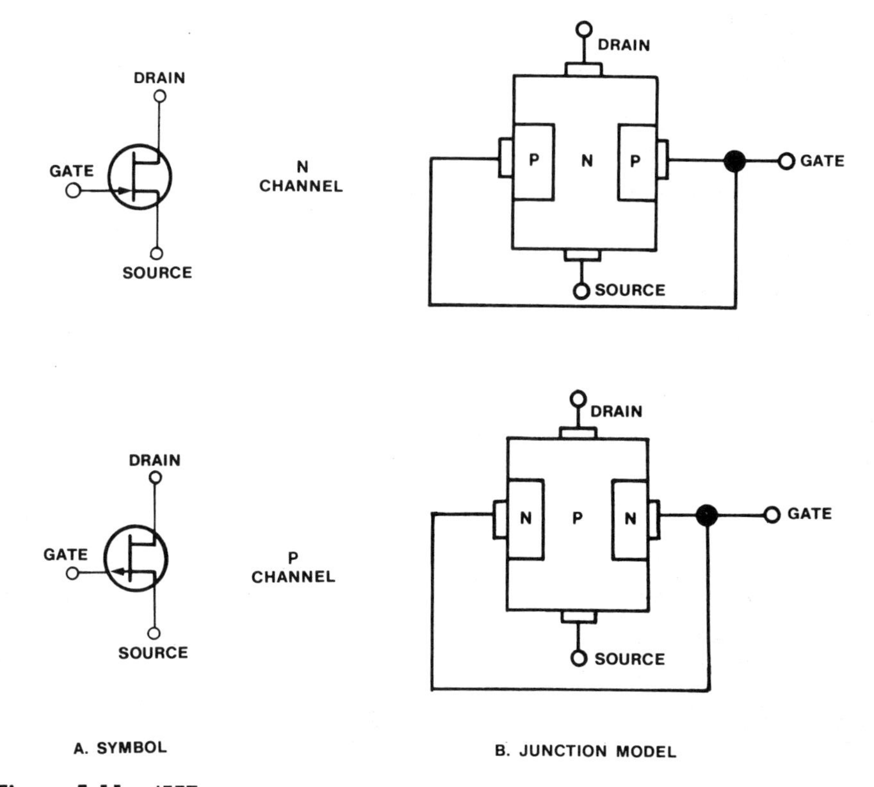

Figure 5-11 JFET

The junction model in Figure 5-11b presents JFET operation. Let's consider the *N*-channel JFET in the upper section of the figure. The N material forms the channel, which extends between two terminals called the *drain* (D) and the *source* (S). Positive voltage polarity is applied to the drain and negative to the source. A slab of *P* material is diffused on both sides of the channel to form a sandwich-shaped structure.

In a no-bias condition (0 bias) on the gate, there is no depletion area at the junction of the gates and the channel. Current flows freely from source to drain. If reverse bias is applied to the gates (−) a depletion area is expanded from the junction. The larger the (−) is, the larger the depletion area. If the (−) is large enough, this area will spread across the channel and stop current flowing from source to drain. This condition is called *pinch-off.* Reverse bias depletes the channel of carriers as in a reverse-biased diode. The extreme conditions are 0 bias and pinch-off. All areas in between these two conditions provide variable current from source to drain. It is in this variable range that amplification takes place.

In the lower section of Figure 5-11, the *P*-channel JFET is modeled. This JFET operates in the same manner as the *N*-channel type. The exception is that pinch-off reverse bias on the gate is (+) and amplification takes place between zero and (+) pinch-off voltage.

The *order of merit* (amplification factor) is a ratio of source to drain current and gate voltage:

$$g_m = \frac{\Delta I_D}{\Delta V_G}$$

JFET as an Amplifier

The circuit shown in Figure 5-12 is typical of the operation of a JFET in an application as a practical amplifier. This particular amplifier is called a common-source amplifier and is used throughout industry. It may be related to the common-emitter amplifier using bipolar transistors (see Figure 5-6). The JFET shown is an *N*-channel type. The resistors R_{G1} and R_{G2} form a voltage divider that sets the gate voltage (V_g).

$$V_G = V_{CC} \frac{R_{G2}}{R_{G1} + R_{G2}}$$

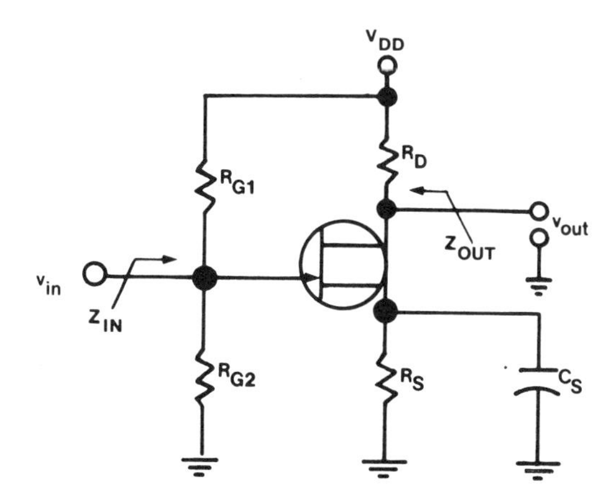

Figure 5-12 JFET as an Amplifier

Circuit input impedance is calculated by looking at the resistance between ground and the gate terminal. This consists of the voltage-divider resistances and the input resistance of the JFET. Since the input resistance of the JFET itself is infinite, we may ignore that value.

Z_{IN} may be calculated as follows:

$$Z_{IN} = R_{G1} \parallel R_{G2}$$

where
Z_{IN} = input impedance
R_{G1} and R_{G2} = voltage-divider resistance

Output impedance is essentially the drain resistance. Output resistance has very little effect on the total output impedance. Therefore,

$$Z_{out} = R_D$$

where
Z_{out} = output impedance
R_D = drain resistance

Voltage gain of the amplifier, as with all amplifiers, is a ratio of voltage out to voltage in.

$$A_v = \frac{V_{out}}{V_{in}} \quad \text{(voltage gain)}$$

We must clarify this further. The mutual inductance (figure of merit) of the circuit, as you may recall, is calculated in the following manner:

$$g_m = \frac{\Delta I_D}{\Delta V_G}$$

If we combine the output impedance and transconductance equations, we create a voltage-gain formula for the voltage change between the gate and the drain.

$$A_v = g_m R_D$$

The reader must realize that the mutual conductance and gain formulas are basic and would be modified with additional or less circuitry. Additions of capacitors and resistance in the bias legs or the source and drain circuits could modify these basic equations.

If the source resistor is not bypassed as in Figure 5-12, the effect is a lowering of voltage gain. It will, however, provide a greater bandwidth.

JFET as a Source Follower

As with the bipolar transistor amplifier, the JFET may be used as an isolation amplifier because of its high input impedance. The source follower is similar to the emitter follower of the bipolar transistor because it also has a very low output impedance. In Figure 5-13, the JFET is used as a *source follower*. It is called a source follower because the output is taken from the source resistance R_S. The circuit is also called a *common-drain* configuration because the drain is connected directly to the dc supply and therefore at ac ground.

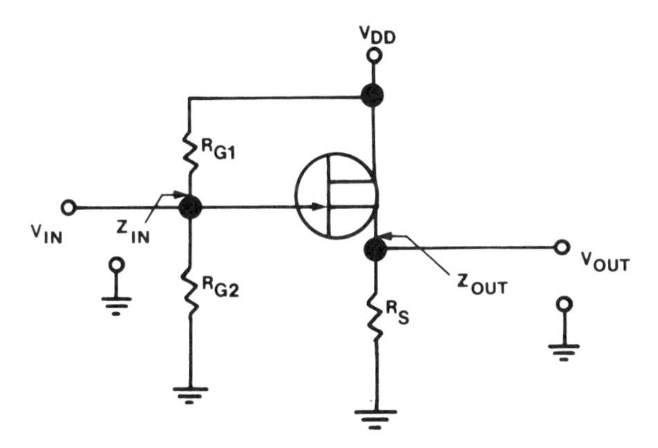

Figure 5-13 JFET as a Source Follower

The resistors R_{G1} and R_{G2} provide a voltage divider that sets the dc gate voltage.

$$V_G = V_{CC} \frac{R_{G2}}{R_{G1} + R_{G2}}$$

The input signal v_{in} sees the gate-to-source resistance $1/g_m$ and Z_{IN}. The output signal v_{out} is taken across the source resistor. The input signal v_{in} physically appears from the gate to ground across the source resistor R_S. The output signal v_{out} essentially sees the source resistor R_S. The only difference between in and out voltages is the small voltage drop across the gate-to-source resistance r_s (1/g_m) of the JFET device.

It stands, then, that input and output voltage are approximately the same:

$$v_{in} \simeq v_{out}$$

The voltage gain is less than

$$A_v \simeq 1$$

Input impedance Z_{IN} is the same as in the common-source configuration.

Output impedance is extremely low because we are looking back at the circuit between the source terminal and ground. This sees the source resistance R_S in parallel to the gate-to-source dynamic resistance r_s (1/g_m):

$$Z_{OUT} = R_S \| r_s$$

The outcome is a very low output impedance.

METAL OXIDE SEMICONDUCTOR FIELD-EFFECT TRANSISTOR (MOSFET)

Another FET device is the metal oxide semiconductor field-effect transistor, or the MOSFET. The MOSFET differs only slightly from the JFET described in the last paragraphs. The principle of operation is the same as the JFET except the gate is made of metal, insulated by a thin layer of glass or oxide. Operation is by current conduction, and control is acheived by varying an electric field.

There are two types of MOSFET, *depletion* and *enhancement*. The depletion type conducts with the gate at zero bias. Conduction increases by applying further negative bias until cutoff. The enhancement type is cut off at zero bias, whereas further positive bias allows it to conduct.

MOSFETs are used in digital logic and memory circuits. Furthermore, MOSFETS are used in op amps and other linear circuits but are not as popular as the bipolar transistor for linear work. The MOSFET is popular for logic circuits for it does not require resistors or diodes. Because of this, integrated circuits may be extremely small. Furthermore, they are simple to manufacture. For these reasons, the MOSFET has become the workhorse of very large scale integration (VLSI) circuits, which include devices such as microprocessors. MOSFETS are packaged in typical bipolar transistor cases.

Depletion-Type MOSFET

In Figure 5-14a the depletion-type MOSFET contains a *P* substrate, an *N* channel, and a metal gate that is insulated from the semiconductor material with an oxide. Source and drain terminals are connected to the drain and source diffused materials. This is an *N*-channel depletion MOSFET. Its symbol is just above the *P*-channel depletion MOSFET symbol in Figure 5-14a. The structural model of the *P*-channel depletion MOSFET is the same as the *N* channel with the substrate made of *N* material and the channel of *P* material.

In operation the depletion MOSFET is much like the JFET. Two heavily doped wells provide a low-resistance connection between the ends of the *N* channel and the source and drain connections. The *N* channel is formed on the *P* substrate. Oxide is grown on the *N*-channel surface. Metal is deposited on the oxide to form the gate. The acronym MOS means metal oxide semiconductor, which describes its structure.

With zero bias, the electrons flow freely through the *N* channel from source to drain. As negative bias is increased, electrons are attracted by holes in the substrate, making them unavailable for conduction. A depletion area will result just below the oxide and will grow larger as negative bias is increased. The result is that the *N* channel will get thinner. Finally, current will be pinched off from source to drain.

Enhancement MOSFET

In Figure 5-14b the enhancement-type MOSFET contains a *P* substrate and a metal gate that is insulated from the semiconductor material with an oxide. Source and drain terminals are connected to the drain and source diffused materials. This is an *N*-channel enhancement MOSFET. Its symbol is just above the *P*-channel enhancement MOSFET symbol in Figure 5-14b. The structural model of the *P*-channel enhancement MOSFET is the same as the *N* channel with substrate made of *N* material and the channel of *P* material.

In operation the enhancement MOSFET is much like the JFET. Two heavily doped wells provide a low resistance connection between the ends of the *N* channel, which exists when the gate is sufficiently positive. Oxide is grown on the top side of the *P* substrate. Metal is deposited on the oxide to form the gate.

You should be able to see that the physical difference between depletion and enhancement MOSFETs is that there is no channel. The channel does not exist if there is no bias. The current path from source to drain is as two diodes in series back to back. No current flows between source and drain without bias.

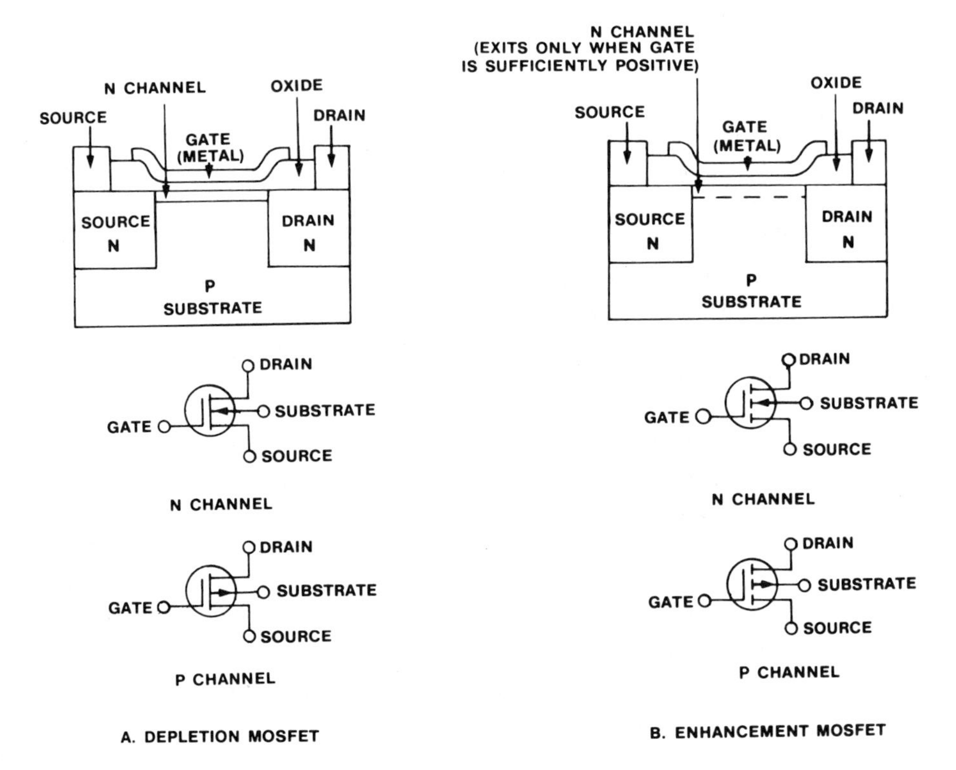

Figure 5-14 MOSFET

If a positive voltage (bias) is applied to the gate, it will attract electrons from the substrate and create an *N*-channel just below the oxide. This channel will become large enough to allow current to flow when a threshold voltage (specified) has been reached. As positive bias is increased, drain current increases.

MOSFETs and Electrostatic Charges

The MOSFET has extremely low leakage current and extremely high input resistance. It is easily destroyed because of these parameters. If a large voltage is applied from the gate to the channel, the oxide will rupture due to electrostatic field stress. This voltage can appear simply by electrostatic discharge from handling. Care must be taken during all handling tasks. When shipping the device, the leads are shorted together. When being handled, the tools and soldering irons are grounded. Technicians even use grounding straps around their wrists to prevent accidental electrostatic voltage applications between the gate and the channel.

MOSFET as an Amplifier

There are many MOSFET amplifier configurations. They cannot all be covered here. The amplifier in Figure 5-15 is typical of a universal circuit using the enhancement MOSFET.

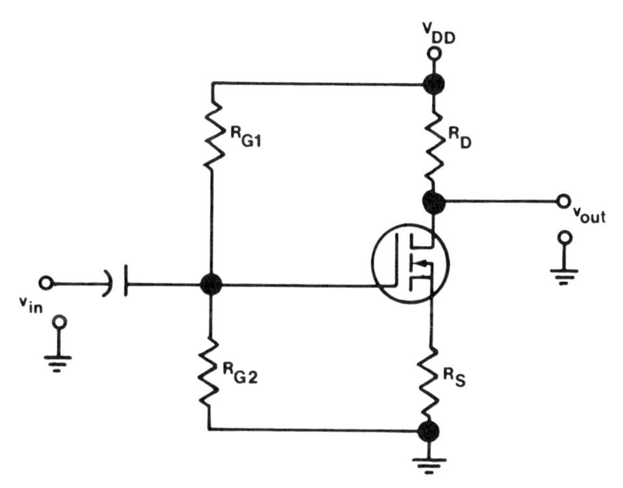

Figure 5-15 MOSFET as an Amplifier

Our analysis begins with the voltage-divider resistors R_{G1} and R_{G2}. The dc voltage drops across these resistors may be easily calculated using the voltage-divider formula for either.

$$V_{RG1} = V_{DD} \frac{R_{G1}}{R_{G1} + R_{G2}}$$

or

$$V_{RG2} = V_{DD} \frac{R_{G2}}{R_{G1} + R_{G2}}$$

Operating drain current is determined by use of the device characteristic curves. Operating drain current is calculated using the following equation.

$$I_{DO} = \frac{I_D}{[1 - (V_{GS}/V_T)]^2}$$

where

I_{DO} = operating (saturation) current of the device

I_D = drain current for specified gate voltage

V_{GS} = gate voltage

V_T = threshold voltage

The voltage drops across the source resistor and the drain resistor are dependent on the operating current.

$$V_{RS} = I_{DO} \times R_S$$

and

$$V_{RD} = I_{DO} \times R_D$$

The figure of merit (voltage gain factor) of the MOSFET amplifier is mutual conductance (transconductance). This value must be decided so that it may be used for further voltage gain calculations:

$$g_m = \frac{2}{V_T}\sqrt{I_{DO}\, I_D}$$

and

$$A_v = g_m R_D$$

where A_v = voltage gain

g_m = mutual conductance

R_D = drain resistance

V_T = threshold voltage

I_{DO} = operating current (saturation)

I_D = drain current at specified gate-to-source voltage (V_{GS})

These equations are very general in form and should be considered as such. If the reader should require in-depth use of the MOSFET rather than generalization of functions, it would be useful to acquire a major book on the subject. The author would recommend *Micro-Electronic Circuits* by Adel Sedra and Kenneth Smith (Holt, Rinehart and Winston, New York, 1982).

INTEGRATED CIRCUIT (IC) AMPLIFIERS

A simple and popular linear integrated circuit is called the *array*. It is several semiconductors on a single IC. The difference between this and other ICs is that the several circuits in the IC may or may not perform a specific function. They may be hooked up in any manner that the designer sees fit. For instance, several diodes and transistors may be on one array. Others may have two differential amplifiers. Still others may have a bridge rectifier and other separate diodes. Linear ICs have evolved from the use of transistors and MOS devices in separate devices to combinations that include both these fabrication techniques. The usual technique in combinations is to use a FET (field-effect transistor) at the input. This supplies the device with an extremely low input current, a notable attribute.

Linear Integrated Circuits

By far the predominant linear integrated circuit is the operational amplifier. In fact, this device is used so often in the electronics industry that many books have been written about it. As you can see by the following list, there is a fairly large variety of linear integrated circuits. Linear integrated circuits perform process and control operations, signal generation, and amplification. The major use is with operational amplifiers. However, the voltage regulators and comparators are in increasing use.

1. Operational amplifiers
2. Voltage regulators
3. Voltage references
4. Instrumentation amplifiers
5. Voltage comparators
6. Analog switches
7. Sample-and-hold circuits
8. Analog-to-digital (A/D) and digital-to-analog (D/A) conversion
9. Telecommunications
10. Audio/radio/television
11. Transistor/diode arrays

Confusion may arise from the word "linear." Linear in electronic devices means that cause and effect are proportional for all values of input and output. Natural devices in real life are not linear. In fact, the linearity across a range of operations is limited. Devices can, however, be nonlinear in operation but still have useful life because the nonlinearity can be well defined, controllable, stable, and available at low cost. Analog relationships such as multiplication, square laws, and log ratios are examples of usefulness. Applications include modulation, power measurement, signal shaping, and correcting for the nonlinearity of measuring devices.

In general, the nonlinear device may be classified according to its smoothness. If the functions are smooth, it may be classified as *continuous* function. If it has one or more discontinuities or jumps, it is classified as a *discontinuous* nonlinear function.

Hybrid Devices

The hybrid device is fabricated by placing devices, including ICs and discrete components, on a single substrate and connecting them electrically to perform electronic functions. There are two basic techniques for fabrication of the hybrid: thin film and thick film. They are similar. *Thin-film fabrication* is accomplished by deposition of a thin layer of metal on a substrate. Substrates are coated and metal layers are selectively removed to form the circuit. Devices such as diodes, resistors, transistors, and capacitors are then attached to complete the circuit on the substrate.

In *thick-film fabrication* a heavy coat is produced by firing a paste deposited on a substrate along with other discrete components. The paste structures are conductors, resistors, and capacitors. The paste is usually stencil screened on a ceramic substrate. Components of the circuit all require a different paste composition. Firing sets the circuit. Thick film is cheaper than thin film, but requires a larger substrate. Thin film is ideal for high frequencies.

Operational Amplifier, an Introduction

An operational amplifier IC is a solid-state integrated circuit that uses external feedback to control its functions. The operational amplifier may be used for a great number of circuits, which include amplifiers, integrators, filters, differentiators, voltage followers, oscillators, regulators, and mathematical circuits. Its greatest characteristic is simplicity in operation.

The operational amplifier can replace entire transistor amplifier circuits in many applications such as audio, radio, analog, and signal-conditioning systems. Its basic operation depends on the use of negative feedback, as described later.

Each operational amplifier has unique qualities that can be determined from the manufacturer's specifications. Some of these qualities are wide frequency range, internal frequency compensation, low power consumption, internal "short" protection, and temperature stability. Operational amplifiers are, in fact, the integrated-circuit evolution to the different amplifier. They are especially noted for their high input impedance, high voltage gain, low output impedance, and zero output signal for zero input signal.

Our description of the operational amplifier will be brief because of the nature of this book. For a complete description of the op amp and its operation, see the author's *Operational Amplifiers,* Reston Publishing Co., Reston, Va., 1983.

Basic Principles of Operation The principle of feedback in the operational amplifier IC is the function that makes it so versatile (see Figure 5-16). This principle is quite simple. A feedback connection is made between the output and inverted input leads, with resistors used as voltage dividers. The voltage divider provides the exact amount of negative feedback for any level of amplification (gain) required or desired.

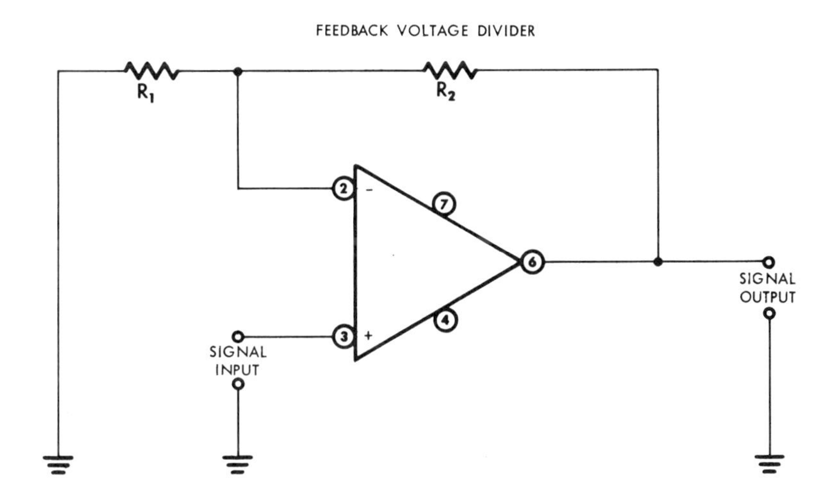

Figure 5-16 Feedback

Closed-loop gain is dependent on the size of the feedback resistance. Open-loop gain (no feedback resistance) can be ignored in most designs, but is used in voltage-comparison and level-detection circuits. The gain of the operational amplifier is controlled externally. The operational amplifier is said to have infinite gain within the restraints of power supplied at pins 7 and 4. It is also said to have infinite input impedance and zero output impedance. When there is no signal applied to either the inverting (pin 2) or noninverting (pin 3) inputs, the output (pin 6) is zero. When a signal is applied to one input, the opposite input strives to approach that signal level so that there is always a differential of zero between the two inputs.

Since feedback is from output to inverting input, the output will always be that amount necessary to drive the two inputs to near null (or an extremely small value). Current is so small at the inputs that it is assumed to be nonexistent. Output polarity is in phase with the noninverting input and out of phase with the inverting input. A more detailed discussion of operation is provided later.

Differences between Operational Amplifiers Each operational amplifier is constructed to perform at its best under certain conditions. Families of operational amplifiers are designed to optimize a set of parameters crucial to particular applications. Usually the titles in the operational amplifier catalogs provide some indication as to the set of parameters that best utilizes the specific "skills" of an op amp. Some of these are listed next.*

1. General purpose (low cost)
2. FET input, low bias current
3. Electrometers (lowest bias current)
4. High accuracy, low drift differential
5. Chopper amplifiers (lowest drift)
6. Fast wideband
7. High output
8. Isolated op amps

Each set of parameters is used under circumstances pertinent to the particular application. For instance, items 1, 2, and 4 are most relevant for transducer applications, while item 7, high output, may be used in power applications.

Ideal Amplifier The search for an ideal amplifier is an exercise in futility. The characteristics of the operational amplifier are good enough, however, to allow us to treat it as ideal. The following are some amplifier properties that make this so. (The reader must realize that these ratings are seldom, if ever, reached.)

1. Gain: infinite
2. Input impedance: infinite
3. Output impedance: zero
4. Bandwidth: infinite
5. Voltage out: zero (when voltages into each input are equal)
6. Current entering the amplifier at either terminal: extremely small

Open-Loop Operation Consider the open-loop response of the amplifier in Figure 5-17. Note that there is no external feedback. Pin 2 is the inverting input and the point at which the signal is input. Pin 3 is the noninverting input.

* From Data-Acquisition Products Catalog, Analog Devices, Inc.

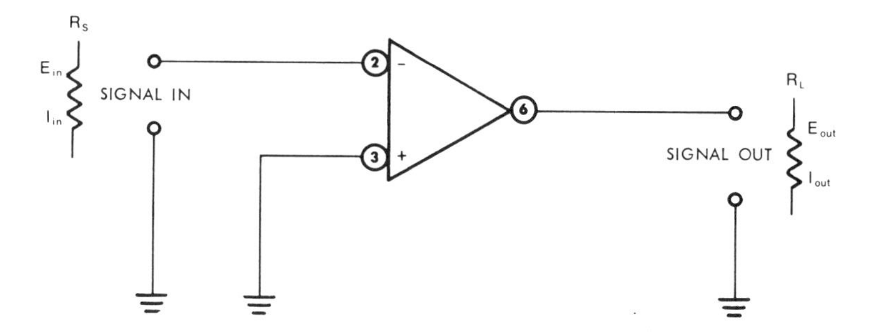

Figure 5-17 Open-Loop Operation

Pin 3 is grounded. Pin 6 is the signal output. Since the gain is extremely high (approaching infinity), the difference in the two inputs must be as near zero as possible. In the ideal amplifier this is the case.

Open-loop gain in some general-purpose operational amplifiers is usually around 10^5. The analysis of open-loop gain is simple. It is the ratio of output to input voltage.

$$\frac{\text{Signal out}}{\text{Signal in}} = \text{open-loop gain}$$

With no signal, there is no input. Therefore, at pins 2 and 3 the voltage is zero. At pin 6 the voltage is also zero. If pin 2 or 3 takes any signal value, voltage at pin 6 increases to saturation.

A small amount of signal, such as 3 mV, will cause the operational amplifier to traverse to the active region. This is the same effect as in a switching transistor circuit; the output is either high (1) or low (0).

Open-loop operation is used in voltage comparators. Other applications are control-loop-gain scheduling, tracking monitors, and level detection. However, open-loop operation is not nearly as versatile as closed-loop operation.

Input Impedance (Z_{IN}) Input impedance of an operational amplifier is said to approach infinity. In actuality, each device does have a typical input impedance, such as 1 MΩ and above. Some of the more sophisticated devices may reach an input impedance of as much as 100 MΩ. This value may be determined from the manufacturer's specifications. The input impedance (Z_{IN}) of an op amp is the resistance at either the inverting or the noninverting inputs with the opposite input at ground.

Output Impedance *(Z_{OUT})* The output impedance of an operational amplifier is said to approach zero. In actuality, each device is different, but for the sake of design this impedance can be assumed to be zero. In specifications, output impedances are said to be 25 to 50 Ω. However, this value can be ignored for most applications.

Because of low output impedance, the output will function as a voltage source and will provide required current for a wide range of loads up to the current limit level (25 mA, typically). With this capability and high input impedance, the operational amplifier becomes an ideal impedance-matching device.

Gain Probably the most important parameter/specification involved with an amplifier is gain. In fact, that is what the amplifier is all about. An output is derived from some input. The amplifier provides an output that is X amount larger than the amount placed into the input.

Voltage gain with operational amplifiers approaches infinity for the ideal amplifier. For nonideal amplifiers, the voltage gain is adjustable with external resistances. In either event, the following voltage gain ratio exists:

$$A_v = \frac{V_{OUT}}{V_{IN}}$$

where

A_v = voltage gain

V = voltage

This may seem rather basic; however, it is essential to most amplifier situations. Two other gain ratios, A_i (current gain) and A_p (power gain), should be stated as fundamental to the amplifier cause:

$$A_i = \frac{I_{OUT}}{I_{IN}}$$

where

A_i = current gain

I = current

$$A_p = \frac{P_{OUT}}{P_{IN}}$$

where

A_p = power gain

P = power

All three of these equations are simply ratios of one value to another and are dimensionless. All other parameters and specifications evolve around modifications to the gain ratios. The modifications are either desirable or undesirable.

Open-Loop Voltage Gain Open-loop gain is the voltage gain of the op amp without a feedback but with a load. These values are around 10^5 (not infinity) and are dependent on frequency. The open-loop dc power gain A_p is around 100 decibels (dB), again, not infinite.

Open-loop gain may vary with temperature, voltage supply, and configuration load. Indeed, open-loop gain may vary somewhat with each device within a specified part number. Open-loop gain will always vary with frequency.

Closed-Loop Voltage Gain Signal gain is closed-loop gain and is a simple ratio for both inverting and noninverting input signals. For inverting stages, the signal gain is

$$A_v = \frac{R_F}{R_M}$$

where A_v = voltage gain

R_F = feedback resistance

R_M = metering resistance

For noninverting amplifiers, the gain is

$$A_v = \frac{R_M + R_F}{R_M}$$

Op Amp as an Amplifier

Inverting amplifiers, using operational amplifiers, are of two basic designs: dc and ac. Both types of amplifiers provide high voltage gain. They are very versatile and are used in automatic control systems, sound systems, communication systems, and instrumentation. Because of the high voltage gains, the devices are usually protected by a current limiter. Inputs to the amplifiers should typically not exceed supply voltages.

The dc inverting amplifier using an op amp consists of the op amp, a feedback resistance R_2, and a metering resistance R_1 (see Figure 5-18). The

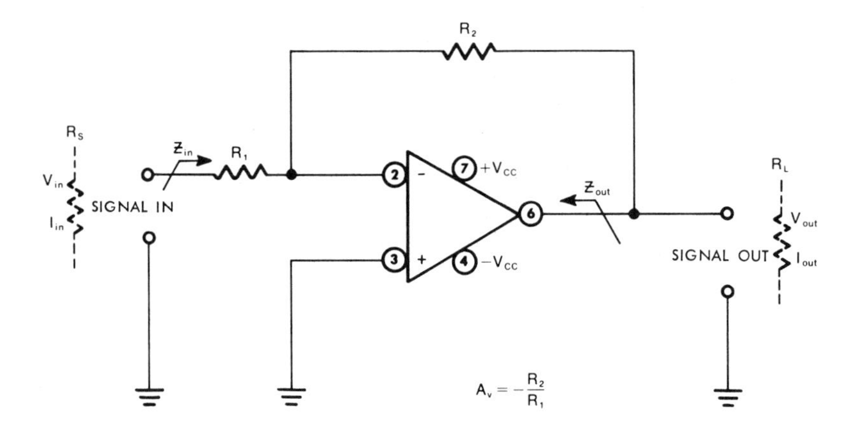

Figure 5-18 Inverting dc Amplifier

author would like to remind the reader of several of the basics of op amps before proceeding:

1. No current flows into the amplifier.
2. The input summing points are at virtual ground.
3. The difference voltage between the inputs, inverting and noninverting, is zero (0).
4. With a feedback attached (closed loop), the negative input signal will be driven to the positive reference input.

In order to match impedances in a discrete amplifier, the typical situation was to ensure that output impedance of the first stage or source was equal to or less than the second stage input impedance. Input impedance of an inverting dc amplifier is not infinite but typically around 40 or 50 times the source impedance, as determined by R_1. This resistance decision is not absolute. It can vary as long as input impedance is much greater than the source resistance. Output impedances are very low (25 to 50 Ω, typically) and therefore can be ignored in design.

Procedures for design are basically as follows:

1. Calculate the voltage gain required.

$$A_v = \frac{V_{\text{OUT}}}{V_{\text{IN}}} = \frac{R_2}{R_1}$$

1. Choose R_2, at typically 50 times the source impedance.
2. Calculate feedback resistor $R_2 = A_v R_1$.

The ac version of the inverting amplifier design is the same procedure with a capacitor added as an input coupling device. The choice of a capacitor is dependent on the lowest frequency the circuit may encounter.

Noninverting amplifiers, using operational amplifiers, provide high voltage gain. Since this is the case, the output is usually protected by a current limiter. Inputs to the noninverting amplifier should typically not exceed the supply voltage. Large signal voltage gain may approach values such as 100,000. Noninverting amplifiers, like inverting amplifiers, are used in automatic control systems, sound systems, communication systems, and instrumentation. Also like inverting amplifiers, the noninverting amplifiers are basically of two types: dc and ac.

The dc noninverting amplifier using an op amp consists of the op amp, a feedback resistance R_2, and a metering resistance R_1. The input impedance at pin 3 is infinite (see Figure 5-19).

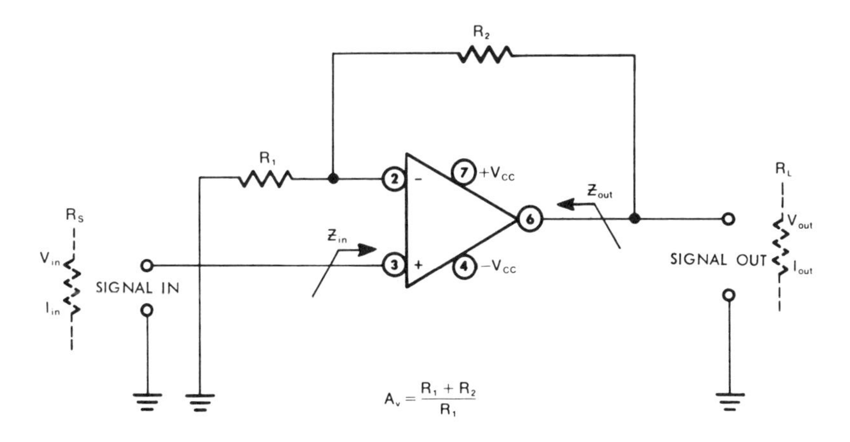

Figure 5-19 Noninverting dc Amplifier

The noninverting amplifier has the advantage of near infinity input the same as the voltage follower. It therefore is used as a buffer for impedance matching. Its output impedance can be said to be zero.

Low-output impedances are generally attributed to a common-collector (emitter-follower) transistor configuration in the output of the amplifier. In Figure 5-19, the R_1 resistor is chosen to match the source impedance and is placed between the inverting input pin 2 and ground. Feedback is arranged between output pin 6 and inverting pin 2. Power, as shown, is applied to pins 4 and 7.

If one of the power supplies becomes disconnected or open, a large voltage may appear at output pin 6. This is dependent on the device used. Voltage gain is

$$A_v = \frac{R_1 + R_2}{R_1}$$

Procedures for design are basically as follows:

1. Calculate the voltage gain required.

$$A_v = \frac{V_{OUT}}{V_{IN}} = \frac{R_1 + R_2}{R_1}$$

1. Choose R_1 to equal the source resistance.

2. Calculate feedback resistor $R_2 = A_v R_1 - R_2$.

The ac version of the noninverting amplifier design is the same procedure with a capacitor added as an input coupling device. The choice of a capacitor is dependent on the lowest frequency the circuit may encounter.

Afterthoughts The op amp is an extremely versatile device and should be studied because of its usefulness in industrial and commercial applications. One should acquire specifications from major IC companies along with other technical books.

Other Operational Amplifier Circuits and Topics As you may have determined, the variety of circuits using the operational amplifier is unending. However, a list of some other applications is provided for reference.

1. Integrator
2. Differentiator
3. Astable multivibrator
4. Bistable multivibrator
5. Monostable multivibrator
6. Comparator
7. Norton amplifier
8. Pulse, triangle, and sine-wave generators
9. Active filters
10. D/A and A/D conversion
11. Voltage regulators
12. Log and antilog amplifiers
13. Sample-and-hold circuits
14. Peak detection
15. Phase-locked loops

6

Electronic Digital Circuits

Digital electronics is usually associated with computers and this is understandable. Its sister is analog electronics, which is associated with linear circuits, usually operational amplifiers and discrete circuit amplifiers using transistors. Analog circuits, which were covered previously, are circuits whose output is usually an amplified version of its input, or whose output is directly related to its input. Digital circuits have two states, either on or off. This chapter will concentrate on the basics of digital electronics and its primary functions both in discrete and integrated form.

We shall begin this section by discussing computer jargon and then the binary numbering system and its relationship to digital circuitry. We shall then cover the logic gates, flip-flops, and other computer-related circuitry.

INTRODUCTION TO DIGITAL CIRCUITS

Computer Jargon

The work done by digital electronics is accomplished with or in the computer. It thus makes sense that persons involved with the software of the computer should understand some of the jargon related to digital electronics.

Bit Binary numbers are 0 and 1. These numbers are often called *bits.* The word bit comes from two words, BInary digiT. A number such as 27 is a two-digit number in the decimal system. That same number in the binary (base 2) system is 11011. The base 2 number is a 5-bit number.

Least Significant Bit (LSB) The binary digit that has the lowest value is called the LSB (e.g., $1101\underline{1}$).

Most Significant Bit (MSB) The binary digit that has the most or highest value is called the MSB (e.g., $\underline{1}1011$).

Nybble A nybble is 4 bits long (e.g., 1101).

Word A word is one or more bits that are related to each other. Words are usually 8, 12, 16, 24, and 64 bits long. The length of the word depends on the computer size. Computer memories are discussed in terms of the number of words they can handle. As an example, an 8K (8000) memory means that the computer will retain approximately 8000 words. These words could be any length, such as 8 or 16 bits long.

Computer Numbering Systems

The operational states of digital electronics are off or on. This is basically why the relationship between the binary numbering system and digital electronics is so compatible. The binary numbering system is based on two digits, 0 and 1. We in the United States and most of the rest of the world depend on the binary system to serve as a vehicle for computer technology. Other numbering systems are used in computer systems. The hexadecimal system and the octal system are two of the more prevalent systems used. The binary numbering system (like any other technical data) is easy to learn once basics are understood. Let's look at some of the basics and attempt to unravel any mystery.

Binary Numbering System A decimal number is one such as 258. We count inclusively the numbers 0 through 9, which are digits, then move to 10, which means that we have now completed one count through the available digits. We then count through these numbers again to 19, then move to 20, which means that we have completed a second count through the available digits. And so on.

The binary numbering system has two digits, 0 and 1. Binary numbers are expressed in terms of these base 2 numbers. If we were to count in binary, the first count would be from 0 to 1, therefore, 1. We then would be at the end of the available digits and would move to 10, which is the second count. The third count is 11 and we would have again reached the end of the available digits and move to 100. If we continue to count, we very soon have a number such as 10,000, which is actually only 16 counts. Table 6-1 provides us with several of the conversions from digital to decimal for comparison purposes.

TABLE 6-1
BINARY TO DECIMAL CONVERSIONS

Decimal No.	*Binary No.*
0	0
1	1
2	10
3	11
4	100
5	101
6	110
7	111
8	1000
16	10000
32	100000
64	1000000
128	10000000

To analyze the structure of a decimal number is rather simple. Let's consider the decimal number 258. This number is made up of 8 units, 5 tens, and 2 hundreds.

$$8 \times 1 = 8$$
$$5 \times 10 = 50$$
$$2 \times 100 = 200$$
$$\text{Total} = 258$$

The number 258 can also be analyzed using powers of 10.

$$8 \times 10^0 = 8$$
$$5 \times 10^1 = 50$$
$$2 \times 10^2 = 200$$
$$\text{Total} = 258$$

The method used to analyze binary (base 2) numbers is very similar in function. Let's consider the base 2 number 11011. Each position from the right digit is represented by a base 2 exponent as follows:

Base 2 No.	*Base 2 Exponents*		*Base 10 Value*	
1	1×2^0	=	1	
1	1×2^1	=	2	
0	0×2^2	=	0	
1	1×2^3	=	8	
1	1×2^4	=	16	
			27	Total

Therefore, 11011 base 2 = 27 base 10.

Binary Coded Decimal (BCD) System Most digital numbers are extremely large in comparison to their decimal equivalent. For instance, a digital number such as 11011011 is equal to decimal number 219. The unwieldy digital number is difficult to read and more difficult to convert. Computer software personnel have found that the binary coded decimal system is much easier to cope with. This system uses four binary numbers to represent each decimal digit. For instance the decimal number 219 is represented by binary numbers as follows:

0010	0001	1001
2	1	9

Binary Coded Hexadecimal System The hexadecimal system is base 16. It is used in digital computers because it is easy to work with and because it can be closely related to the binary system. The first 10 digits of the hexadecimal system are 0 through 9, which is the same as the decimal system. The last 6 digits in hexadecimal are letters A through F. Table 6-2 provides a list of corresponding binary and hexadecimal numbers. These are available on periphery equipment of computers. The hexadecimal system is related to the binary system using four digits. Table 6-3 is a structural analysis of the hexadecimal system.

Binary Coded Octal (BCO) System The binary coded octal system is associated with the binary numbering system. The octal system itself is not as satisfactory as the BCO because the octal has only 8 digits as compared to the

TABLE 6-2
BINARY CODED HEXADECIMAL SYSTEM

Binary No.	*Hexadecimal No.*
0000	0
0001	1
0010	2
0011	3
0100	4
0101	5
0110	6
0111	7
1000	8
1001	9
1010	A
1011	B
1100	C
1101	D
1110	E
1111	F

TABLE 6-3
HEXADECIMAL STRUCTURE

Binary Position	3	2	1	0
Exponent	16^3	16^2	16^1	16^0
Decimal Value	4096	256	16	1

Conversion Examples:

Hexadecimal Numbers

138 $= 8 \times 16^0 + 3 \times 16^1 + 1 \times 16^2 = 8 + 48 + 256 = 310_{10}$

283 $= 3 \times 16^0 + 8 \times 16^1 + 2 \times 16^2 = 3 + 128 + 512 = 643_{10}$

A29 $= 9 \times 16^0 + 2 \times 16^1 + A(10) \times 16^2 = 9 + 32 + 2560 = 2601_{10}$

2CB3 $= 3 \times 16^0 + B(11) \times 16^1 + C(12) \times 16^2 + 2 \times 16^3 = 3 + 176 + 3072 + 8192 = 11443_{10}$

decimal system's 10 digits. Table 6-4 provides a list of binary to octal systems. Table 6-5 is a structural analysis of the binary coded octal (BCO) system.

TABLE 6-4
BINARY CODED OCTAL SYSTEMS

Binary Number	*Octal Number*
000	0
001	1
010	2
011	3
100	4
101	5
110	6
111	7

TABLE 6-5
BINARY CODED OCTAL STRUCTURE

Binary Position	3	2	1	0
Exponent	8^3	8^2	8^1	8^0
Decimal Value	512	64	8	1

Conversion Examples:

Octal
Numbers

$12 = 2 \times 8^0 + 1 \times 8^1 = 2 + 8 = 16_{10}$

$258 = 8 \times 8^0 + 5 \times 8^1 + 8 \times 8^2 = 8 + 40 + 512 = 560_{10}$

$3456 = 6 \times 8^0 + 5 \times 8^1 + 4 \times 8^2 + 3 \times 8^3 = 6 + 40 + 256 + 1536$
$= 1838_{10}$

THE GATES

Logic Gates

There are five basic logic gates: NOT, AND, NAND, NOR, and OR. The logic gate is the reasoning or decision component of digital electronics. The logic gate simply makes a decision. It decides between functions such as yes or no, true or false. Each input to a gate is considered high or low, HI or LO, 1 or 0. Each output has these same two states. These logic functions are called *binary variables.*

Each logic gate has a mathematical expression from Boolean algebra that describes its function.

NOT Gate The NOT gate is the simplest of the logic gates. Its mathematical expression is thus. If A is high, then X is not high.

$$A = \overline{X}$$

The input A is either high or low (1 or 0). The output $\overline{X}$ is inverted. The mathematical symbol for this inversion is the bar over the symbol $\overline{X}$.

AND Gate The AND gate is called the logical product. Its mathematical expression is thus. If A and B are high, then X is high.

$$A \bullet B = X$$

Its mathematical symbol is $\bullet$. The inputs A and B can be either high or low (1 or 0). Only if both inputs are high (1) can there be a high output. The $\bullet$ is often left out.

OR Gate The OR gate is called the logical sum. Its mathematical expression is thus. If A or B is high, then X is high.

$$A + B = X$$

Its mathematical symbol is $+$. The inputs A or B can be either high or low (1 or 0). If either is high (1), the output is high (1).

NAND Gate The NAND gate is the inverse or negation of the AND gate. Its mathematical expression is thus. If A and B are high, then X is not high.

$$A \bullet B = \overline{X}$$

Its mathematical symbol is $\bullet$. The inputs A and B can be either high or low (1 or 0). Only if both inputs are high (1) can there be a low output. The $\bullet$ is often left out.

NOR Gate The NOR gate is the inverse or negation of the OR gate. Its mathematical expression is thus. If A or B is high, then X is not high.

$$A + B = \overline{X}$$

Its mathematical expression is +. The inputs A and B can either be high or low (1 or 0). If either input is high (1), then X is low (0).

Truth Tables

The truth table provides a list of relationships between possible inputs and their outputs. Inputs are either 1 or 0 (high or low). Outputs are either 1 or 0. Inputs are in binary order. Each of the basic gates has fixed truth tables. These tables are given in Table 6-6.

To find the number of binary input combinations for gates with more than one input, powers of 10 are used. For a two-input gate (A, B), there are four binary input combinations, that is, two inputs with two possibilities (1, 0), or 2^2. For a three-input gate (A,B,C), there are eight binary input combinations, that is, three inputs with two possibilities (1, 0) or 2^3. Some of these combinations are listed in Table 6-7.

TABLE 6-6
BASIC GATE TRUTH TABLES

A	*X*	*A*	*B*	*X*	*A*	*B*	*X*	*A*	*B*	*X*	*A*	*B*	*X*
0	1	0	0	0	0	0	0	0	0	1	0	0	1
1	0	0	1	0	0	1	1	0	1	1	0	1	0
		1	0	0	1	0	1	1	0	1	1	0	0
		1	1	1	1	1	1	1	1	0	1	1	0
NOT GATE		AND GATE			OR GATE			NAND GATE			NOR GATE		

Discrete Logic Circuits

Discrete circuits are made up of individual components, whereas integrated circuits are constructed entirely on one chip. It should be pointed out that logic discrete circuits are seldom used in today's computer technology. However, a discussion of logic would not be complete without a functional analysis of discrete circuits. The circuits we shall discuss are functional. They do not, however, represent the circuits within the integrated circuits except by their functions.

TABLE 6-7
BINARY INPUT COMBINATIONS

No. of Inputs	*Power of 10*	*Binary Combinations*
1	2^1	2
2	2^2	4
3	2^3	8
4	2^4	16
5	2^5	32
6	2^6	64
7	2^7	128
8	2^8	256
9	2^9	512
10	2^{10}	1024

NOT Gate Discrete Circuit The NOT gate is the simplest of the gates. It has one input and one output. These are inverted. Consider the circuit in Figure 6-1. The collector is reverse biased by the positive applied voltage V_{CC}. The base–emitter junction is forward biased whenever there is a high (1) input. The load resistor R_L has a voltage drop of the applied voltage V_{CC} during forward bias. The transistor is on and saturated. The collector of the transistor is at ground potential; therefore, the output is low (0).

When there is a low (0) input, the transistor is off, the load resistor has no current, and therefore no voltage drop. The collector is high (1). Therefore, the output is high (1).

NOR Gate Discrete Circuit The NOR gate is similar to the NOT gate in construction. It has two or more inputs and one output. We shall discuss the circuit in Figure 6-2. The collector is reverse biased by the positive applied voltage V_{CC}. The base–emitter junction is forward biased whenever there are any high (1) inputs. The load resistor R_L has a voltage drop of the applied voltage V_{CC} during forward bias of the junction. The transistor is on and saturated. The collector of the transistor is at ground potential; therefore, the output is low (0).

When there are all low (0) inputs, the transistor is off and the load resistor R_L has no current, and therefore no voltage drop. The collector is high (1). Therefore, the output is high (1).

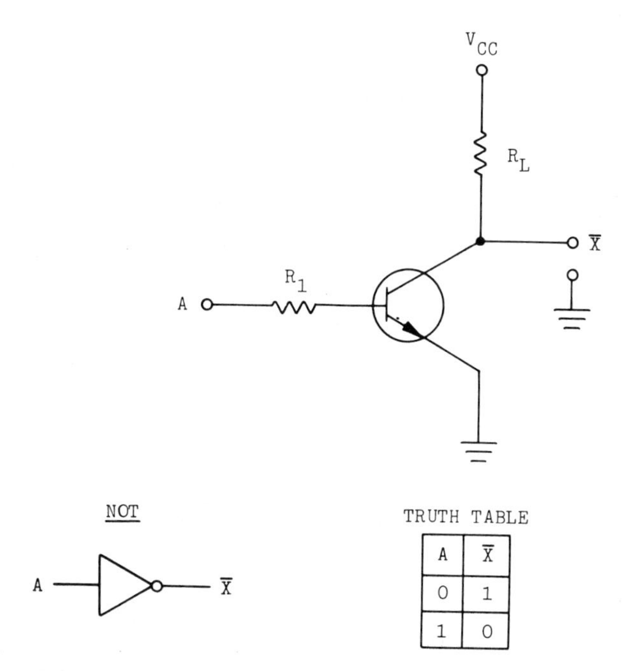

A	$\overline{X}$
0	1
1	0

Figure 6-1 NOT Gate

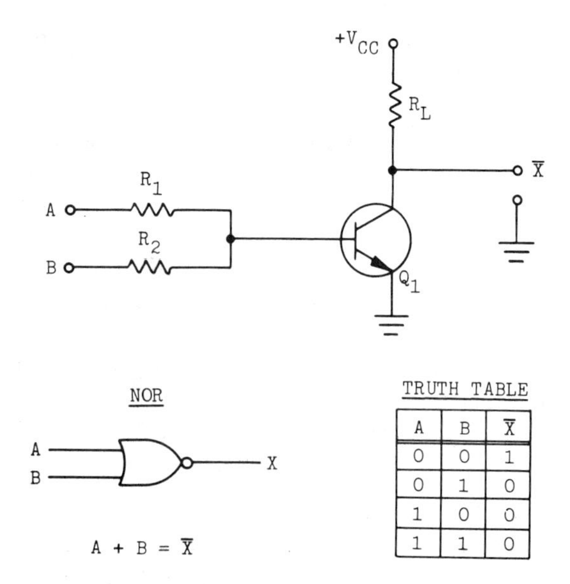

A	B	$\overline{X}$
0	0	1
0	1	0
1	0	0
1	1	0

Figure 6-2 NOR Gate

NAND Gate Discrete Circuit The NAND gate has two or more inputs and one output. Note in the discrete circuit shown in Figure 6-3 that the NAND gate has a pair of diodes at its input. These diodes are forward biased with a low (0) input. The collector is reverse biased by the positive applied voltage V_{CC}. The diodes are reverse biased by two high (1) inputs. A voltage divider, R_1 and R_2, forward biases the base–emitter junction, turning the transistor on into saturation. The load resistor R_L has a voltage drop of the applied voltage V_{CC} during forward bias of the base–emitter junction. The collector of the transistor is at ground potential; therefore, the output is low (0).

When there are low (0) inputs, one or both of the diodes are forward biased and a low potential appears at the junction between R_{B1} and R_{B2}. This low potential turns the transistor off. The load resistor R_L has no current and therefore no voltage drop. The collector is high (1). Therefore, the output is high (1).

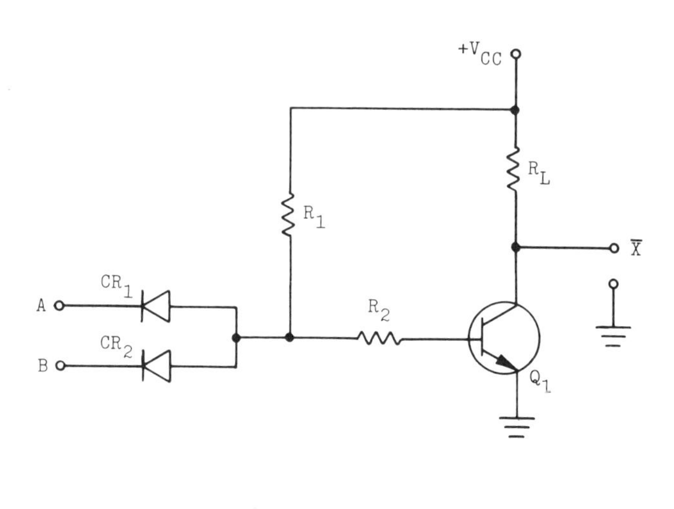

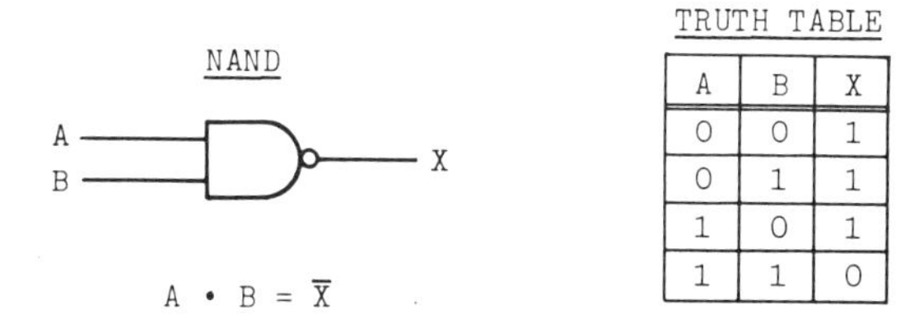

TRUTH TABLE

A	B	X
0	0	1
0	1	1
1	0	1
1	1	0

Figure 6-3 NAND Gate

OR Gate Discrete Circuit The OR gate is the same as the NOR gate with a second stage. Logically, it is a combination NOR gate and NOT gate. It has two or more inputs and one output. In the circuit of Figure 6-4, the collectors of transistors *Q*1 and *Q*2 are reverse biased by the positive V_{CC}. The base–emitter junction of transistor *Q*1 is forward biased whenever there are any high (1) inputs. The load resistor R_{L1} has a voltage drop of the applied voltage V_{CC} during forward bias of the junction. The transistor *Q*1 is on and saturated. The collector of *Q*1 is at ground potential. This ground potential turns off transistor *Q*2. The load resistor R_{L2} has no current and therefore no voltage drop. The collector of *Q*2 is high (1); therefore, the output is high (1).

When there are all low (0) inputs, the transistor *Q*1 is off. Therefore, the transistor *Q*1 can be assumed to be inactive and essentially out of the circuit. With this condition, a voltage divider, R_{L1} and R_{B3}, exists to forward bias the transistor *Q*2 on into saturation. The load resistor R_{L2} has a voltage drop of the applied voltage V_{CC} during forward bias of the base–emitter junction. The collector of the transistor *Q*2 is at ground potential; therefore, the output is low (0).

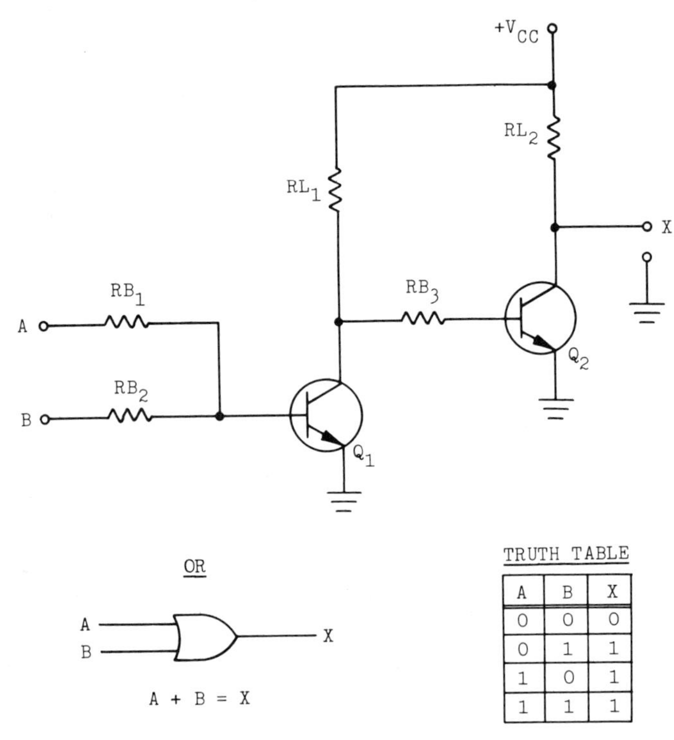

A	B	X
0	0	0
0	1	1
1	0	1
1	1	1

Figure 6-4 OR Gate

AND Gate Discrete Circuit The AND gate is the same as the NAND gate with a second stage. Logically, it is a combination NAND gate and NOT gate. It has two or more inputs and one output. In Figure 6-5, the collectors of transistors $Q1$ and $Q2$ are reverse biased by the positive V_{CC}. The diodes C_{R1} and C_{R2} are reverse biased by two high (1) inputs. With the diodes in reverse bias, a voltage divider R_{B1} and R_{B2} forward biases the base–emitter junction of transistor $Q1$, turning it on to saturation. The load resistor R_{L1} has a voltage drop of the applied voltage V_{CC} during forward bias of the base–emitter junction. The collector of transistor $Q1$ is at ground potential. This ground potential turns off transistor $Q2$. The load resistor R_{L2} has no current and therefore no voltage drop. The collector of $Q2$ is high (1); therefore, the output is high (1).

When there are any low (0) inputs, one or both of the diodes are forward biased and a low potential appears at the junction between resistors R_{B1} and

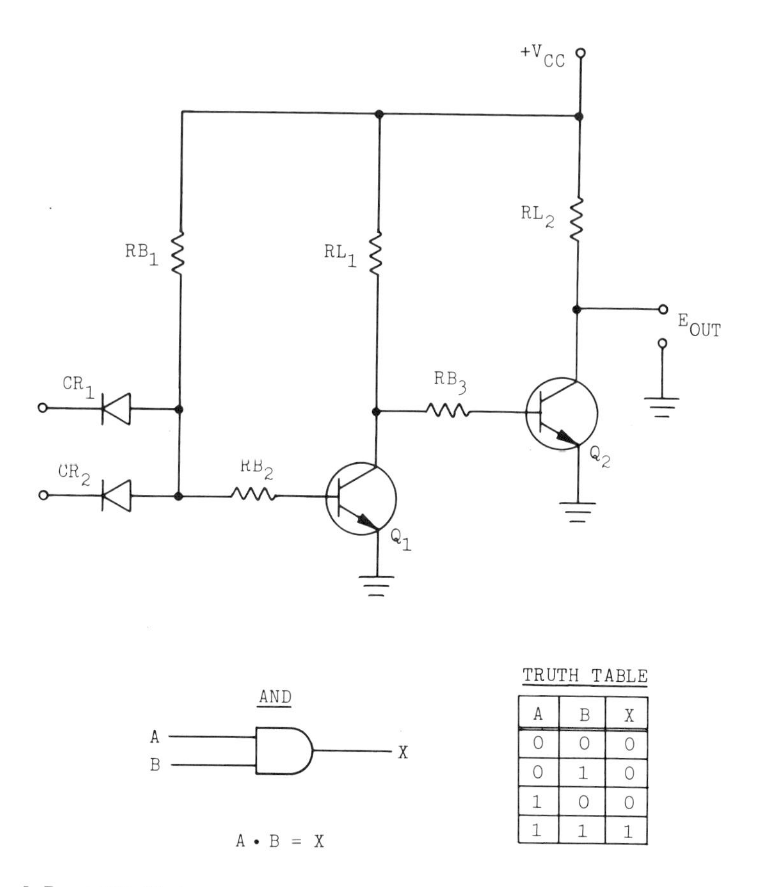

A	B	X
0	0	0
0	1	0
1	0	0
1	1	1

Figure 6-5 AND Gate

R_{B2}. This low potential turns transistor $Q1$ off. Therefore, the transistor can be assumed to be inactive and essentially out of the circuit. With this condition, a voltage divider, R_{L1} and R_{B3}, exists to forward bias the transistor $Q2$ on into saturation. The load resistor R_{L2} has a voltage drop of the applied voltage V_{CC} during forward bias of the base–emitter junction. The collector of transistor $Q2$ is at ground potential; therefore, the output is low (0).

Exclusive OR Gate

The exclusive OR gate is not included in the five basic logic gates. However, it is used extensively in logic circuitry. The symbol for the exclusive OR gate is illustrated in the Figure 6-6. The symbol is exactly like the OR gate with an extra arc across the inputs.

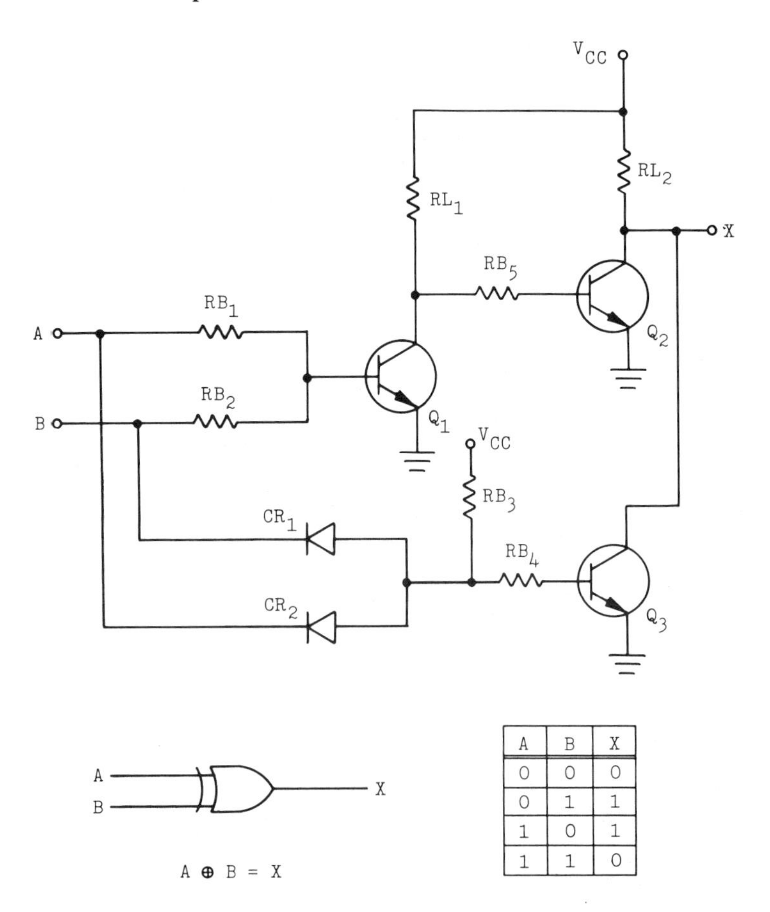

A	B	X
0	0	0
0	1	1
1	0	1
1	1	0

Figure 6-6 Exclusive OR Gate

The mathematical expression for the exclusive OR gate is thus. If A or B is high (1), then X is high (1), but not if both inputs are high (1).

$$A \oplus B = X$$

The truth table in Figure 6-6 is also similar to the OR gate with the last input (1,1) giving a low (0) output.

Let's consider the discrete circuit for the exclusive OR gate. The circuit in Figure 6-6 is a functional circuit consisting of an OR gate at the top and a NAND gate underneath. The operation of the OR and the NAND gate is as described in the preceding paragraphs. With any low (0) inputs one of the diodes is forward biased and the junction between resistors R_{B3} and R_{B4} is low, turning off transistor $Q3$. The circuit simply acts as a NAND gate. With both inputs high (1), the load resistor R_{L2} becomes part of the NAND gate circuit, and a ground appears at the collector of both $Q2$ and $Q3$. The resistor R_{L2} has a voltage drop of the applied voltage V_{CC}, and the output is low (0).

Integrated Logic Circuits

An integrated logic circuit contains several to several hundred transistor configurations on one chip. Its logic functions may be arithmetic or memory. Many of the circuits are gates, while others are multivibrators. The entire integrated circuit is usually less than ½ cm^2. The circuit is usually made from a silicon chip. Logically, the digital integrated circuit has one or several outputs and one or several inputs.

Types of Integrated Logic Circuits There are basically two types of integrated logic circuits: bipolar and unipolar. These types are named from their fabrication techniques.

The process in creating integrated or discrete solid-state devices is the planar process. The process is called planar because the leads of the device are brought to the top surface of the solid-state wafer to be connected, interconnected, or bonded. A physics or electronics text will describe this process in detail.

The two fabrication processes, bipolar and unipolar, are essentially the same. The differences between the two processes are subtle. Unipolar devices are simpler to manufacture because they require fewer steps. Transistor action takes place closer to the surface of the devices. Diffusion of the layers in the device is less critical. Unipolar devices are less expensive, smaller and consume less power than bipolar devices. Unipolar devices are used for integrated circuits that require a large number of gates, say 100 to 10,000. Unipolar devices are slow in comparison to bipolar devices.

Bipolar integrated devices switch at higher speeds, use more power, are generally larger, and are more expensive than unipolar devices. They are generally used where 100 or so gates are required.

The real difference between bipolar and unipolar devices is in their operation not their construction. Bipolar devices use *both* (*N* and *P*) (negative and positive) carriers in operation. Unipolar devices use *either* (*N* or *P*) (negative or positive) carriers in operation.

Classification of Digital Integrated Circuits All integrated circuits are classified by their order of complexity. The complexity is tied to the amount of logic circuits in their makeup. Digital integrated circuits are made by both the bipolar and unipolar fabrication methods.

Small-Scale Integration (SSI). Small-scale-integration (SSI) integrated circuits are those that have 10 to 12 logic circuits in their makeup on one chip.

Medium-Scale Integration (MSI). Medium-scale-integration (MSI) integrated circuits are those that have 10 to 100 logic circuits in their makeup on one chip.

Large-Scale Integration (LSI). Large-scale-integration (LSI) integrated circuits are those that have 100 to 10,000 or more logic circuits on one chip.

The Logic Family of Integrated Circuits In the development of integrated circuits, it was soon found that different approaches for IC production led to many circuit forms. Some of these forms are described in this section. Note that the circuits are simplified for explanation purposes.

Diode Logic (DL). Diode logic (Figure 6-7) is the simplest of the logic circuits within integrated circuits. Diodes are used at the input of the circuit. They are placed in a manner so that an output will only have an effect on the output when a diode is forward biased.

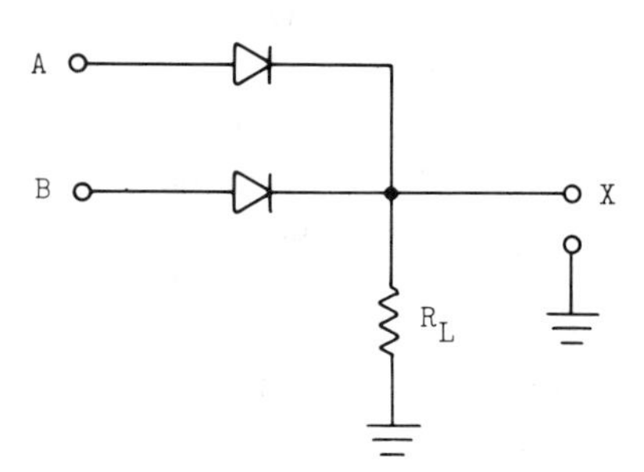

Figure 6-7 Diode Logic (DL)

Resistor–Transistor Logic (RTL). In Figure 6-8, inputs are placed on two bias resistors and the output is taken from transistor collectors with the common load R_L. Base resistors limit the current during high (1) inputs.

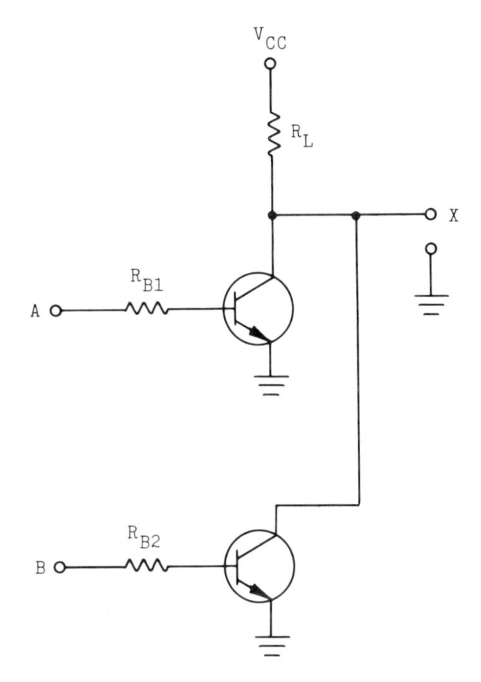

Figure 6-8 Resistor–Transistor Logic (RTL)

Diode–Transistor Logic (DTL). The DTL integrated circuit was developed because of its speed. In Figure 6-9 the input is placed on a pair of diodes while the output is taken from the collector of a transistor with the load resistor R_L.

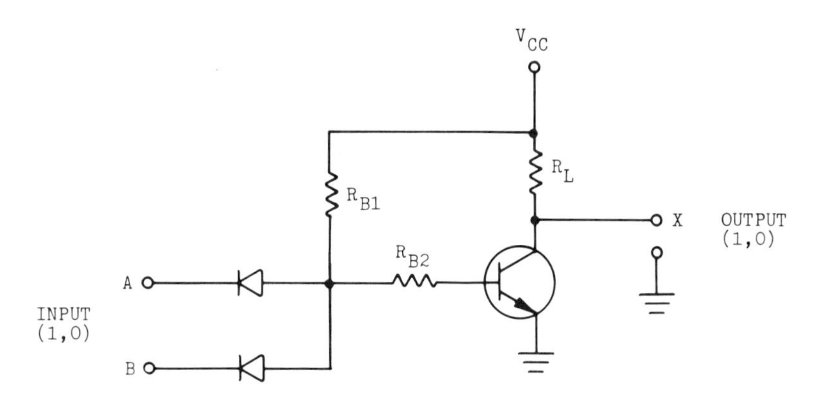

Figure 6-9 Diode–Transistor Logic (DTL)

Transistor–Transistor Logic (TTL). TTL logic (Figure 6-10) is an outgrowth from DTL logic. Inputs are applied to a multi-emitter junction. The output is taken from the collector of a second transistor. The 5400/7400 TTL integrated circuits are the most popular digital logic family used today. There are, of course, persons who would disagree with this because of advancements in manufacturing technology. We will look at the pad layouts for these devices later in the chapter.

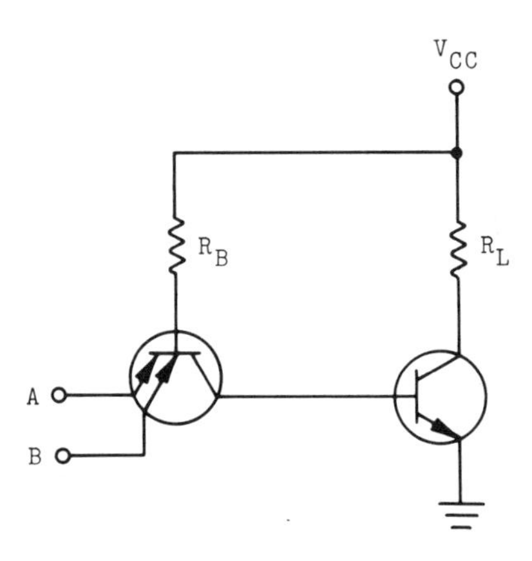

Figure 6-10 Transistor–Transistor Logic (TTL)

Schottky TTL Logic. Schottky TTL logic increases the operating speed of TTLs with some increase in power. A diode is connected between the collector and the base of the logic transistor. This action prevents the transistor from operating in saturation and thereby increases its operating speed.

Integration Injection Logic (I^2L). The I^2L logic uses transistors throughout the integrated circuit. Transistors are the active component in integrated circuits. Inactive components such as resistors are not used in I^2L circuits. Transistors are installed in their place to supply base current and other functions. I^2L logic has low power and medium speed. Passive components such as resistors take up more room on the chip than transistors. Therefore, the I^2L logic can be manufactured with more parts.

Emitter-Coupled Logic (ECL). Emitter-coupled logic is extremely fast, much faster than its counterparts, RTL, DTL, and TTL. These logic family members are defined as saturation-mode logic. Their fabrication method ensures that they are turned on to saturation, then turned off to cutoff. To saturate a transistor, the semiconductor material must be filled with charge carriers, thereby taking time to switch. Emitter-coupled logic ECL biases the transistors above cutoff and below saturation. This places their operation between the two extremes; thus ECL transistors are never saturated or cut off. ECL transistors operate in a current mode (under current at all times).

Small-Scale Integration (SSI)

You may recall that SSI integrated circuits are those that have between 10 and 12 transistor circuits on one chip. The typical integrated circuit that defines SSI are the base gates and flip-flops. The discrete gates we have discussed are AND, OR, NOT, NAND, NOR, and exclusive OR. The flip-flops are utilized as memory devices to store 1 bit of information (high or low) (1 or 0).

We shall feature the bipolar 5400/7400 series for this explanation as it is commonly used in industry. The bipolar 5400/7400 series has faster speed and greater power than the unipolar devices (MOSFETS). Therefore, the emphasis placed by industry on TTL is more significant (at least to this author) than the MOS devices.

Integrated Gate Circuits In Figure 6-11 the basic gates are shown in pad layout looking at the top of the integrated circuit. The pad layout that you see is for the dual-in-line package (DIP) integrated circuit. There are other packages such as the flat pack and the can.

The 7400 series gates are typical of logic ICs available in today's market. The 7400 is a two-Quad NAND gate. This means that each gate has two inputs and there are four gates within the one DIP. Note that pin 14 is designated for applied voltage V_{CC} and pin 7 is common ground.

The 7402 is the two-Quad NOR gate. The identification of gates is the same as in the 7400 except the input and output pins make the individual gates point to the left, rather than the right. There is nothing significant about this except for the statement. Each gate is manufactured in a way that allows the simplest extension of pin leads from the circuit to the gate external pins.

The 7404 is a hex inverter, colloquially called a six-pack inverter. The 7408 is a two-Quad AND gate. The 7432 is a two-Quad OR gate. The 7486 is a two-Quad exclusive OR gate.

The package style in Figure 6-11 is 14 pin. Other dual-in-line packages are 8 pin and 16 pin. All TTL devices require a 5-V power supply (V_{CC}). Their logic levels are high (1) and low (0). Logic voltage levels are 5 V and 0 V, respectively. All TTL family members are compatible with one another and may be used without regard to impedance matching.

Other families such as CMOS are compatible as to the power supply but have different propagation times. Therefore, the two families (CMOS and TTL) are seldom used concurrently unless there are special adaptations used in coupling the devices, such as an optoisolator.

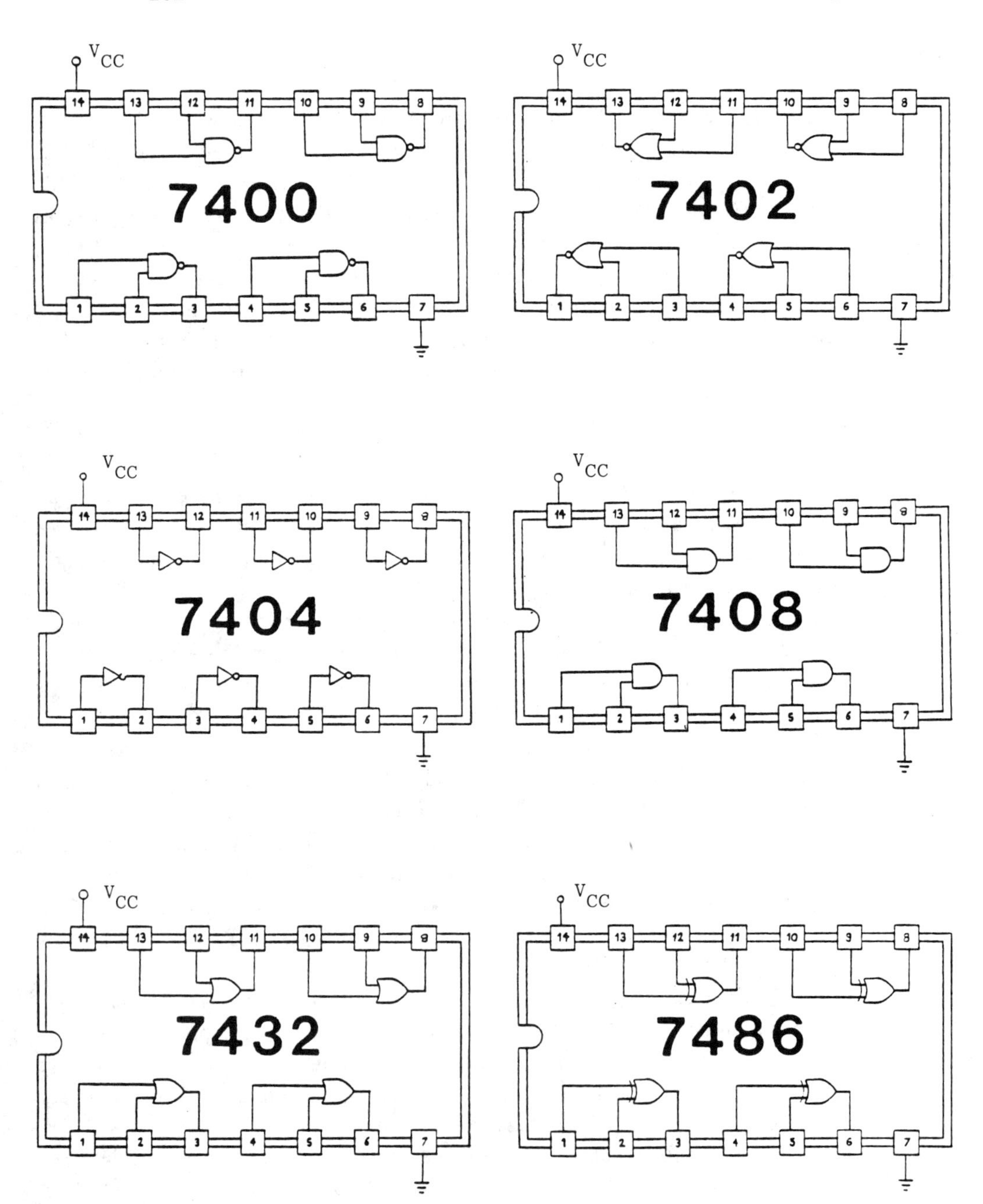

Figure 6-11 Integrated Gate Circuit Pad Layouts

Flip-Flop Circuit The basic gates and flip-flop circuits comprise the bulk of all logic circuitry. The flip-flop is constructed of NAND and NOR gates. The flip-flop is a memory circuit that stores 1 bit of information either high or low. The basic flip-flop, when turned on by switching action, remains in that state until told to change states. Flip-flop circuits are often made with one of the basic gate ICs or are on an integrated circuit of their own. They may function individually to set or reset, be clocked to change states, or operate synchronously.

RS Flip-Flop Most fundamental of the flip-flop circuits is the set–reset (RS). Figure 6-12 shows this basic circuit made from a pair of NAND gates. The truth table for the RS flip-flop and its logic symbol are also included in the illustration. Noting the logic symbol, the $\overline{Q}$ output is the inversion of the Q. Inputs are called set and reset (S and R). Thus the name RS flip-flop. In the first state condition as shown in the truth table, two lows are not allowed, for this will cause the flip-flop to go to an unknown state. A low on set (S) with the reset (R) position high will make $\overline{Q}$ go to high and Q go to low. The logic high at $\overline{Q}$ is fed to the NAND gate input of reset, and the gate output Q is driven low. The set (S) input now may be changed to high, but the state of the flip-flop will not change. Q and $\overline{Q}$ will remain in their last state. This action is called *latching*. Often the RS flip-flop is called a *latch*.

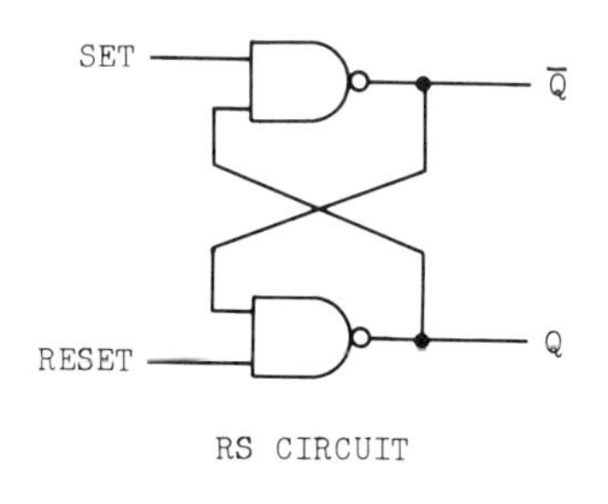

RS CIRCUIT

R	S	$\overline{Q}$	Q
0	0	NOT ALLOWED	
0	1	0	1
1	0	1	0
1	1	$\overline{Q}$	Q

TRUTH TABLE

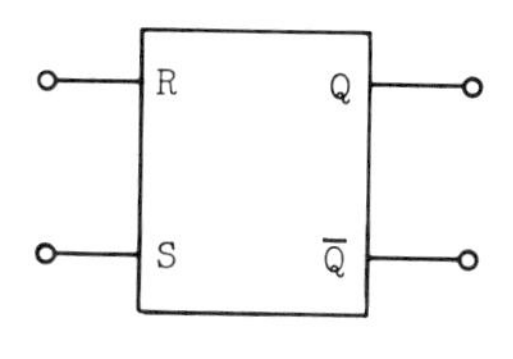

LOGIC SYMBOL

Figure 6-12 RS Flip-Flop

A low on reset with set high will cause Q to go high. In each condition the two outputs will be inverted from one another.

In some applications the set position may also be called *preset* and the reset position *clear*. It must be noted that the truth table may be expanded to eight possible conditions. The switching involved in the truth table explanation assumes that the states of the flip-flop are as shown. They could indeed be just opposite.

Clocked RS Flip-Flop The RS flip-flop is an asynchronous circuit that is not related in time. The inputs are placed in the gates at random. The clocked RS flip-flop is timed, and therefore called *synchronous*. The clock decides in timed sequence when the flip-flop will change states. Figure 6-13 is representative of the synchronous RS flip-flop. The circuit uses a basic flip-flop with a pair of NAND gates and a clock input added. The clock signal is NANDed with the set and reset inputs. When the clock is high (1), the outputs of either of the input gates are dependent on the set and reset input functions. If, for instance, the set input was high (1) and the reset low (0), the set gate will have a low (0) output. The reset gate in turn will have a high (1) output. If the clock changes to low (0), the set and reset gates both will then be high (1). The second set of NAND gates operate in the same manner as the basic RS flip-flop. The logic symbol is also the same as the RS flip-flop with the added clock (Ck) input.

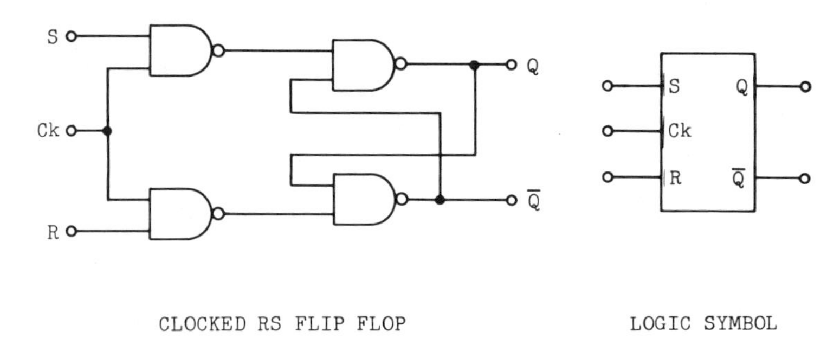

Figure 6-13 Clocked RS Flip-Flop

Toggle (T) Flip-Flop In Figure 6-14 each input to the AND gates causes the flip-flop to toggle (switch). The first input pulse (high) sets Q to high and $\overline{Q}$ to low. The second input pulse (high) returns Q back to low. Note that it took two pulses (high) to provide this counting action. If the T flip-flop is used as a counter with the circuit shown, counting by 2 is accomplished.

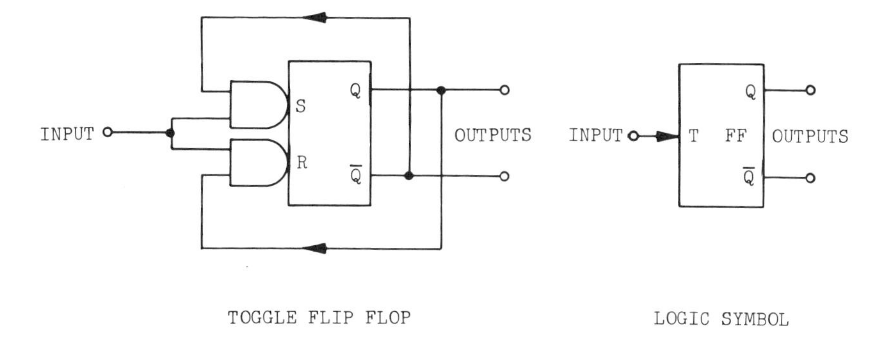

Figure 6-14 Toggle (T) Flip-Flop

D Flip-Flop Another synchronous flip-flop is the D type (Figure 6-15). Note that the D flip-flop is an RS-type flip-flop. This circuit is also called a *latch*. Data are placed in the flip-flop as high (1) or low (0) at D, whenever a trigger pulse occurs at Ck. Most D flip-flops also have set and reset functions, which react like that circuit. The D flip-flop may be used as a toggle type by feeding the $\overline{Q}$ output back to the D terminal and applying a clock pulse to the Ck terminal.

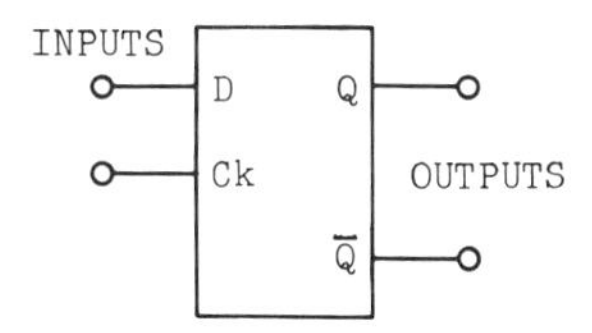

Figure 6-15 D Flip-Flop

JK Flip-Flop The basic JK flip-flop utilizes a pair of AND gates at the input. Each AND gate has a data input (J and K). A clock input makes this flip-flop a synchronous one. The outputs are standard (Q and $\overline{Q}$). Essentially, the JK flip-flop is a clocked flip-flop with preset and clear functions. The letters JK mean absolutely nothing. As far as anyone can determine they were simply selected. The JK utilizes the functions of the RS, the T, and the D flip-flops.

Figure 6-16 is a JK flip-flop circuit and its logic symbol. In the first step of the truth table, J and K inputs are both low (0) and will be unchanged by a trigger from the clock (Ck).

If J is high and K is low when the clock is low, Q goes high or stays high, whichever may be the case, and $\overline{Q}$ will be low. If K is high and J is low when the clock is low, $\overline{Q}$ goes high, whichever may be the case, and Q will be low. If J is high and K is high, the flip-flop toggles at each clock pulse.

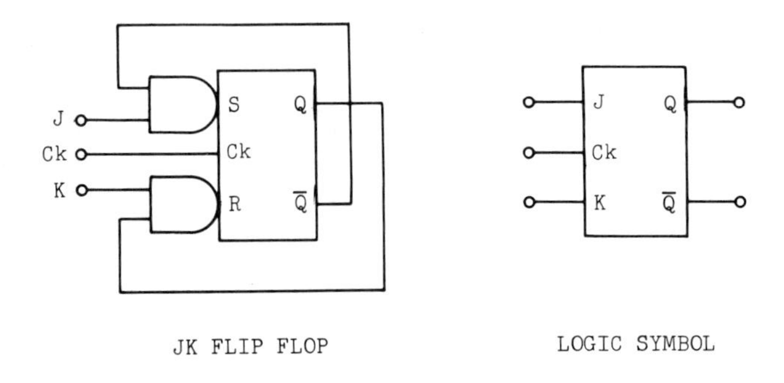

TRUTH TABLE

J	K	Q	$\overline{Q}$
0	0	NO EFFECT	
0	1	RESET	
1	0	SET	
1	1	TOGGLE	

Figure 6-16 JK Flip-Flop

Medium-Scale Integration (MSI)

The medium-scale integrated circuits you may recall are those that have between 10 and 100 transistor circuits. Most of these devices are indeed combinations of gates and flip-flops.

Ripple Counter The fundamental counter circuit (Figure 6-17) counts and remembers clock pulses. Some counters are small-scale integration (SMI) devices; others are medium-scale integration (MSI) devices. The circuit is called a ripple counter. It is usually constructed with JK flip-flops. In the initial stage, all the JK flip-flops are set so that all outputs are low. A clock pulse applied to the first flip-flop causes the $Q1$ output to flop to high from low. The second pulse into the first flip-flop causes $Q1$ to go low and triggers the second flip-flop. The second flip-flop output goes from low to high. The clock input continues in this manner with the following sequence:

$Q1$: changes every second pulse
$Q2$: changes every fourth pulse
$Q3$: changes every eighth pulse
$Q4$: changes every sixteenth pulse

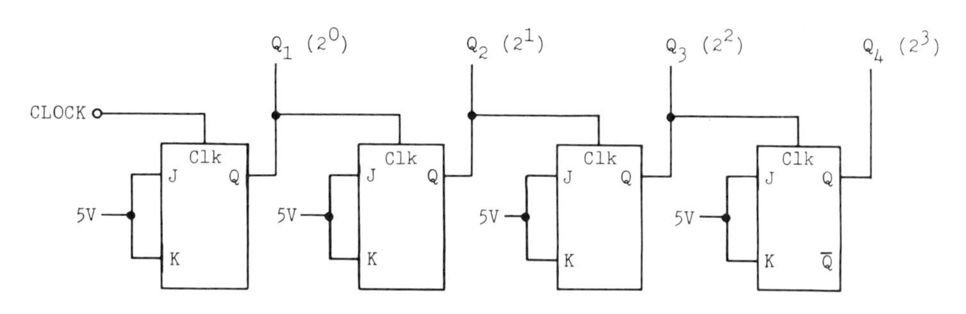

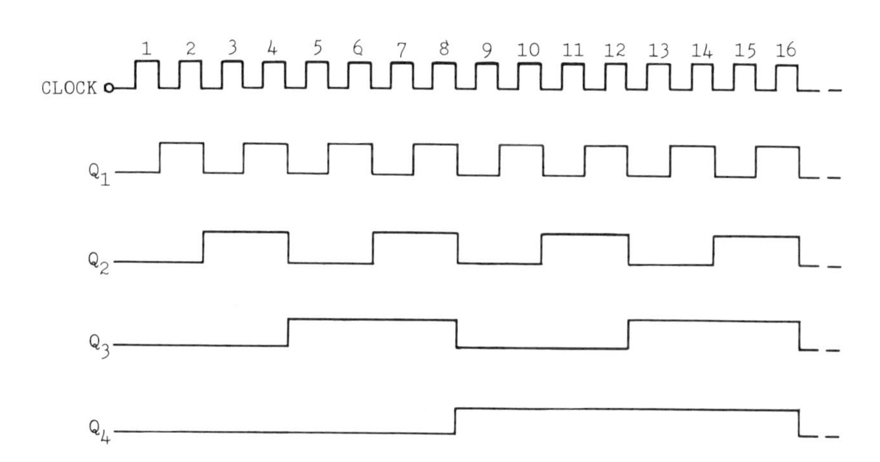

Figure 6-17 Ripple Counter

This ripple effect is where the counter gets its name. The output would be 1111 (*Q*1, *Q*2, *Q*3, *Q*4) at the fifteenth pulse and 0000 at the sixteenth. That is, all four of the flip-flops must change state. The clock pulse must ripple through all flip-flops.

The lower part of Figure 6-17 shows the pulse states of the flip-flops at different clock periods. Wave forms at *Q*1, *Q*2, *Q*3, and *Q*4 are presented in time and pulse periods.

The maximum count of a counter is $2^N - 1$, where N equals the number of flip-flops. Two flip-flops will provide $2^2 - 1$ or a count of 0 through 3. Three flip-flops will provide $2^3 - 1$ or a count of 0 through 7. Four flip-flops will provide $2^4 - 1$ or a count of 0 through 15.

Typical of the TTL counters are the 7490, 7493, 74160, and 74190.

Register The register is a memory circuit. You may recall that flip-flops are devices that store a single bit of binary information. When flip-flops are used in combinations to store multiple bits of information, the combination is called a *register*. Some registers are set simultaneously. These are called *parallel* registers. Other registers operate 1 bit at a time. These are called *shift* registers. Registers are characterized by their shift direction and their data out–data in capabilities.

In Figure 6-18, a typical register shifts to the right. That is, the data enter at the left flip-flop and move through stages of flip-flops to the right. Other flip-flops move data from right to left. Still others are universal in operation. Data can be placed in either end and move through stages in either direction. We therefore have three types: right, left, and right–left.

In Figure 6-18, a serial in–parallel out register is shown. Each bit of information must be placed into the register one at a time and then shifted to make room for the next bit of information. This process continues until the register is at its capacity. A serial-in register is one that accepts inputs one at a time. In a parallel-in register, all flip-flops are loaded simultaneously. A serial-out register shifts the bits of information out one at a time. A parallel-out register may shift all the information out simultaneously.

The following are the register in and out combinations:

Serial in–serial out
Parallel in–serial out
Serial in–parallel out
Parallel in–parallel out
Serial in–serial/parallel out
Parallel in–serial/parallel out

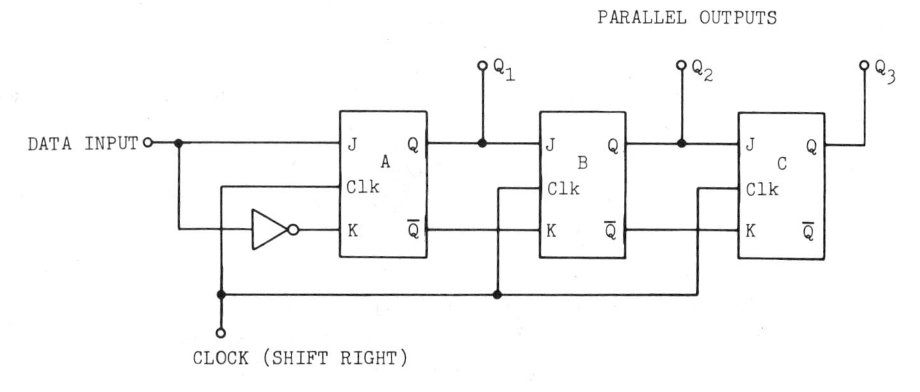

Figure 6-18 Register

Let's consider the serial in–parallel out register in Figure 6-18. Data are placed in the input 1 bit at a time and are taken from the flip-flops in parallel simultaneously. The K input is inverted from the J input. A high is on the first flip-flop, providing a high on $Q1$. The next pulse transfers this high to $Q2$, and so on through the register.

Typical of the TTL registers are the 7494, 7495, 74194, and 74195.

Decoder The purpose of the decoder is to convert digital information from one format to another. Let's consider the block diagram shown in Figure 6-19. The output of a counter is binary. This output is placed into a decoder whose output is coded and drives a 7-segment display. Decoder outputs may be decimal, 16 states for a four-stage counter, or as a driver for seven-segment displays. Typical of TTL decoders are the 7447 and the 74141.

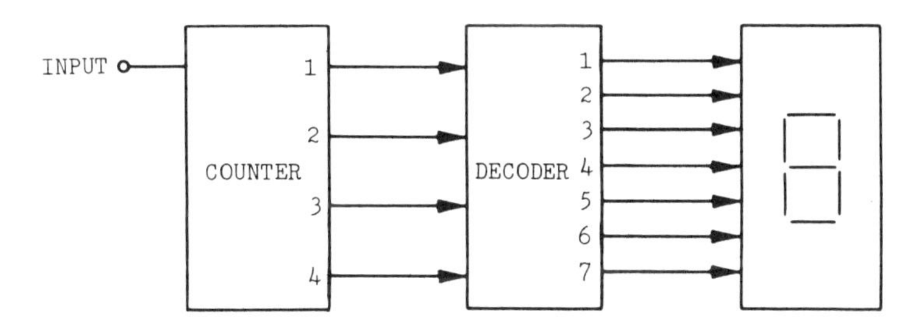

Figure 6-19 Decoder

Adder Binary addition, when you understand it, is rather like child's play. There are as a total only four possible combinations of ones and zeros to perform addition on.

$$\begin{array}{r} 0 \\ +0 \\ \hline 0 \end{array} \qquad \begin{array}{r} 0 \\ +1 \\ \hline 1 \end{array} \qquad \begin{array}{r} 1 \\ +0 \\ \hline 1 \end{array} \qquad \begin{array}{r} 1 \\ +1 \\ \hline 0 \end{array} \text{ and carry 1}$$

The answers here are the sum. The carry is either zero (0) or carry 1. A truth table can be created, which may define this action in another way.

A	*B*	*Sum*	*Carry*
0	0	0	0
0	1	1	0
1	0	1	0
1	1	0	1

A logic equation may be written to describe this truth table. The equation would be defined as all cases in which the output is high (1).

$$\text{Sum} = \bar{A}\,B + A\,\bar{B}$$
$$= A + B$$

The carry output has only one output and can be written

$$\text{Carry} = AB$$

Figure 6-20 illustrates the half-adder. It is called a half-adder because it does not have the provisions for adding a carry from a previous set of additions. In actuality, it is an exclusive OR gate and a NAND gate.

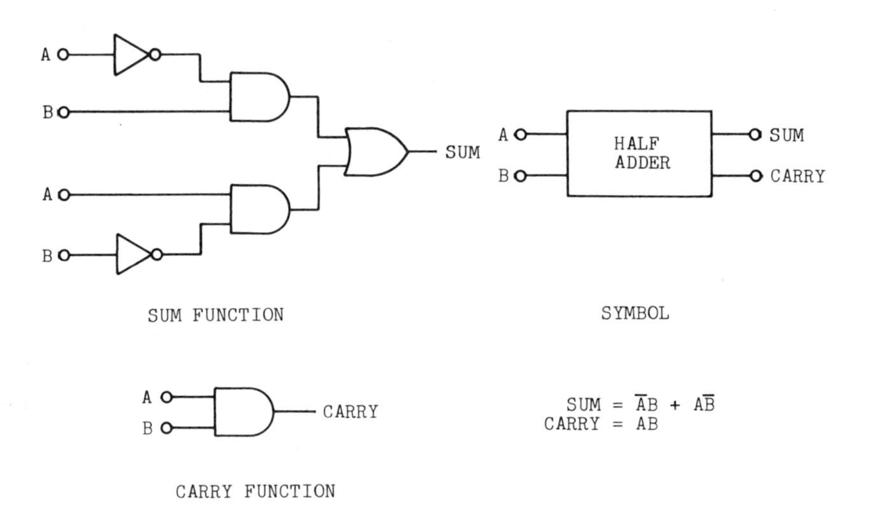

Figure 6-20 Half-Adder

The full adder takes up where the half-adder leaves off. As you may recall, the half-adder does not have the capability of adding a carry from a previous addition set. The full adder accomplishes this using two half-adders and a NOR gate. In Figure 6-21 a full adder schematic is shown with its logic symbol. In the schematic it is shown that the inputs to the full adder are two letters and a carry in. Its outputs are a carry out and a sum. The first half-adder calculates the sum and carry inputs from A and B inputs. The second half-adder calculates the carry in from a previous adder and the carry from the first half-adder. The

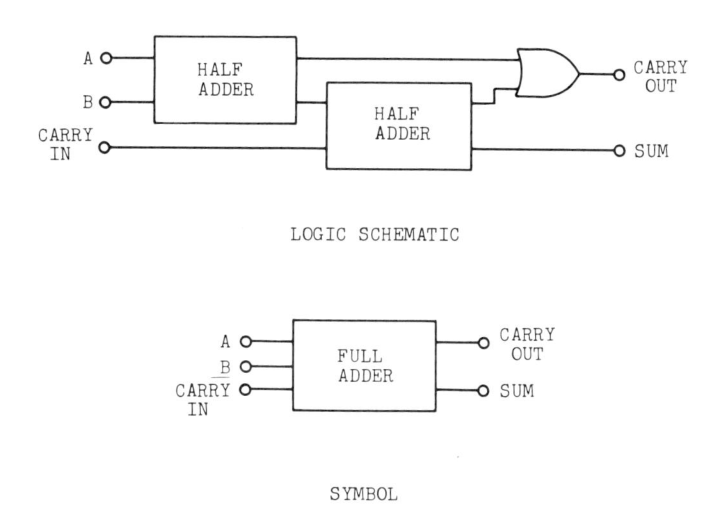

Figure 6-21 Full Adder

full adder outputs are carry out and sum. Therefore, the full adder provides summing and carry value functions both in and out.

The 4-bit adder in Figure 6-22 is typical of the devices that add numbers consisting of several bits. These are usually combinations of single-bit full adders. The least significant bits are A_o, B_o, and C_{io}. The most significant bits are A_3, B_3, and C_{i3}. Each adder in turn calculates and produces an output into the next adder as a carry out. Typical of TTL 4-bit adders is the 7483.

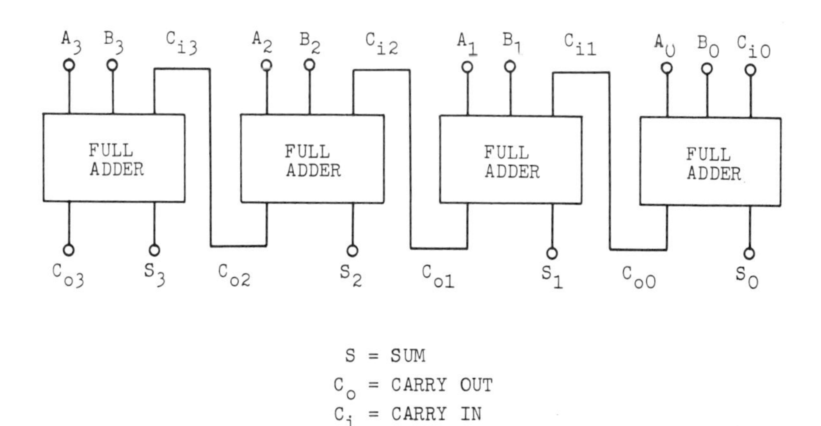

Figure 6-22 Four-Bit Adder

Large-Scale Integration (LSI)

Large-scale integration components have 100 or more transitor logic circuits on one chip. LSI components are largely dominated by unipolar devices (metal oxide semiconductors, MOS). These devices include random-access memory (RAM), read-only memory (ROM), digital clocks, calculators, and microprocessors. LSI devices are complex combinations of logic circuits. We shall leave these combinations to books dedicated in their entirety to digital electronics. The scope of this book does not allow the detail required to develop an understanding of these devices.

7

Microprocessors and Microcomputers

The *microprocessor* is a part of a microcomputer. It is a single integrated circuit chip containing a part of the computer. The part it usually contains is called a central processing unit (CPU). Some microprocessor chips contain arithmetic units, control registers, and some memory. The microprocessor may indeed be the most thrilling and innovative electronic device ever conceived. The *microcomputer* is an extremely small (miniature) digital information processing system consisting of integrated circuits and a microprocessor. The two names, microprocessor and microcomputer, are used synonymously.

It's rather interesting to know that the word computer comes from the Latin *computare* meaning to reckon or to think. A digital computer is one that reckons or calculates. Humans first reckoned with fingers, toes, and stones, all of which helped their minds to compute. The abacus evolved from the use of pebbles around 4000 B.C. In 1615, John Napier devised what was called "Napiers bones," numbered squares to calculate with. Blaise Pascal, a 19-year-old French mathematician, invented the first accounting machine. It was composed of a series of gears each with 10 teeth to represent the numbers 0 to 9. The gears turned using a pencillike stylus. When 9 was reached, the gearing made an automatic change to 10. This same principle is used in modern-day digital electronics. In 1673, a German named Gottfried Leibnitz developed a similar machine that also multiplied. A Frenchman, Basile Bouchon, in 1725 invented punched paper tape for the operating of weaving machines. Still another Frenchman, Joseph Jacquard, in 1891, used punched coded cards for controlling looms. An American, Thomas, invented a desk calculator in 1820 that used the same basic principles as the Leibnitz machine. The "Thomas" calculator was used in business for the next hundred years.

The first analytical machine was invented by Charles Babbage, an Englishman, in 1833. He was a brilliant mathematician, but not very popular. His design was sound, but technology was not advanced enough to build the machine even with British government funding. Parts of it were completed, and worked, but the whole machine failed because the concept was far in advance of the engineering technology.

George Boole, an English mathematician, published an article in 1854 called "Investigation of the Laws of Thought on Which are Founded the Mathematical Theories of Logic and Probabilities." These laws are what are known today as Boolean algebra, used extensively in computer programming.

Dr. Herman Hollerith, head of the U.S. Census Bureau, in 1886 invented a machine using the punched card that saved 60% to 70% of the time for completing the census of 1890. The card was followed by a sorting device built by Hollerith. This opened up a new wide field of computing aids.

Business and industry grew and by the time of World War I, a mechanical keypunch, a gang punch, a vertical sorter, and a tabulator were meeting the accounting needs of the nation and the world. All these machines were, as you may have determined, mechanical. The first electrical machine was designed by Dr. Howard Aiken of Harvard in 1939. It was called the Mark I and was built by IBM Corporation. This was the first true electric computer and was constructed of electromagnetic relays and punched cards.

The first electronic computer was designed and built by J. P. Eckert, Jr. and J. W. Mauchly of the University of Pennsylvania in 1945. It was called the Electronic Numerical Integrator and Calculator (ENIAC). Programs were hard wired. To change a program meant soldering thousands of circuits. The ENIAC used 18,500 tubes (estimate) and weighed over 30 tons. It covered 15,000 square feet of floor space and consumed 200 kilowatts of power. It could add two numbers in two-tenths of a millisecond. Compare that with modern-day computers, which can perform millions of additions per second.

After ENIAC, the next computer of significance was the Selective Sequence Electronic Calculator (SSEC) built by IBM Corporation in 1948. It was a large-scale digital computer and was used in the atomic energy field. The success of this computer led many manufacturers into the development of digital computers for commercial use. Also in 1948, M. V. Wilks of Cambridge University, England, built the first stored-program computer, called EDSAC.

The UNIVAC, or Universal Automatic Computer, was produced in 1951 by the Remington Rand Corporation. Eckert and Mauchley had left the University of Pennsylvania and established a company of their own which became a division of Remington Rand Corporation. UNIVAC I along with the IBM 700 series were the nucleus for the first-generation computers, the tube type. MANIAC was built by Princeton. SEAC was built by the Bureau of Standards. WHIRLWIND was built by the Massachusetts Institute of Technology.

Second-generation computers were of the transistor type. These were designed and developed in the late 1950s. They were smaller, faster, and many times more reliable.

Integrated circuits were the basis for third-generation computers in the 1960s. IBM 360 and UNIVAC 1108 were standards.

Fourth-generation computers came about in the late 1960s with the advent of solid-state medium-scale integration (MSI) and large-scale integration (LSI). Processing systems were simply built on single chips.

These several paragraphs provide the reader with a touch of history to set the stage for the microcomputer and microprocessor chip. This chapter will provide you with a cursory look at these miracles of modern science. Detailed descriptions will be left to computer books. In these few pages we will provide the reader with a foundation of their function.

THE CALCULATOR

The calculator is an electronic machine used to perform mathematical functions such as add, subtract, multiply, and divide. The basic calculator circuit in Figure 7-1 illustrates the components involved in this work.

The calculator consists of a power supply, a calculator integrated circuit, a digital driver, a numerical display, and a keyboard. These are all packaged together in a small case that is usually designed for its usefulness and charisma. The calculator integrated circuit (IC) is typically a 24-pin MOS/LSI. Two pins are connected to a 9-V battery. MOS devices require little power, and therefore the calculator will work for long periods of time.

The numerical display is a package of light-emitting diodes (LEDs) placed in seven-segment displays. Each segment will light up when signaled to create different numbers from 0 through 9. There are as many as 10 or more seven-segment displays on modern-day calculators.

Power is applied to the calculator IC and the keyboard. The keyboard provides control and signal inputs to the calculator IC and control to the digital driver. Keys on the keyboard are arranged in a matrix and are electronically scanned and debounced by the calculator IC. The digital driver buffers the calculator IC inputs and drives the display segments.

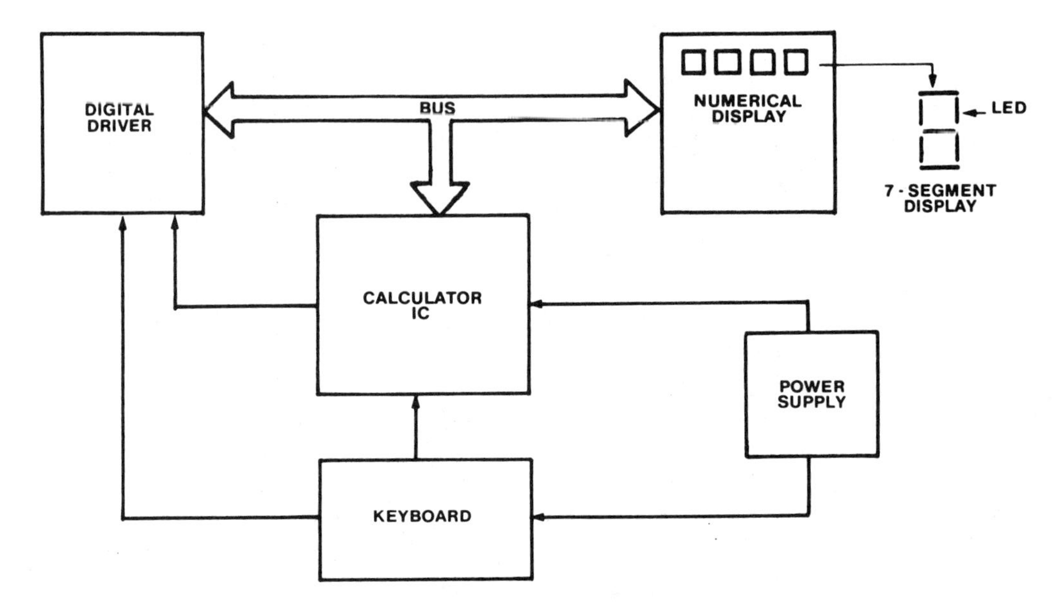

Figure 7-1 Calculator Schematic

Calculators have evolved into units that provide for math functions such as trigonometry, logarithms, percentages, and just about any other professional need.

THE BASIC COMPUTER

The basic digital computer consists of three fundamental subsystems (see Figure 7-2). These are the input and output (I/O), the control processing unit (CPU),

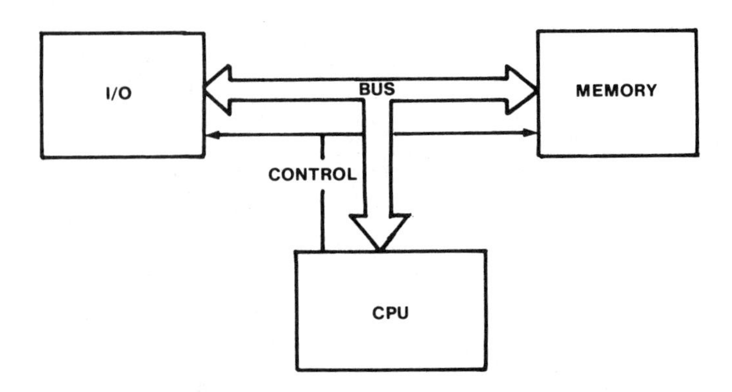

Figure 7-2 Fundamental Computer

and the memory. These are all tied together by electrical control wiring and signal data buses.

The I/O unit is an input/output device that allows the operator to interface with the computer. It may input data and/or programs and achieve results from a readout or a printer. The memory unit stores the input, the program, and other data, and gives it up when told to do so. The central processing unit (CPU) performs mathematical and logical operations and controls the operation of all units. The data bus carries signal data between components. Control wiring carries commands to direct the system to function in its various modes.

Some Computer Vernacular

All computers have similar components and circuitry. Some standard terms can be applied to all computer systems.

Programming Instructions input to the computer to carry out a sequence of operations are called programs. These are applied by keyboard, tape, punched cards, and other methods. Programming may be directed to the processing unit or the memory.

Hardware The physical parts of the computer are called hardware. These parts may be major components such as the CPU, memory, I/O, and keyboard. Other hardware may be the entire package or wiring. Connectors and printed circuits also fall into this category.

Software Instructions to the computer are called software. The instruction may be in the form of keyboard inputs to memory or processor, magnetic tape, punched cards, paper tape, and so on. Programmed instructions are loaded into the computer memory called random-access memory (RAM).

Firmware These are programs that are fixed within the computer. These programs are called read-only memory (ROM).

ROM Read-only memory (ROM) is permanent programmed memory. It cannot be erased or modified. It may be called up (read out) but not written into. ROMs are not affected by shutoff of power to the computer.

RAM Random-access memory is that memory which may be read out of, written into, erased, and reused. In other words it may be modified by almost any means.

Volatile The word volatile has to do with loss of memory. If memory is lost when power is turned off, the memory is termed volatile.

Nonvolatile If memory is retained when power is turned off, the memory is termed nonvolatile.

External Memory Memory that is added to the basic computer and interconnected to the CPU is external memory.

Language

There are many language types used with a computer. Each has its place in computer operation and understanding. Operation and programming are accomplished in computer language. The lowest-level computer language is *machine language.* Machine language is a string of 0's and l's used to trigger the gates and multivibrators that control the computer operation. Machine language consists of two sets of instructions that instruct the operation to be used and the address to be operated on.

Assembly language is an instruction from machine language translated into a mnemonic code such as CLR, which means clear. It is a memory aid and allows the programmer to construct easily remembered instructions.

COBOL is an acronym meaning COmmon Business Oriented Language. FORTRAN is another acronym meaning FORmula TRANslation. Obviously, this is a mathematical-directed language. BASIC (Beginner's All-Purpose Symbolic Instruction Code), along with FORTRAN and COBOL, are procedure-directed languages. That is, they are used to direct the computer to perform some logical task rather than to deal with machine operation. Such operations as retrieving data from memory or loading data are basic and typical functions of these higher-level languages.

Microprocessor

The microprocessor chip (Figure 7-3) may have as many as 40 pins or more. Typical of the microprocessor chip is the 8080. The 8080 microprocessor pin designation is shown in Figure 7-3. The processor's internal function diagram is illustrated in Figure 7-4.

The primary functional sections of the microprocessor are as follows:

1. The accumulator
2. Arithmetic logic unit
3. General register
4. Instruction register
5. Instruction decoder

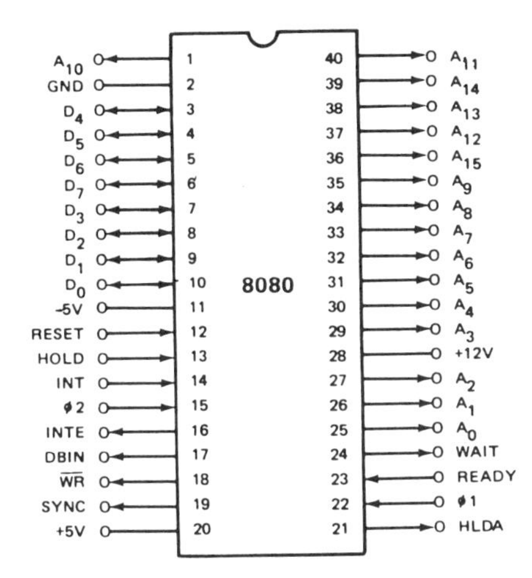

Figure 7-3 Pin Connection Diagram for 8080 Microprocessor

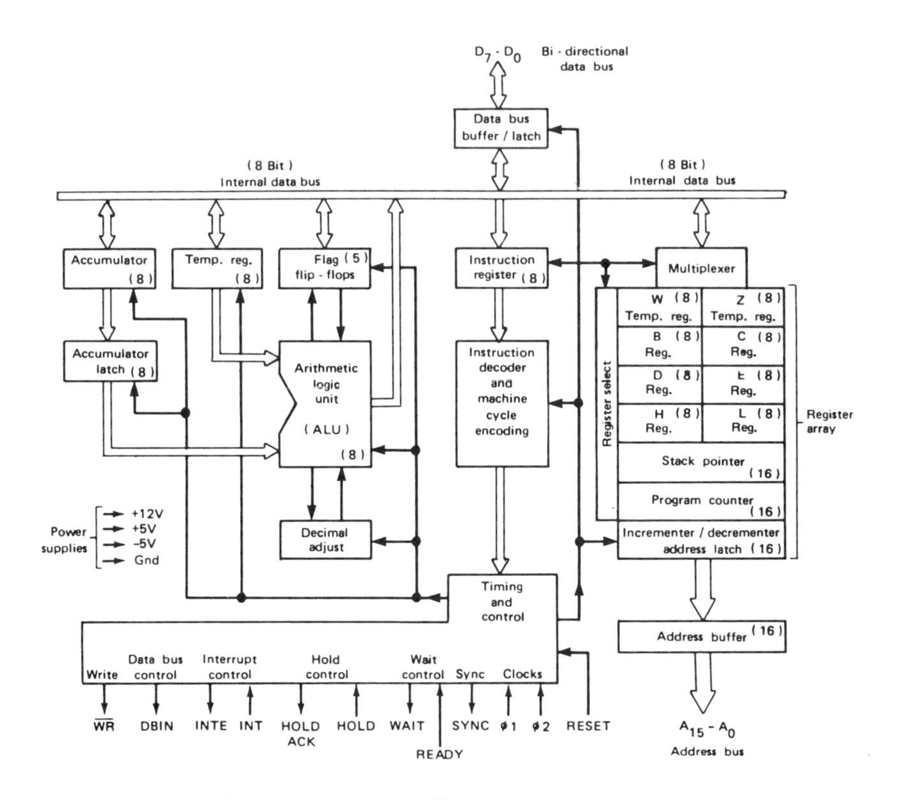

Figure 7-4 Internal Function Diagram for 8080 Microprocessor

Each of these major units is accompanied by a bus driver. The bus driver controls the direction of data flow. The block diagram in Figure 7-4 will provide an understanding of the 8080 function. The 8080 interfaces with an I/O and a memory. Internal to the microprocessor, data are carried on an 8-bit bus. External to the microprocessor, data are carried on a 16-bit bus between the memory and the I/O and the CPU. The internal bus is on lines D_0 thru D_7. The external bus uses lines A_0 through A_{15} to address the memory and the I/O. The bus is bidirectional and is controlled by the instruction decoder.

Incoming instructions are temporarily stored in the instruction register and decoded to prepare for the next operation. Information may be stored in the register array prior to being operated on or before being sent to the I/O.

Mathematical operations are performed by the arithmetic logic unit (ALU). Tied directly to the ALU are registers and accumulators that temporarily store information while it is being operated on. Data are input and output by way of the 8-bit internal bus.

The program counter locates the CPU's next instruction and automatically increments itself after each instruction. A clock pulse provides accurate timing for simultaneous operation of all components. Twelve control signals provide timing and control. The microprocessor is powered by $+12$ Vdc and $\pm$ 5 Vdc. The 8080 is a complete central processing unit used for general-purpose digital computer systems.

Microprocessors are best known for their work with digital computers. However, these versatile devices are used in many industrial applications. The microprocessor may be interfaced with special-purpose machinery to provide what are known as dedicated machines. Dedicated machines use external memory (ROM) to perform very specific functions. Some of these functions are machine shop cutting operations, home appliance controls, electronic games, automatic drafting equipment, and communication controllers.

8

Transducers, Sensors, and Detectors

A *transducer* is a device that converts one form of energy or physical quantity to another. The energy or stimulus determines the quantity of the signal.

A *sensor* is a device used to detect, measure, or record physical phenomena such as heat, radiation, and the like, and to respond by transmitting the information, initiating changes, or operating controls.

A *detector* is a device used to sense the presence of an entity such as heat, radiation, or other physical phenomena.

As you can readily see, any difference between these three devices is an extremely thin line. There are other devices such as the gage and the pickup that essentially have the same definitions. For the purposes of this book, we shall attempt to understand the transducer as the basic signal-gathering and transmitting device. However, we may digress at times. By developing our comprehension of the functions of a transducer, we will provide a broader definition of these devices.

MEASUREMENT FUNCTIONS

The functions of a transducer or sensor are deeply involved in measurement or control. In Figure 8-1a, a sensor is attached directly to an indicator. This is the most basic of measurement systems. The sensor is acting as a sensing unit and also as a driver for an indicator pointer.

In Figure 8-1b, a sensor responds to some physical quantity such as heat. Wires, called extension wires, made from the same material as the sensor, extend to an indicator. The indicator pointer responds to a change in the physical quantity felt at the sensor.

In another simple system, shown in Figure 8-1c, a transducer is used. The sensor responds to some physical quantity. The response is coupled to a compatible transducer, which converts the sensor signal to an electrical signal. The electrical signal is transmitted by wire to an indicator pointer. The pointer responds to a change in the physical quantity felt at the sensor.

In the more complex system of measurement shown in Figure 8-1d, a signal conditioner is used. The sensor responds to some physical quantity. The

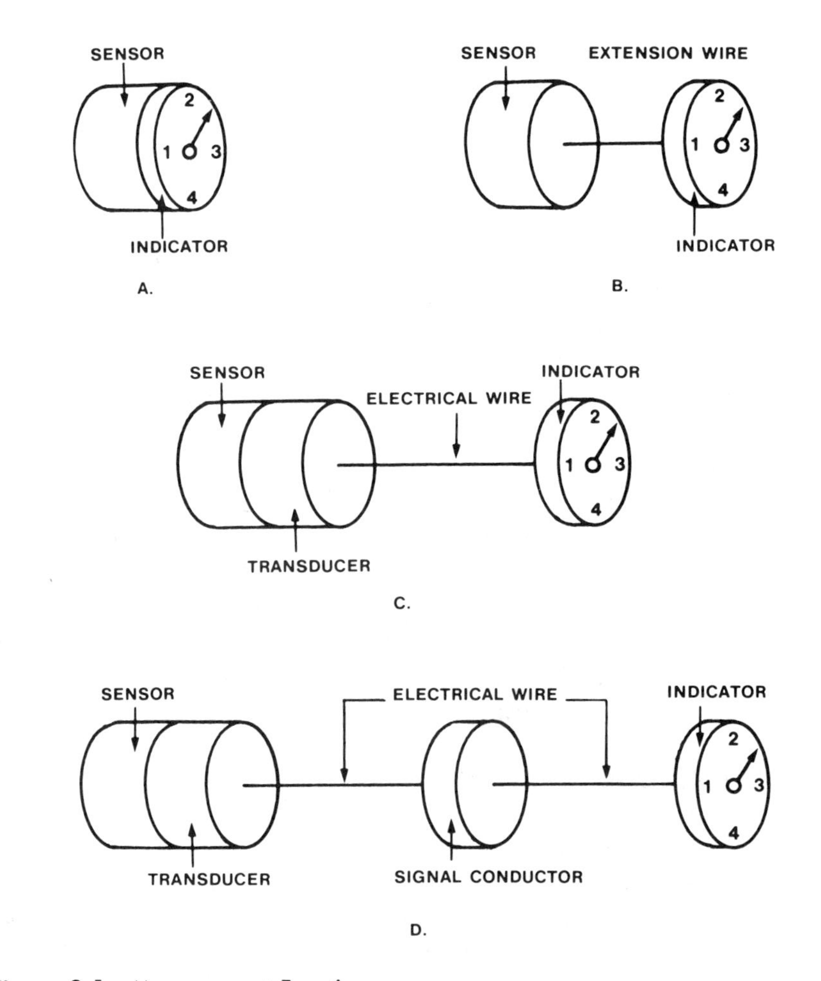

Figure 8-1 Measurement Functions

response is coupled to a compatible transducer, which converts the sensor signal to an electrical signal. The electrical signal is transmitted to a signal conditioner whose job is to modify the signal for display. The modified signal is transmitted to the display for readout. The readout responds to a change in the physical quantity felt at the sensor.

Figure 8-1 provides a basis for measurement system organization.

SIGNAL ANALYSIS

The purpose of the transducer is to detect signal phenomena. There are three major areas of signal phenomena. The first is *radiation*; here the transducer must be capable of detecting the emission of radiation or reacting to radiation or the effects of it. The second major area of signal phenomena is *electrical*. This area involves detection of an electrical signal such as current or voltage. The third area is *physical* signal phenomena, such as mass or volume. In Table 8-1, a list of signals is provided along with the science that deals with the method of measurement of each. This table could be expanded to include all the variations of light wavelengths, along with numerous electrical parameters and new study sciences in physical phenomena. Signal results are either qualitative or quantitative. Qualitative results are those whose molecular, atomic, physical, or functional features are found to be different from some standard. Quantitative results are measurable in numerical quantities.

TABLE 8-1. SIGNAL PHENOMENA AND THEIR SCIENCES

Signal Phenomena	*Science Study*
Radiation	
Emission	Spectroscopy
Scattering	Turbidimetry
Refraction	Refractometry
Absorbtion	Spectrophotometry
Electrical	
Potential	Potentiometry
Current	Polarography
Resistance	Conductimetry
Physical	
Mass	Gravimetric analysis
Volume	Volumetric analysis
Rate	Kinetic methods
Temperature	Thermal conductivity

Transducer Output Functions

Outputs from transducers may take the form of an electrical voltage or current or be nonelectrical in nature. Readout devices include such instruments as indicators, tape recorders, digital displays, or oscilloscopes. The signal, in the event the output signal is electrical, must be compatible with the readout device. The signal may be subjected to amplification, integration, or differentiation. It may be added, subtracted, multiplied, divided, or in some other way modified.

Transducer Types

One function of a transducer is to detect and convert a measurand into an electrical quantity. The measurand is the item being measured. The measurand need not be energy, but could be material quantities, such as the amount of coal in a hopper.

There are basically two wide areas into which transducers are categorized. These are active and passive. *Active transducers* generate a voltage or current as a result of some form of energy or force change. For instance, a thermocouple, when heated, generates an electrical signal that is proportional to the amount of heat applied. The *passive transducer* changes its properties when exposed to light energy. Therefore, the current through the photoresistor changes. In this event, we can again say that the amount of electrical output change is dependent on the amount of energy (light) to which the photoresistor was exposed.

There are other special transducer elements that may embody the active or passive elements for some combination of special interest. The electrokinetic transducer, for instance, operates with polar fluid maintained between two diaphragms to change the difference of potential between faces of its electrical interface. This element does not quite fit into the categories of active and passive transducers.

ACTIVE TRANSDUCER ELEMENTS

Active transducer elements are usually defined as those elements that generate a voltage as a result of some energy or force change. The generation of signals is accomplished by six major methods. The methods include electrostatics and chemical. These two are not in general use as transducer element types. The other four methods are electromechanical, photoelectrical, piezoelectrical, and thermoelectrical. These are covered in some depth in the next several paragraphs.

Electromechanical Element

Relative motion through a magnetic field will produce a voltage at the ends of the conductor (Figure 8-2). That is, if a conductor passes through a field or the field is moved across a conductor, the motion will produce a voltage at the ends of the conductor. Such is still the case if both the conductor and the magnetic field are moving. All these statements are representative of Faraday's *law of induction*. In the 1800s Michael Faraday, an English scientist, developed a simple machine in which a conductive disk was rotated in a magnetic field. Sliding contacts (brushes) were used to pick off the voltage from the conductive disk. The mode of operation is called a single-pole generator. In Figure 8-2a, a fixed-level conductor is moved through a field. The motion produces a constant level of voltage across the conductor. In Figure 8-2b, the conductor is rotated in the magnetic field, thereby producing an alternating voltage. The magnitude of the voltage generated (E) is dependent on the flux density (B) of the magnet, the length of the conductor (L), and the speed of the conductor (V).

$$E = BLV$$

Faraday's rotational conductor was somewhat different. An equation was developed as a result of the Faraday law of induction. The induced voltage (V) produced in a conductor increases as the number of conductors (N) moving through the field increases:

$$V \propto N \frac{d\phi}{dt}$$

where

V = induced voltage

$\propto$ = proportional to

$\frac{d\phi}{dt}$ = rate at which the magnetic flux crosses the conductor

The equation suggests that induced voltage increases whenever the number of poles increases or the rate of motion of the conductor through the magnetic field increases.

The reader will note that the Faraday law of induced voltage changes somewhat with a loaded circuit. That is, when a load resistance is added to the conductor, as in Figure 8-2c, a current is induced through the conductor and

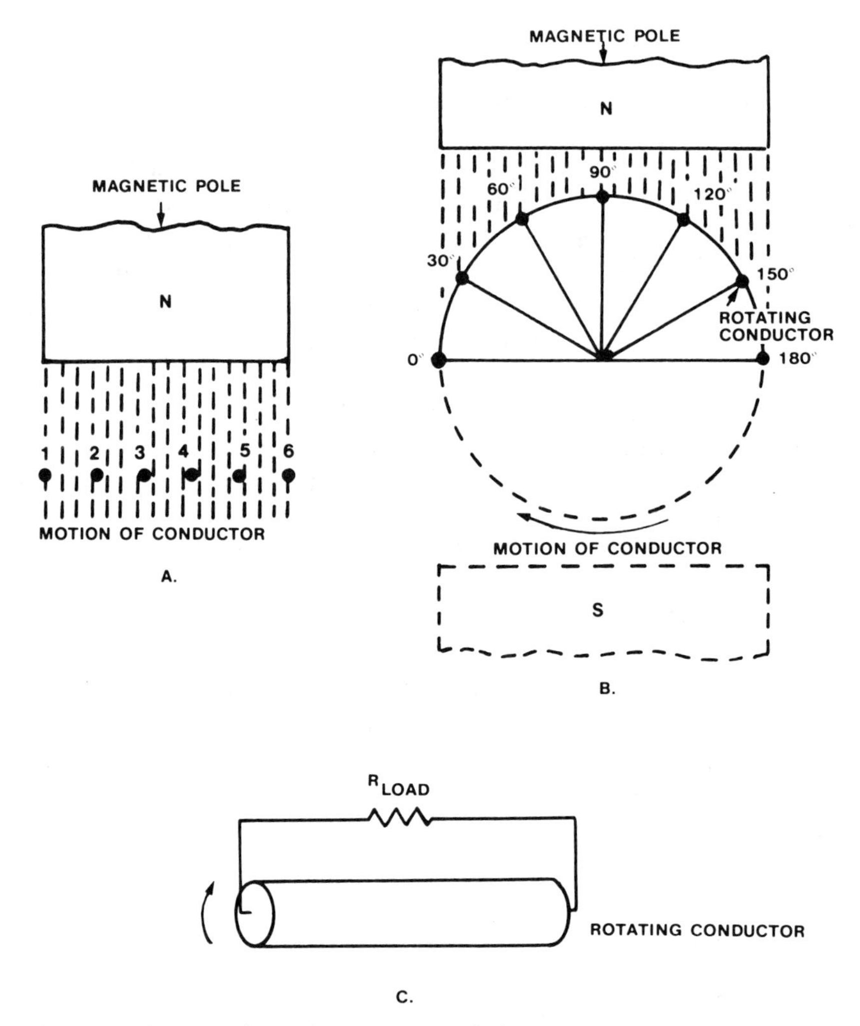

Figure 8-2 Electromechanical Generation of Electricity

load that opposes the motion of the conductor in the magnetic field. Without the load resistance, the conductor moves easily through the field. With the load resistance installed, the conductor does not move easily. The reason for this is that the current through the conductor produces a magnetic field that opposes the motion that generated the voltage in the first place. Lenz's law modifies the Faraday equation and can be stated mathematically as

$$V = -N\frac{d\phi}{dt}$$

Note the similarity to the induced voltage equation previously stated. The minus (−) sign denotes the opposition developed by the load resistance.

Photoelectric Element

Light striking on conductive material produces an effect called the *photoelectric effect*. The sensing of light is also called *photodetection*. Photodetection is defined around three phenomena. These are photoemission, photoconduction, and photovoltaic actions. All quantum photodetectors respond directly to the action of incident light. The first phenomenon, *photoemission*, involves incident light that frees electrons from a detector's surface (see Figure 8-3). The cathode of this tube is coated with radiation-sensitive material. Electrons are released as radiation is applied. This usually occurs in a vacuum tube. Note that electron current flows from negative to positive. With *photoconduction*, the incident light on a photosensitive material causes the material to alter its conduction (see Figure 8-4). Photoconductive material decreases resistance as radiation is applied. The third phenomenon, *photovoltaic action*, generates a voltage when light strikes the sensitive material of the photodetector (see Figure 8-5). Radiation causes a difference of potential between the two terminals. Note that there is no power source such as a battery involved with photovoltaic action.

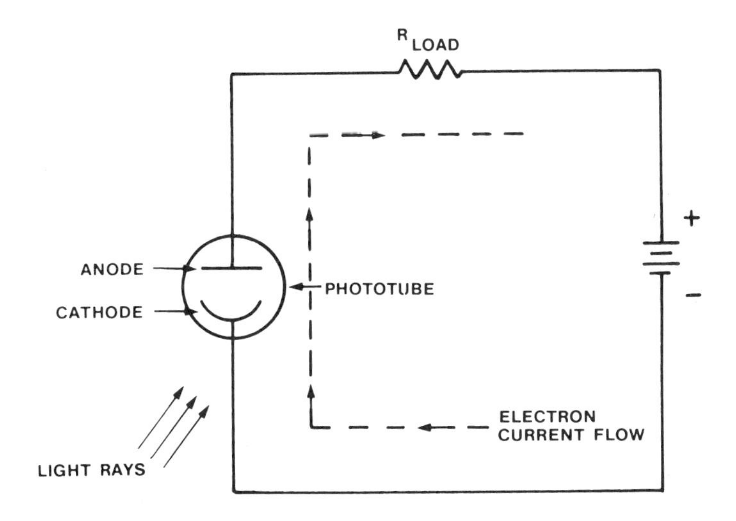

Figure 8-3 Photoemission

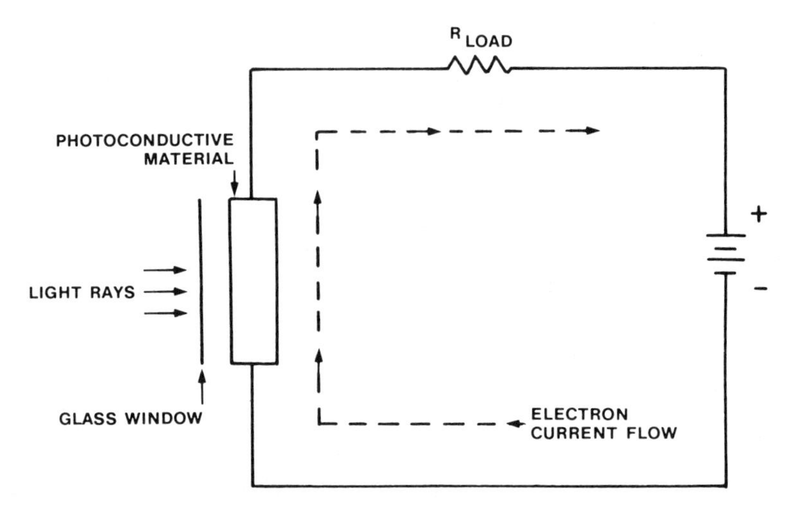

Figure 8-4 Photoconduction

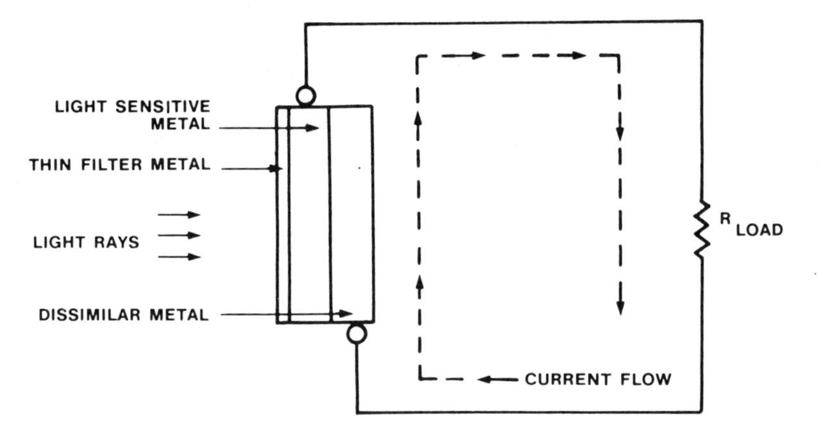

Figure 8-5 Photovoltaic Cell Operation

Piezoelectric Element

Synthetic or natural crystals such as quartz have special physical qualities. When crystalline material is stressed (pressure causes the crystal shape to change or distort), a voltage is produced at the surface. If a material performs in this manner, it is said to be *piezoelectric*. Piezoelectric material also has another quality that is the direct reverse of piezoelectric. This other quality is called *electostriction*. Electrostriction involves the application of an electrical field to the crystal substance. The electrical field alters the crystal shape. Probably the most common example of piezoelectric material is the ceramic material in the stylus of a disk record playback. As the record turns, the ceramic crystal flexes in the grooves, causing a voltage that represents the amount of flex. The voltage is then amplified and sent to a loudspeaker. Another common use of the piezoelectric effect is the control of radio frequencies in transmitters with quartz crystals.

A most basic piezoelectric transducer function is illustrated in the Figure 8-6. In the figure a quartz crystal with a beam is tied to a supporting base. The crystal and its beam are mechanically connected to a pressure-sensing diaphragm that acts as a force summing device. The diaphragm, in turn, is open to a pressure port. A change in pressure causes a mechanical change on the crystal's beam which causes the crystal to oscillate at a specific frequency or generate an electrostatic charge signal. This description is, of course, extremely simplified.

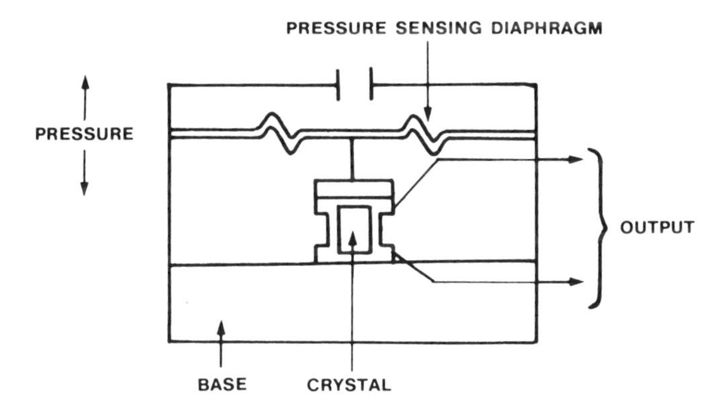

Figure 8-6 Fundamental Piezoelectric Transducer

These devices have high-frequency response, are self-generating, and are small and rugged. They are, however, sensitive to temperature changes and cross-accelerations, have high-impedance outputs, and do not recover very quickly after extreme shock.

Thermoelectric Element

The sensing of temperature is usually accomplished with the aid of a thermoelectric device such as the thermocouple, the resistance temperature detector (RTD), or the thermistor. Whichever device is used, it plays a major role in the monitoring of heat energy, such as in conversions from coal to steam and air conditioning.

Three thermoelectric effects play a large part in the generation of electricity by way of temperature. The first is the *Seebeck effect* (see Figure 8-7). Thomas Seebeck, a German physicist, fused two dissimilar metal wires together on both of their ends. He then heated one of the junctions and found that electrical current flowed from one wire to the other. He caused electrons to flow from a copper wire to an iron wire. This effect developed into what we know now as the *thermocouple*.

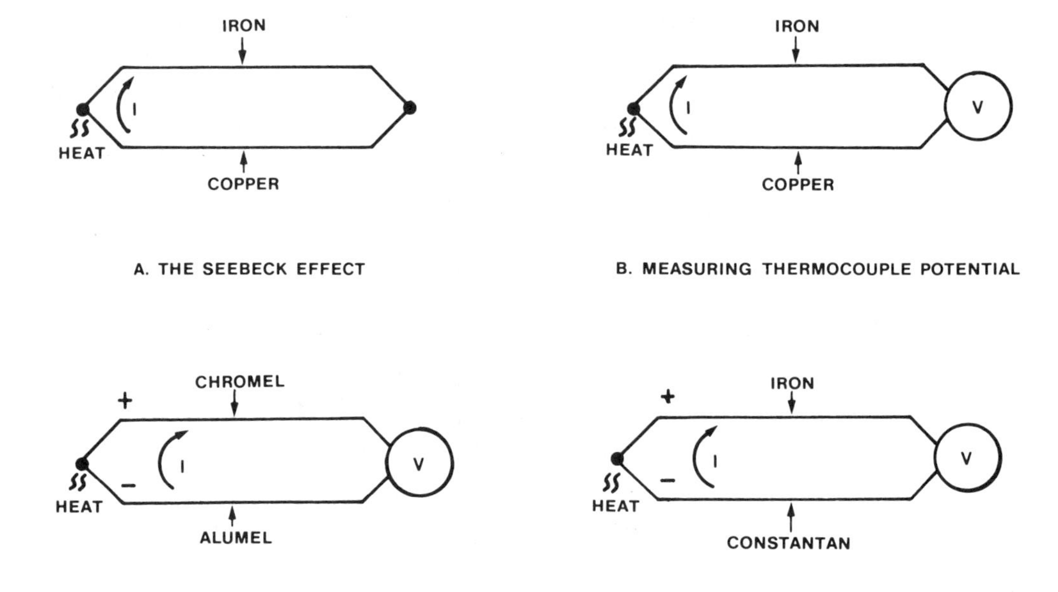

Figure 8-7 Thermocouple Effects

The second thermoelectric effect is called the *Peltier effect* (see Figure 8-8). Jean Peltier, a French physicist, applied current to two dissimilar metal wires attached together at their ends. As electrons moved from the copper wire to the iron wire, the junction became warm (hot). As the electrons moved from the iron wire to the copper wire, the junction became cool. The reasoning is that electrons moving from a higher energy state (iron) to a lower energy state (copper) create an excess of energy, and therefore heat.

The third thermoelectric effect is called the *Faraday effect* (see Figure 8-9). Michael Faraday found through experimentation that in certain materials resistance is decreased as temperature increases; that is, they have a negative

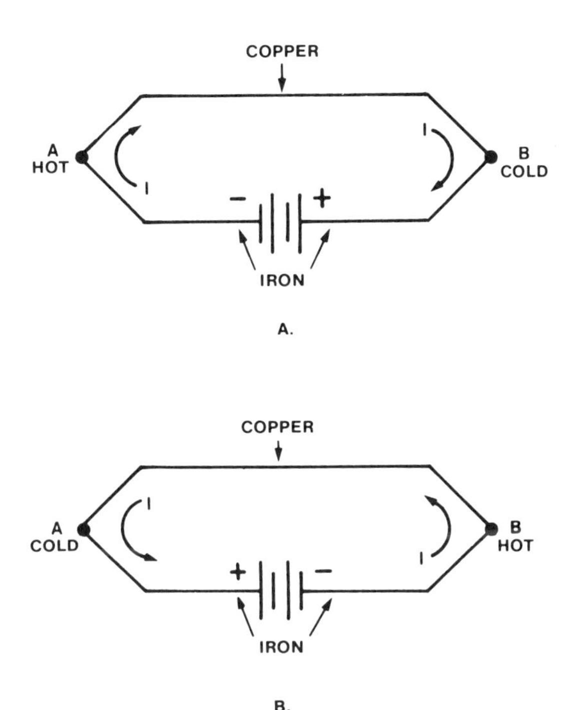

Figure 8-8 Peltier Effect

temperature coefficient. The resistance of oxides such as cobalt, manganese, and nickel is decreased when they are heated.

Thermoelectric materials are chosen for their ability to provide a uniform voltage–temperature relationship.

PASSIVE TRANSDUCER ELEMENTS

Each passive transducer has an element that, under some force, responds by moving mechanically to cause an electrical change. Let's discuss some of these elements in a cursory manner.

Capacitive Element

The capacitive transducer element consists of two fixed conductive plates that are isolated from a housing by insulated standoffs (see Figure 8-10). A pressure

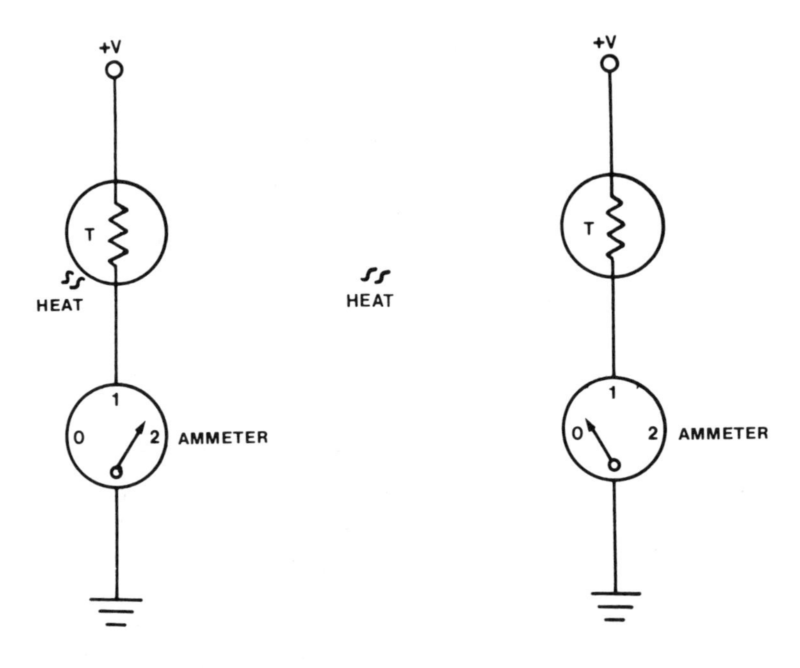

Figure 8-9 Faraday Effect

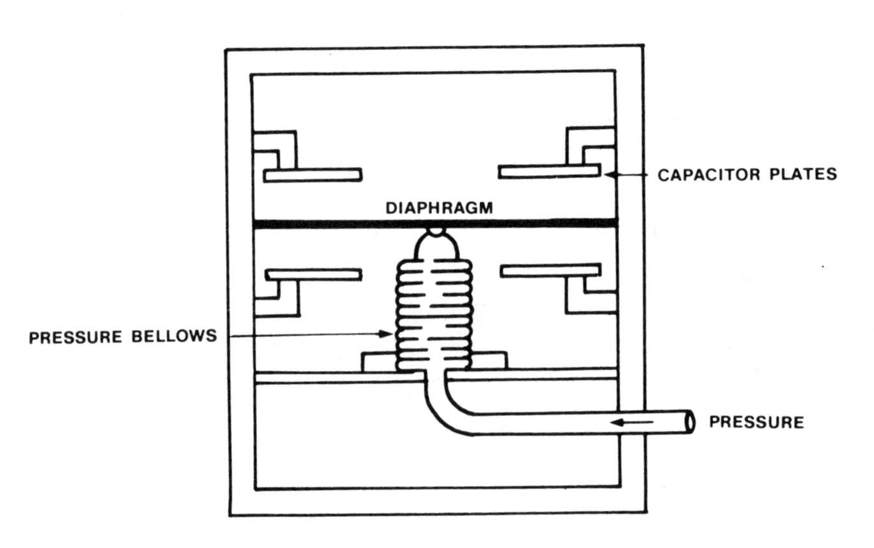

Figure 8-10 Capacitive Element

port directs pressure into a bellows. Attached to the bellows is a diaphragm. As pressure changes, the bellows expands or retracts, changing the position of the diaphragm that serves as one capacitor plate. This causes a capacitance change in two separate capacitive circuits. In some transducer designs, only a single capacitive element is employed. The change in capacitance is used to vary the frequency of oscillators or to null a capacitance bridge. The dual concept allows two oscillators to operate in a highly linear mode. Small displacement of the capacitor diaphragm is a major advantage. Capacitance sensors may be used for pressure transducers in ranges from 100 pounds per square inch (psi) and down to medium vacuum ranges.

The capacitive transducer has excellent frequency response, standard hysteresis, repeatability, and stability with excellent resolution. Its disadvantages include a high-impedance output with additional electronics. Capacitor leads from the sensor must be short to eliminate stray pickup. Finally, the capacitive transducer is extremely sensitive to temperature variations.

Inductive Element

The inductive transducer element (Figure 8-11) consists of a diaphragm or core called an *armature* that is driven by force from a measurand. The armature is either displaced or rotated near the coils of a C-shaped pickoff (Figure 8-11a). The displacement or rotation of the armature causes a change of inductance of the magnetic flux in the coil by varying the air gap in the flux path. In an

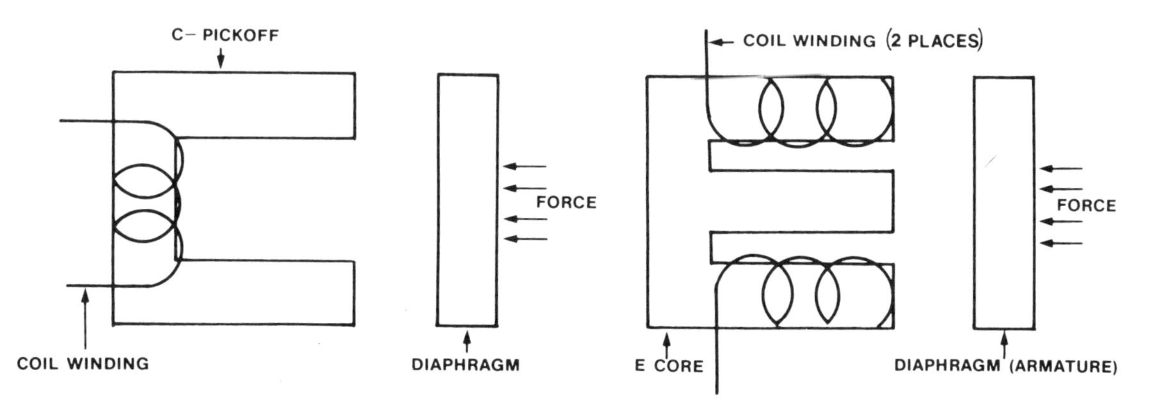

Figure 8-11 Inductive Element

excited circuit this would cause a change in inductive reactance, which, in turn, would modify current flow.

In Figure 8-11b, the armature is either displaced or rotated between a pair of coils of an E-type pickoff. The displacement or rotation of the armature causes a change of inductance between the two coils.

Let's consider what may happen in the event of a measurand change. The parameters that are involved in the physical makeup of the coil can be modified by the force of the measurand. The inductance of a coil or inductor may be changed by the core material, the relationship between the armature and the E pickoff, the number of coil turns, the diameter of coil, and the coil length.

An inductor operates on the principle of induction. Current through an inductor lags voltage by 90° in a pure inductive circuit and by some angle between 0° and 90° in other R_L circuits. As current increases through an inductor, a magnetic field is built up around the coil. The strength of this field is dependent on the physical makeup of the inductor and the core permeability. The magnetic field produces a counterelectromotive force that opposes a change in current.

The core of the inductive pickup has some properties that cause some error. The most prominent of these losses are in *eddy currents* and *hysteresis.* Eddy currents are prevalent in iron cores. Alternating current induces voltage in the core, causing currents to flow in a circular path through the core. This causes a loss in signal power. Eddy currents occur more often at high-frequency ac operation. The second type of loss is hysteresis. Hysteresis also occurs most prominently at high frequencies; it is loss of power as a result of switching or reversing the magnetic field in magnetic materials.

Differential Transformer Element

The differential transformer element (Figure 8-12) consists of a primary and secondary of a transformer with a magnetic movable core. One end of the core is attached to a push rod. The push rod is mechanized by a pressure-sensing diaphragm, bellows, or Bourdon tube. As pressure changes, the push rod displaces the magnetic core within the transformer. As the core is displaced, an unbalance is produced within two secondary windings. AC excitation is usually 50 to 60 Hz, but 10,000 Hz is employed to reduce the size and mass of sensor components. The sensitivity of the transducer is a tradeoff in design between the core and the turns ratio. Owing to the relatively large core, the transducer cannot be used in areas of high vibration or acceleration.

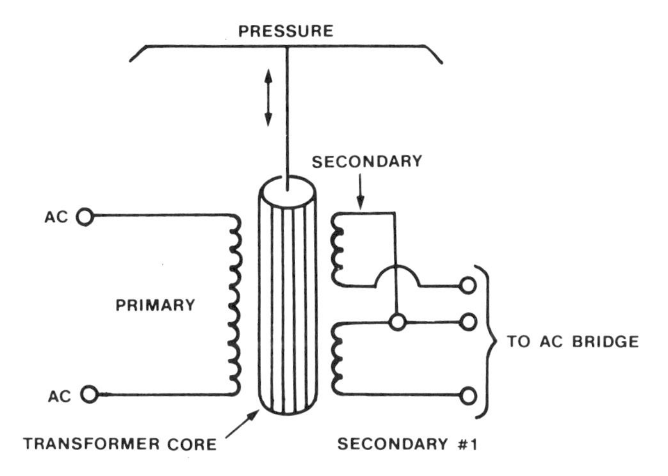

Figure 8-12 Differential Transformer Element

Potentiometric Sensing Element

A basic potentiometric pressure transducer element (Figure 8-13) has a force-summing pressure bellows attached to the linkage. The linkage is mechanically connected to a potentiometer wiper. The wiper travels over a multiturn wire coil or deposited resistor, as the sensor reacts to pressure changes. Usually some motion amplification between the force-summing element (pressure sensor) and the wiper track across the resistance element is employed to minimize acceleration error.

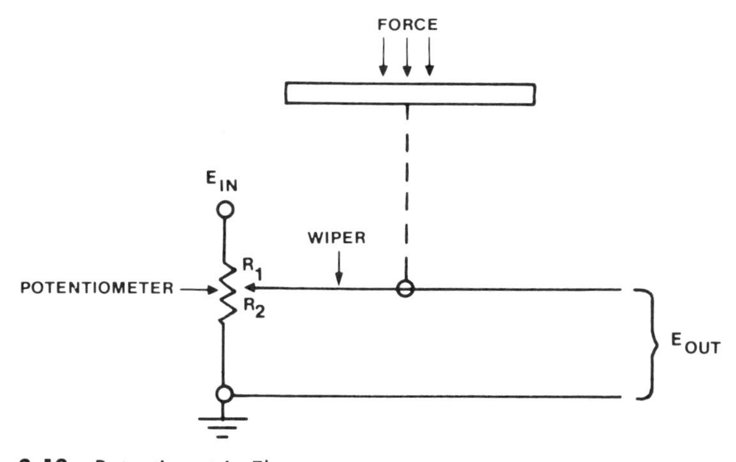

Figure 8-13 Potentiometric Element

The resistance of the potentiometer can provide specialized output linearities such as linear, sine, cosine, and exponential. As a voltage ratio device, close regulation of the excitation voltage is not required. High-level output is inherent in the potentiometer concept. Output load impedance must be kept high to limit loading effects.

SOME SPECIAL TRANSDUCER ELEMENTS

Some transducers employ the basic action of active or passive transducers. Others may employ some special combination of characteristics that set it somewhat aside from the standards of active or passive. We shall take at least a cursory look at these transducers.

Electrokinetic Elements

An electrokinetic transducer element (Figure 8-14) consists of polar fluid contained between a pair of diaphragms. A porous disk is inserted in the polar fluid between two partially porous plates called diaphragms. A change in force from the measurand causes the diaphragm to deflect and allow a very small amount of polar fluid to flow through the porous disk partition to effect a deflection on the second diaphragm (left in the illustration). This flow causes a difference of potential between the plates of the porous plug. A second effect may be obtained in reverse operation. An electrical potential may be applied to plates of the porous plug, which in turn causes polar fluid flow and diaphragm deflection.

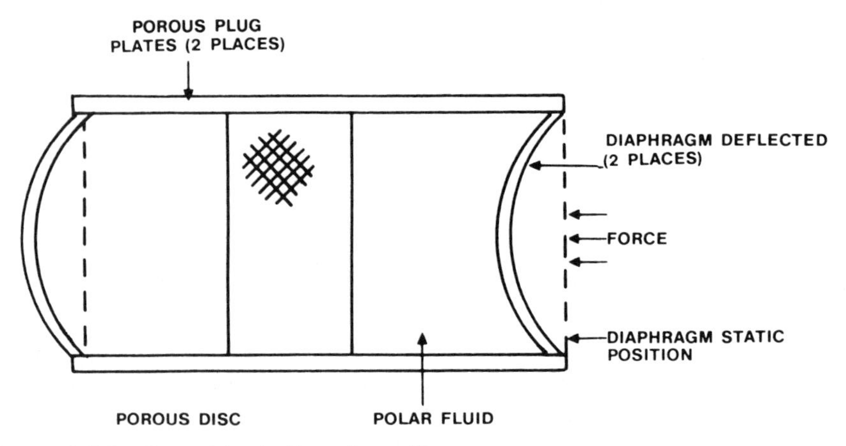

Figure 8-14 Electrokinetic Transducer Element

The electrokinetic transducer element is self-generating with relatively high frequency response and high output. It does have disadvantages in that it cannot monitor static pressures or linear accelerations. The element cannot be calibrated without flow. A further disadvantage is that the polar fluid is often volatile.

Force-Balance Element

In Figure 8-15 the actual sensing element is the capacitor in a force-balance element. However, this element could just as well be an inductive element. Operation of this device is rather simple. A change is effected by a measurand change, and force is applied to a force-summing capacitor. The physical properties of the capacitor change, and therefore its capacitance also changes. The variable signal is routed to an amplifier and in turn to a servo mechanism. A feedback signal equal to the variable output of the capacitor is fed back to the force-balance element to return it to its original state before a measurand force change. The feedback may be mechanical and/or electrical and may be mechanized by a servo system or electromagnetism. Actual motion of the mechanical linkage and force-summing devices is hard to detect. The device has a fairly high output and is accurate and stable. It does, however, have a low-frequency response and is sensitive to acceleration and shock. The element is heavy and may be expensive.

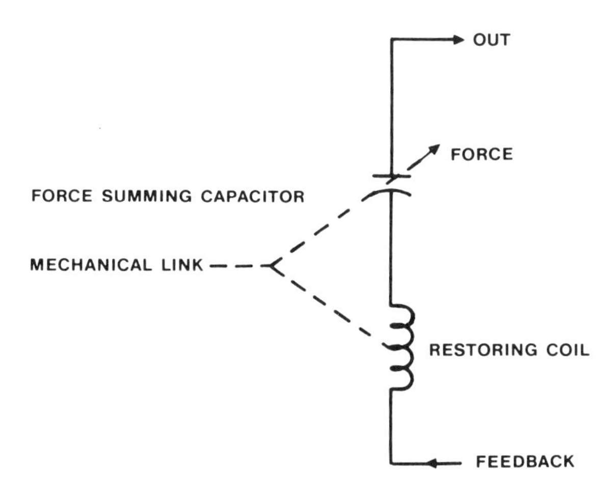

Figure 8-15 Force-Balance Element

Oscillator Element

In Figure 8-16 a fixed capacitor and a variable inductor (coil) are used as a frequency-resonant circuit input to an oscillator. The variable unit may be either of these components. The force element may be water or air pressure, for example. As the measurand changes, the force-summing bar causes the inductor to change inductance, which alters the frequency at which the circuit is resonant. Any change in the resonant circuit will be felt at the oscillator amplifier input.

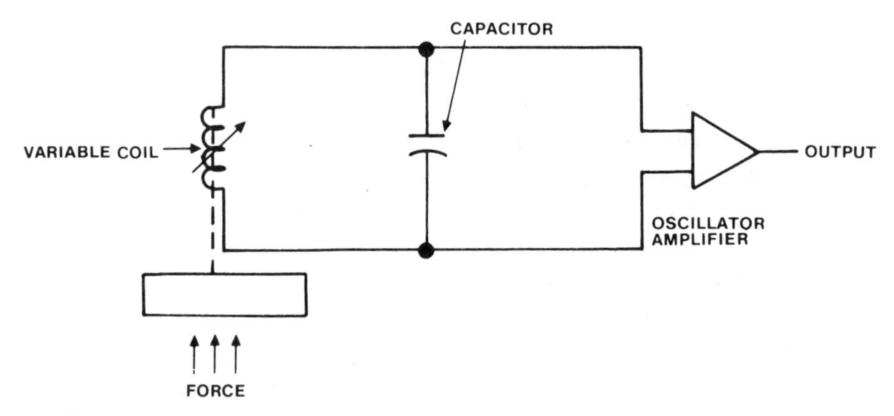

Figure 8-16 Oscillator Element

The oscillator element may be very small with a fairly high output. Temperature may restrict the operating range, and the devices used may play a part in poor sensitivity. The accuracy of the device may be poor unless expensive components are utilized.

Photoemission–Photodetection Combination

The photoemission–photodetection combination (Figure 8-17) consists of a force-summing diaphragm, a window, a light source, and a detector. Some measurand such as air or water causes the diaphragm to deflect. The window opens or closes as the diaphragm modulates the amount of light from the light source through the window opening. A photodiode on the window's opposite side detects the light and feeds the varying amounts as current flow to some receiving or amplifying component. The output then is dependent on the force applied by the measurand. This is a linear system and can be made extremely accurate with a strong and sensitive light-emitting diode (LED) as the light source. The unit is simple and can provide a high output.

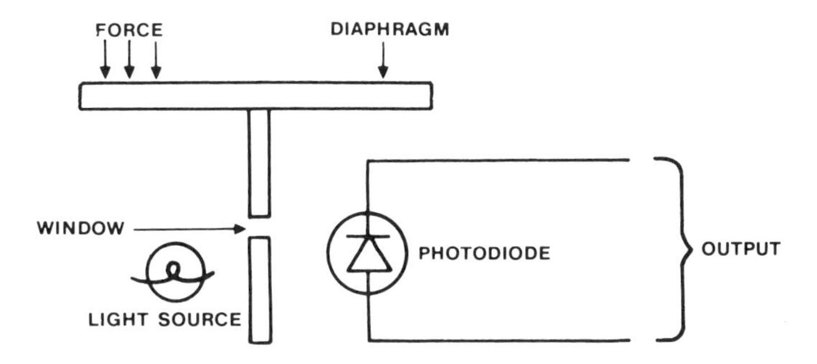

Figure 8-17 Photoemission–Photodetection Combination

Frequency response may be low, and long-term stability is not always achieved. Temperature, as in other transducer elements, is a factor in efficiency.

Vibrating Wire Element

The vibrating wire element (Figure 8-18) consists of a fine tungsten wire strung through a strong magnetic field so that the lines of force are maximum crossing the wire. As the measurand causes a force change, the wire vibrates at a frequency that is determined by the length of the wire and the tension applied. As the wire vibrates in the magnetic field, an electrical signal is generated at the output (taken from the wire) and fed to an amplifier. A feedback signal is fed back to the wire to maintain oscillation at the force frequency. Modification of the wire vibration frequency is then dependent on the measurand change. The output of the vibrating wire element can be high and can be transmitted over long distances

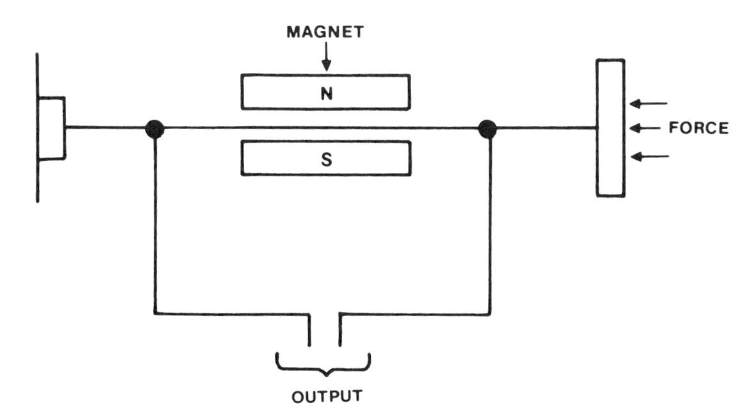

Figure 8-18 Vibrating Wire Element

without much loss. The device is sensitive to acceleration and shock and is not considered a stable element. Temperature affects the wire and its hookup.

Velocity Element

The velocity element (Figure 8-19) consists of a moving coil within a magnetic field. One end of the coil is attached to a pivot, while the other end is free to move within a restricted area. Output is taken from the moving coil. A voltage signal is generated by the coil moving within the magnetic field. The output is proportional to the velocity of coil movement. Electrical feedback may be used for dampening purposes.

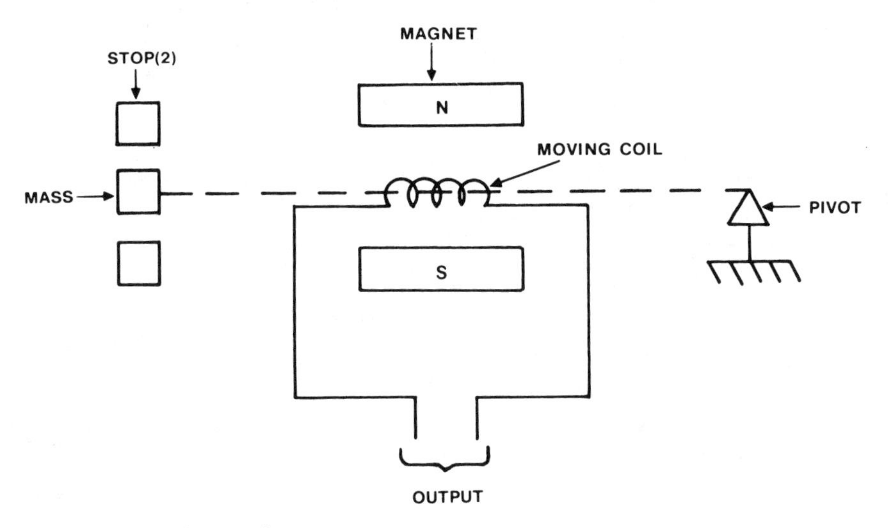

Figure 8-19 Velocity Element

Strain Gage Element

The purpose of a strain gage element (Figure 8-20) is to detect the amount or length displaced by a force member. The strain gage produces a change in resistance that is proportional to this variation in length. Each strain gage has a property known as a *gage factor,* which provides this function. Gage factor is a ratio of resistance and length.

$$\text{GF} = \frac{\Delta R/R}{\Delta L/R}$$

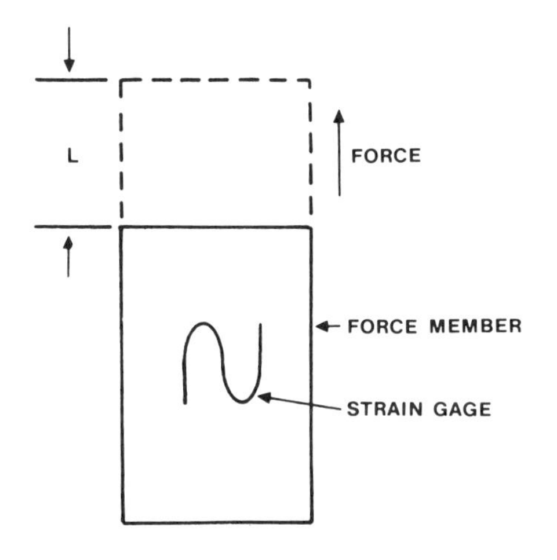

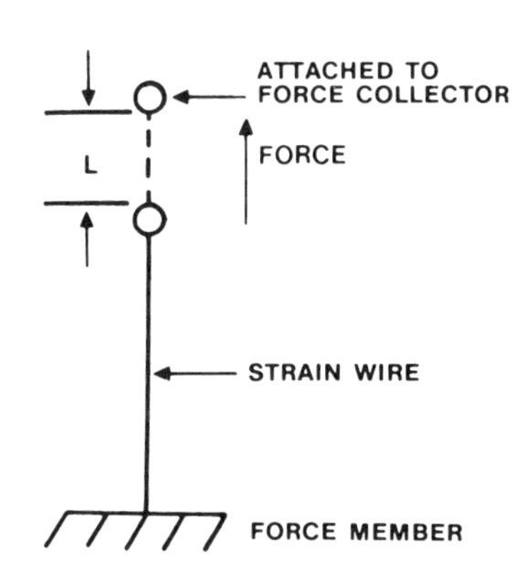

A. BONDED GAGE

B. UNBONDED GAGE

Figure 8-20 Strain-Gage Element

where GF = gage factor

ΔR = change in resistance

R = original resistance

ΔL = change in length

L = original length

The gages are electrically installed as part of a Wheatstone bridge for circuit applications.

There are two basic strain gage types: bonded and unbonded. The *bonded gage* (Figure 8-20a) is entirely attached to the force member by an adhesive. As the force member stretches in length, the strain gage also lengthens. The *unbonded gage* (Figure 8-20b) has one end of its strain wire attached to the force member and the other end attached to a force collector. As the force member stretches, the strain wire also changes in length. Each motion of length either with bonded or unbonded gages causes a change in resistance. Strain gages are made of metal and semiconductor materials. Strain gages are very accurate, may be excited by alternating or direct current, and have excellent static and dynamic response. The signal out is small, but this disadvantage may be corrected with good periphery equipment.

Other Transducers Other specialized transducers are made in this country and the rest of the world that do special tasks and are unique. Throughout this chapter the author will attempt to provide at least cursory coverage of most of today's transducer types.

NATURAL FREQUENCY

Most transducers that are involved with static acceleration or vibration have what is called an expected natural frequency. The natural frequency of a seismic mass possessing one degree of freedom is:

$$f_n = \frac{1}{2\pi}\sqrt{\frac{K}{M}}$$

where

f_n = natural frequency

K = maximum stiffness

M = minimum mass

The sensitivity of acceleration and vibration transducer devices increases in proportion to mass (M) and decreases in inverse proportion to stiffness (K).

Transducers that do respond to vibration or acceleration usually have a point at which they freely oscillate without being forced. Some transducers respond with a ringing noise. Ringing may also occur with a step change in the measurand. Others simply vibrate or oscillate at their own frequency. Manufacturers will freely provide information regarding the transducers' natural frequency and/or resonant points.

MEASUREMENT AND ERROR

The concept of measurement evolves around three separate parts. These are the quantity or quality of the physical phenomena being measured, the reference for comparison, and the method of comparing the sensed quantity or quality to the reference. Thus the act of measurement must be based on three fundamental operations: *sensing, processing,* and *utilizing collected information.*

Regardless of the components used in the measurement, there is room for error. Error is usually considered to be static or dynamic. *Static error* can be further separated into predictable errors such as parallax, interpolation, and environmental causes, such as heat, humidity, pressure, and radiation. *Dynamic error* is the difference between the sensed physical quantity and the value indicated after installing the measuring equipment. In other words, dynamic error is due to loading or connecting the measurement system. Error must be determined and either be compensated for automatically by the system or be added to or subtracted from indicated results.

TRANSDUCER TYPES

We cannot possibly cover all the transducer types in a single chapter, nor could we accomplish that task in an entire book. The world of transducers represents an enormous combination of components. With industrial technology growing at its existing rate, transducers will take on an even more important role.

Semiconductor Strain Gages

The recent technologies developed for the manufacture of transistors and integrated circuits have now been applied to sensors for pressure transducers (Figure 8-21). IC manufacturing techniques can be employed to build, calibrate, and test pressure transducers. High-accuracy, stable pressure transducers can be produced in high volume at very low costs.

Within the semiconductor or piezoresistive family of strain gages, two types of sensors are employed: the bar gage and the diffused gage. The *bar gage transducer* is nearly identical to the bonded foil strain gage instrument. Its gages are individually bonded with epoxy to a mechanical structure, such as a beam or diaphragm. The *diffused gage transducer* uses a silicon element for the mechanical structure, and the strain gage is an integral part of the silicon element rather than being bonded to it; the gage is diffused into the structure.

In the bar gage, an impurity such as boron is introduced into the silicon crystal as it is grown. In a bar gage, the entire piece of silicon is an active gage. The individual gages are then electrically matched, as nearly as possible, and four legs of a bridge circuit are cemented to the pressure sensor (either a diaphragm or bending beam).

Load Cell

A load cell is a transducer device that produces an output that is proportional to an applied force. The load cell facilitates the accurate measurement of force

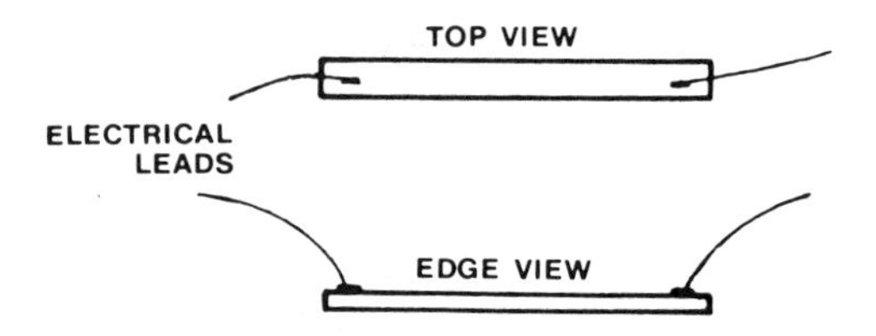

A. BAR GAGE

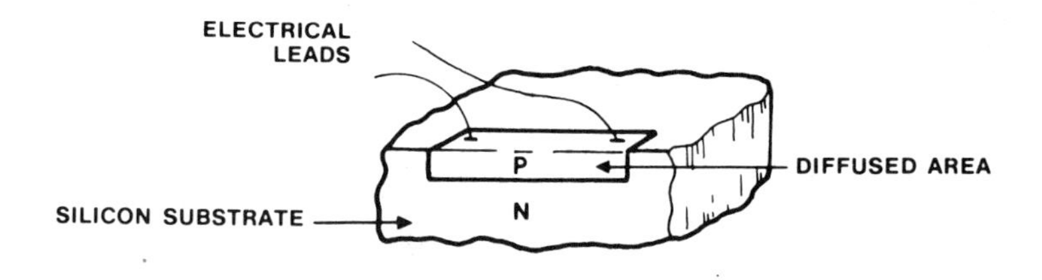

B. DIFFUSED GAGE

Figure 8-21 Semiconductor Strain Gages

and weight. Some load cells are strain-gage types. This section describes the load cell and its loading and weighing function. There are, of course, several types of load cells. The concepts behind all types are similar. A typical load cell type is the variable differential transformer (LVDT). The unit is based on a cantilever beam principle (see Figure 8-22).

The heart of the LVDT load cell is the deflection element, which is a series–parallel arrangement of multiple center-loaded beams. When an axial load is applied to a center-loaded beam (A) in Figure 8-22, it produces a proportional linear deflection (B) that is sensed by the armature (C) of a differential transformer (LVDT). As the armature moves relative to the primary coils (D) of the LVDT, it changes the distribution of the magnetic field and consequently the voltage that is induced into the secondary coils (E). Thus, for every change in applied load, there is a specific deflection and output voltage of the load cell.

A weigh cell is a transducer used in fill by-weight systems, in batch weighing

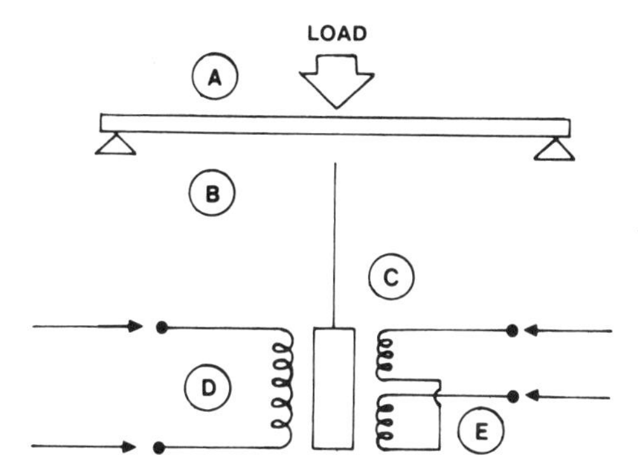

Figure 8-22 Operation of an LVDT Load Cell

systems, and check weighing systems. The weighing system may use several types of transducers, the most prominent of which is the LVDT-type transducer.

Accelerometer

An accelerometer is a transducer that responds to acceleration in one or more axes. Other names for an accelerometer are vibrometer, seismometer, velocimeter, and inclinometer. These devices all involve similar basic principles. They are all vibration pickups consisting of a mass and a spring constant. They are most generally damped and fit into some type of housing. They all differ, however, in their natural frequency and the method by which motion between the mass and the housing produces an electrical signal.

Acceleration is the rate of a velocity change. Velocity is a vector quantity that involves both magnitude and direction. Velocity may change in amount and direction or in both. Acceleration, then, is also a vector quantity.

Acceleration is usually measured in feet per second squared or meters per second squared. If a mass is moving at a constant speed, the acceleration is zero. If a mass is moving in a constant direction, the acceleration results in a continuous change in speed.

There are two fundamental forms of acceleration: linear and angular. *Linear acceleration* occurs when all points of a rigid mass are moving in parallel straight paths. The mass is moving in a linear path. An aircraft, for instance, when it is moving in a straight path, speeds up or slows down. In each case it experiences a linear acceleration. When making a bank, the aircraft changes direction and again experiences a linear acceleration.

Angular acceleration is a vector quantity that represents a rate of angular velocity change of a mass in a rotational motion. A merry-go-round rotates 360°

in one revolution. It has rotated through 2π radians in angular motion. The angular velocity is measured in radians per second or degrees per second. Angular velocity is a vector quantity that has direction and magnitude. Angular acceleration is the rate of change of angular velocity.

Figure 8-23 is representative of a fundamental accelerometer or seismic system. The sensing element involved in most linear accelerometers and other vibration devices may be represented by a spring–mass combination that deviates from a fixed damped position within a frame. This follows Newton's second law of motion, $F = ma$, where F equals force, m equals mass, and a equals acceleration.

The seismic system (short for seismographic) contains a mass (seismic element) suspended from a frame with a spring. The mass is centered (guided) by a pair of spring guides. Damping of the mass is often provided by mechanical or electrical means. An electrical pickoff monitors the position of the mass in relation to the frame.

In Figure 8-23, X represents movement in space of the body being monitored. Y represents the motion or movement of the suspended mass with respect to the frame. K is the spring constant of the mass suspension.

The seismic system responds to an acceleration by producing a force proportional to the applied acceleration. An equal reaction force is developed by the spring. The deflection is a linear function of acceleration. The acceleration

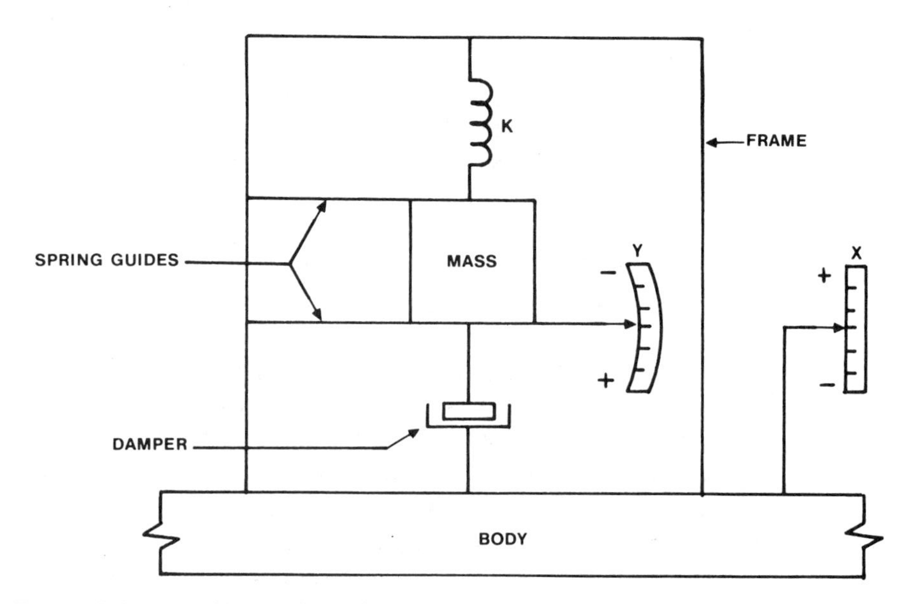

Figure 8-23 Fundamental Accelerometer

is held within the realm imposed by the natural frequency of the seismic system and its damping ratio.

Piezoelectric Transducer

Figure 8-6 shows a piezoelectric element. It is also representative of a piezoelectric transducer. In Figure 8-6, quartz with a beam is tied to the supporting base. The crystal and its beam are mechanically connected to a pressure-sensing diaphragm that acts as a force-summing device. The diaphragm, in turn, is open to a pressure port. A change in pressure causes a mechanical change on the crystal's beam, which causes the crystal to oscillate at a specific frequency or generate an electrostatic charge signal. This description, of course, is extremely simplified.

These devices have high-frequency response, are self-generating, and are small and rugged. They are, however, sensitive to temperature changes and cross-accelerations, have high-impedance outputs, and do not recover very quickly after extreme shock.

The crystal elements in piezoelectric transducers perform a dual function. They act as a precision spring to oppose the applied pressure or force and supply an electrical signal proportional to their deflection. The physical configurations of pressure, force, and acceleration transducers employing transverse elements are shown in Figure 8-24. Note that they differ little in internal configuration. In accelerometers, to measure motion, the invariant seismic mass, m, is forced by the crystals to follow the motion of the base (or structure to which the base is attached). This is an implementation of Newton's law, which states that force and acceleration are proportional.

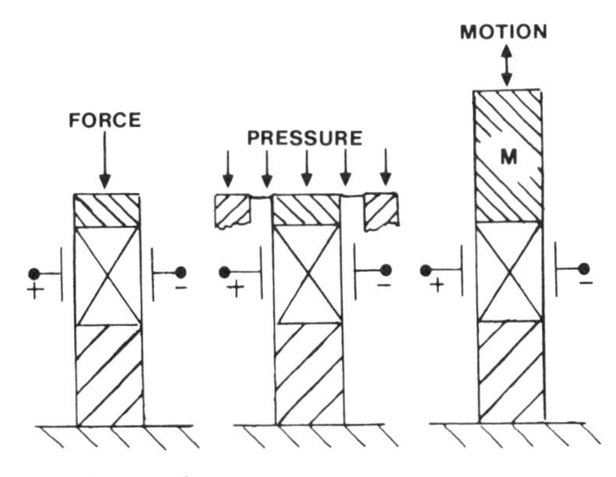

Figure 8-24 Crystal Transducer

An extremely complex manufacturing process with heat and pressure provides the transducer field with synthetic crystals. These crystals, primarily made from barium titanate and lead titanate–lead zirconate, are manufactured as large as 30 tons and small enough to be fed through a tiny vein to the heart.

Piezoelectric Ceramics The piezoelectric ceramic is an electromechanical device. Piezoelectric ceramics are isotropic before poling. Isotropic means that the material has the same properties regardless of the direction of measurement. Poling is a procedure that involves momentary application of a strong direct current. After poling, the ceramic becomes anisotropic (varies according to the direction of measurement). That is, their electromechanical properties differ for electrical or mechanical excitation along different directions. If the crystal's unit cells have no center of symmetry, they will become piezoelectric after poling. If they do have a center of symmetry, they will be inert when they are excited electrically or mechanically. A ceramic is composed of many crystals in random orientation, each unit cell containing a dipole. By application of electrodes and a strong dc field, the dipoles are aligned parallel to the field, thus making the ceramic piezoelectric. Not every domain aligns its dipoles, but enough of them do to achieve piezoelectricity. Once polarized, the ceramic takes on its own personality, exhibiting specific electrical and physical properties that occur in the various transducers to be discussed later in this chapter. Polarization is the last step, other than testing, in the manufacture of ceramics.

Proximity Transducers

Proximity detectors are electrical or electronic sensors that respond to the presence of a material. This electrical or electronic response is utilized to activate a relay or switch and/or to perform automation functions. Proximity detectors act as sensing devices in applications such as limit switches, liquid quantity level controls, speed controls, counters, and inspection tools. There are many types of proximity detectors. The major types are inductive, magnetic, and capacitive. The inductive and magnetic sensors require that the monitored material be metal. The capacitive proximity sensor can monitor nonmetal materials.

For the sake of review, let's take a second look at these transducer types. In the inductive transducer element, the inductance of the element is changed by the proximity of metal materials. In the magnetic transducer element, the magnetic field is changed by the presence of metal materials. In the capacitive transducer element, the capacitance of the transducer element can be changed by the presence of nonmetallic materials. The eddy-current transducer uses the principle of impedance variation. An eddy current is induced into a conductive metal target and monitored by a sensor. The temperature transducer monitors temperature with a sensitive conductive thermopile.

Proximity may be defined as the state or quality of being near. In transducer applications, it is this nearness that allows the change in electrical or electronic function to take place. In Figure 8-25a, typical proximity distances for metallic materials are shown. In Figure 8-25b, typical proximity distances for nonmetallic materials are shown.

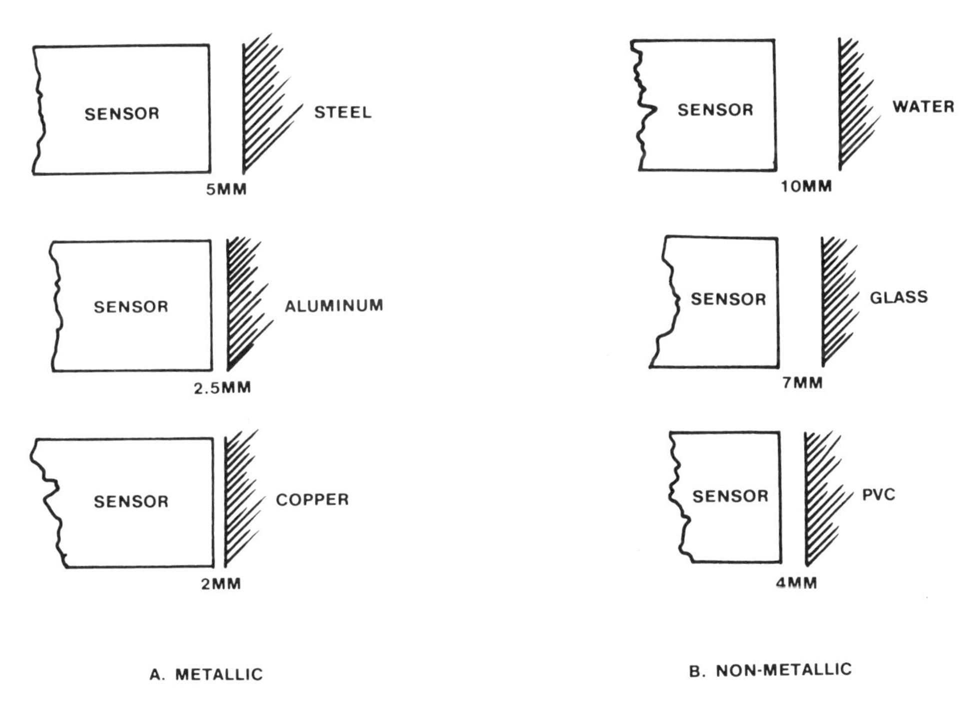

Figure 8-25 Typical Proximity Distance

Proximity detectors have no moving parts and therefore do not wear out from constant use. Since they do not contact their target materials, they are not destroyed by rough or abrasive parts. Proximity detectors can sense small, lightweight parts without touching them and delicate, painted, or polished surfaces without marring them. They can detect irregular-shaped objects regardless of the direction of entry into the sensing field. Control functions are possible at electron speed.

Power-Measuring Transducers

The problems of measuring power are considerable. The variables involved are complex and numerous. Power variables have to do with voltage, current, resistance, and frequency. Each of these measurable components of electricity is also complex. Each has terms that, when monitored, may provide meaningful information to an engineer or a user. This part of the chapter is dedicated to several transducers that monitor power and/or its terms.

The Hall effect was discovered by E. H. Hall in 1879. The Hall effect, for purposes of this discussion, is that characteristic of a certain crystal such that, when it is conducting current (control current) and is placed in a magnetic field, a potential difference is produced across its opposite edges. The potential difference is proportional to the product of the control current, the strength of the field, and the cosine of the phase angle between the control current and magnetic flux. By putting the crystal in a magnetic structure such that an ac current generates the flux and an ac voltage produces the control current, we have a multiplying device. It produces an output proportional to IC Cos $< \theta$ (power). This is in essence the Hall generator watt transducer. In Figure 8-26, the load current I_{ac} produces a proportional flux through the crystal in the air gap. The

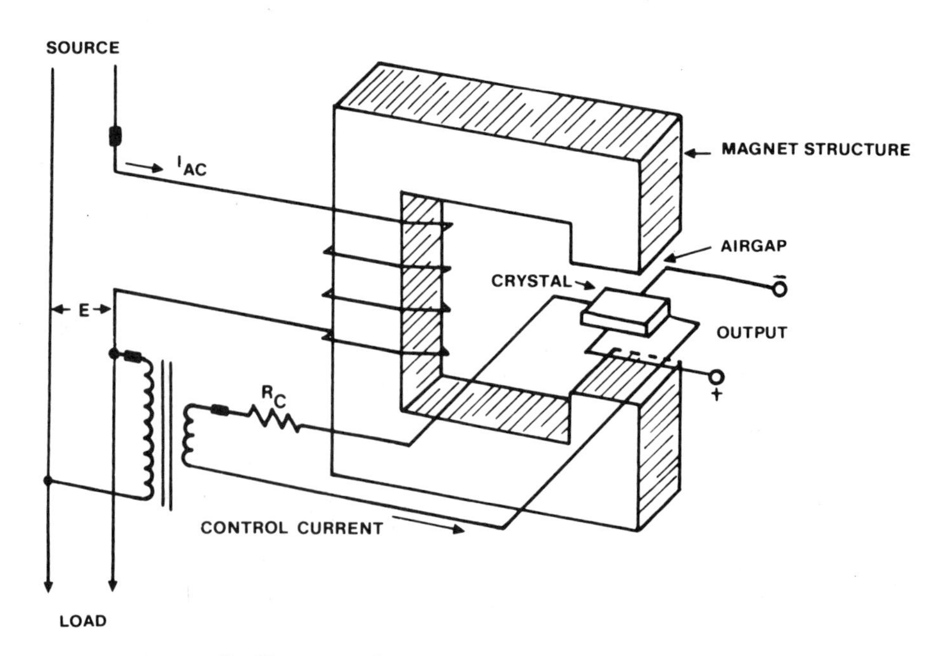

Figure 8-26 Hall Effect Transducer

load voltage E produces a proportional control current through the crystal. The output is proportional to the product of the two and the phase angle between them.

Actually, the output of the Hall generator consists of a dc voltage proportional to true power (watts) plus a double-frequency ac voltage proportional to the volt-amperes in the circuit. When the transducer is used with devices that do respond to ac, the double-frequency component must be filtered out. Filters are provided especially for this purpose.

The output of a Hall element decreases with increasing temperature. The watt transducers are temperature compensated with a thermistor–resistor network in the output circuit. The load resistance is part of the network; therefore, it is fixed for a particular transducer.

Other power monitoring devices are modifications of the ammeter, voltmeter, and frequency counter.

Fluid Transducers

One of the broadest areas in the transducer field is that of fluid transducers. When we speak of fluid parameters, we must consider pressure, flow, and levels of both hydraulic and pneumatic fluid (liquid and gas). Other parameters are also involved, such as temperature and viscosity.

The orifice transducer (meter) is shown in Figure 8-27. When a fluid flowing through a closed duct encounters a restriction, a local pressure drop is developed. The magnitude of the pressure drop is related to the flow rate at which the fluid flows through the duct.

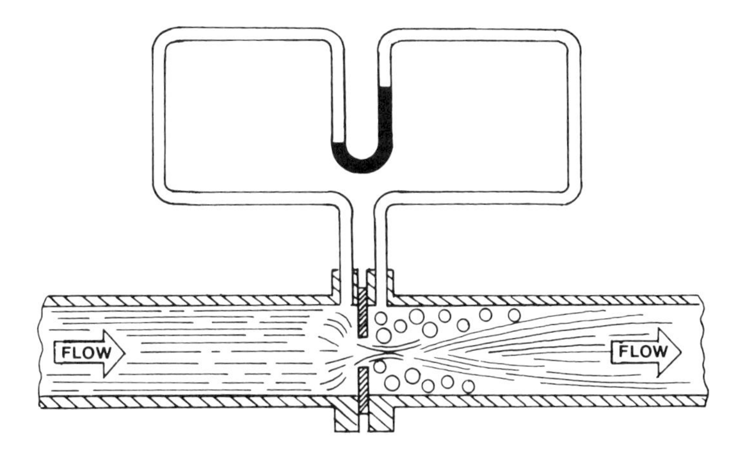

Figure 8-27 Orifice Meter Operation

Note that the pressure drop is related to flow rate, not to volume. Timing devices are used to record the time–flow rate data, and the integration of these elements is used to produce volumetric readings. Thus, the orifice meter is basically a velocity meter, and the volumetric measurement is derived or inferred from the observed fact of velocity versus time.

Gas turbine meters are velocity-sensing devices like orifice meters. The direction of flow through the meter is parallel to a turbine rotor axis, and the speed of rotation of the turbine rotor is nominally proportional to the rate of flow. Gas volumes are derived or inferred from the rotations of the turbine rotor.

The essential function of a gas meter is to provide an accurate readout of total volume throughout. It is in the fulfillment of this function that the turbine meter excels. Direct readouts can be displayed on any of a wide variety of mechanical devices. Additionally, gas turbine meters adapt readily to electromechanical fully electronic measurement systems.

Theoretically, a turbine rotor mounted in a friction-free atmosphere would rotate when a single gas molecule impinged on a rotor blade regardless of the velocity of the gas molecule (see Figure 8-28a). As a practical matter, mechanical friction in the supporting bearings and gears of a gas turbine meter requires a minimal amount of kinetic energy to overcome this mechanical friction and cause rotor rotation at a speed directly proportional to the gas flow rate (see Figure 8-28b). In addition to this mechanical friction, there is a fluid friction caused by the gas flowing through the passages of the meter, which also adds to the base kinetic energy requirement of the meter.

Pressure Transducer

A pressure transducer is a device that responds to liquid or gas pressure and converts the pressure into an electrical variable. Within the pressure transducer is a unit called a pressure-force summing device. This device detects the pressure and responds by changing the pressure to a physical displacement. The physical displacement, in turn, causes an electrical transduction.

In Figure 8-29, several of these summing devices are illustrated. The diaphragm device reacts to pressure applied to a simple diaphragm. The convoluted (also called corrugated) diaphragm and the capsule are variations of the simple flat diaphragm, but require a change in pressure from the source or measurand.

The entities involved in pressure-force summing are the mass, spring constant, and natural frequency. The frequency response of a pressure device is an

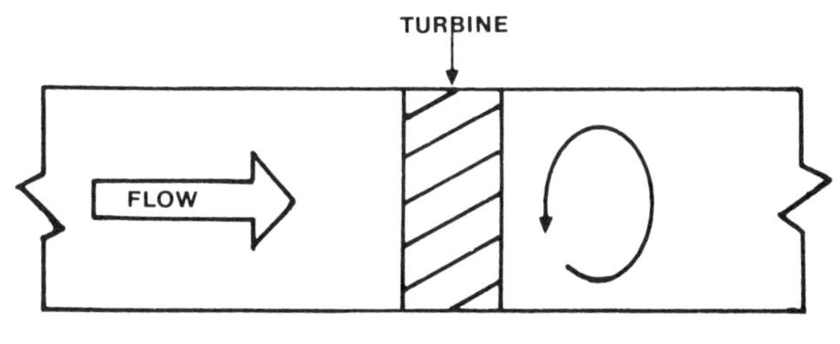

A. THEORETICAL TURBINE METER

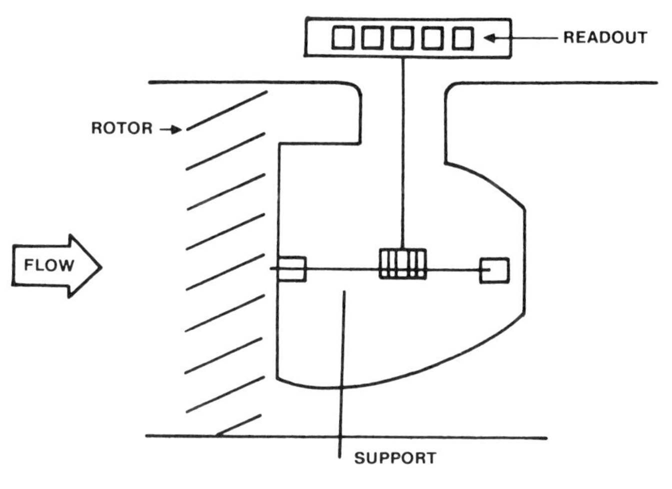

B. ACTUAL TURBINE METER

Figure 8-28 Turbine Meter

extremely important parameter used in the selection of a pressure transducer. This selection revolves around the pressure (measurand) environment. The selector must be aware of what effects will occur before, during, and after measurement. There are essentially three direct pressure applications. These are illustrated in Figures 8-30 through 8-32. Figure 8-30 shows an absolute pressure transducer. This transducer is referenced to a vacuum. Typically, the interior of the transducer case is evacuated and sealed. Some absolute pressure transducers are referenced to a bellows or a capsule whose exterior is acted upon by the measurand (pressure). In the Figure 8-31, the case is vented to atmosphere rather than being evacuated. Other than the vent, the gage pressure application is the same as absolute pressure transducer. In Figure 8-32, sealed gage pressure application is presented. This transducer is designed to prevent ambient media from entering the transducer case. Usually these instruments are of a range that

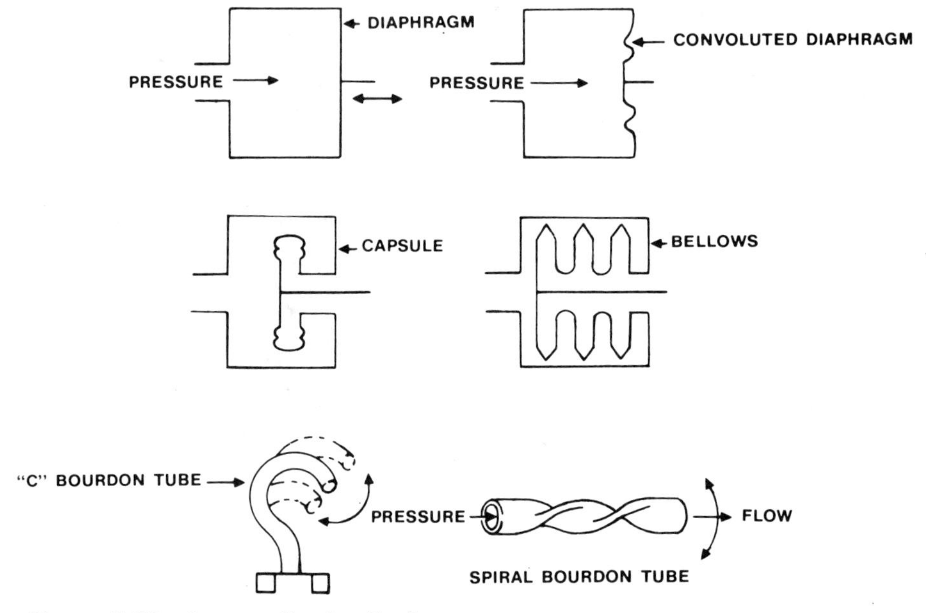

Figure 8-29 Pressure-Sensing Devices

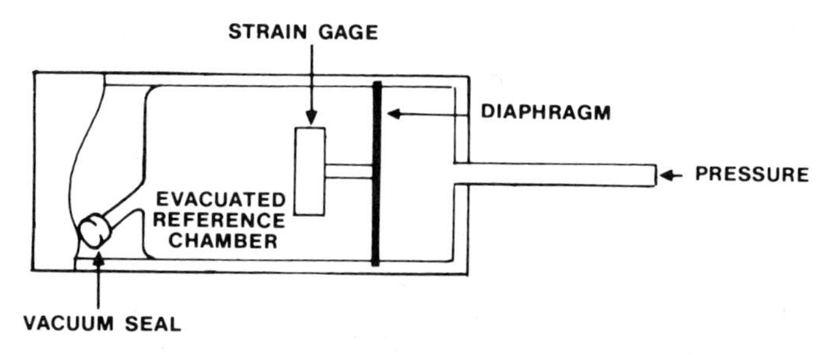

Figure 8-30 Absolute Pressure Transducer

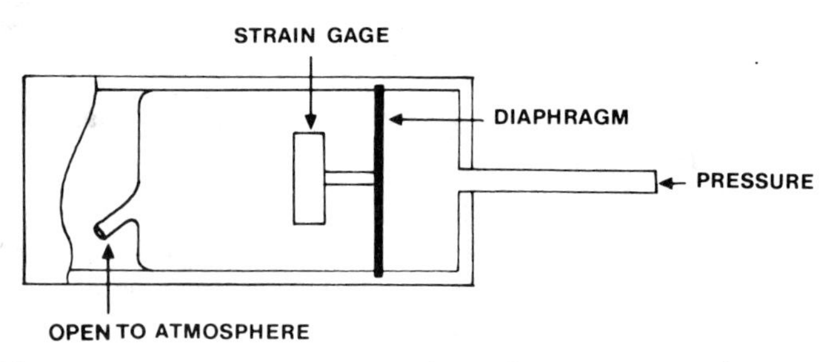

Figure 8-31 Strain-Gage Pressure Transducer Open to Atmosphere

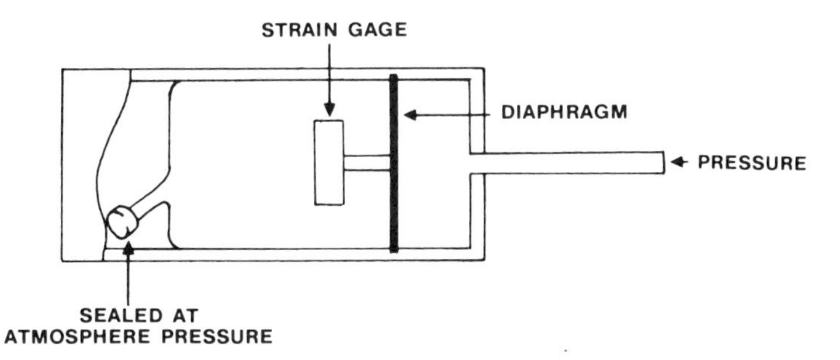

Figure 8-32 Sealed Strain-Gage Pressure Transducer

atmospheric pressure does not affect the measurement. The application is generally built with a partial atmosphere of helium sealed within. The application is called a sealed gage pressure transducer.

The capacitive pressure transducer consists of two fixed conductive plates that are isolated from a housing by insulated standoffs (see Figure 8-10). A pressure port directs pressure into a bellows. Attached to the bellows is a diaphragm. As pressure changes, the bellows expands or retracts, changing the position of the diaphragm in relation to the capacitor plates. This causes a capacitance change in two separate capacitive circuits.

The LVDT pressure sensing system does not utilize any levers, gears, or other linkages. It consists simply of a magnetic core fastened directly by a short stem to a metal chamber. Pressure entering through the "HI" pressure port causes expansion of the capsule and corresponding movement of the core in the LVDT, producing a known change in the LVDT electrical output (see Figure 8-33). Depending on the particular electronic circuitry supplied internally, the unit provides voltage or current outputs precisely proportional to the pressure input.

Pressure applied through the "LO" pressure port enters a metal isolating chamber surrounding the capsule and acts to compress the capsule (opposes "HI" pressure core motion). Thus, when pressures enter both "HI" and "LO" ports simultaneously, the unit's output becomes precisely proportional to the pressure difference. In the case of low range differentials, proper operation requires careful filling procedures to eliminate trapped air.

A basic potentiometric pressure transducer has a force-summing pressure bellows attached to linkage (Figure 8-34). The linkage is mechanically connected to a potentiometer wiper. The wiper travels over a multiturn wire coil or deposited resistor, as the sensor reacts to pressure changes. Usually some motion amplification between the force-summing element (pressure sensor) and the wiper track across the resistance element is employed to minimize acceleration error.

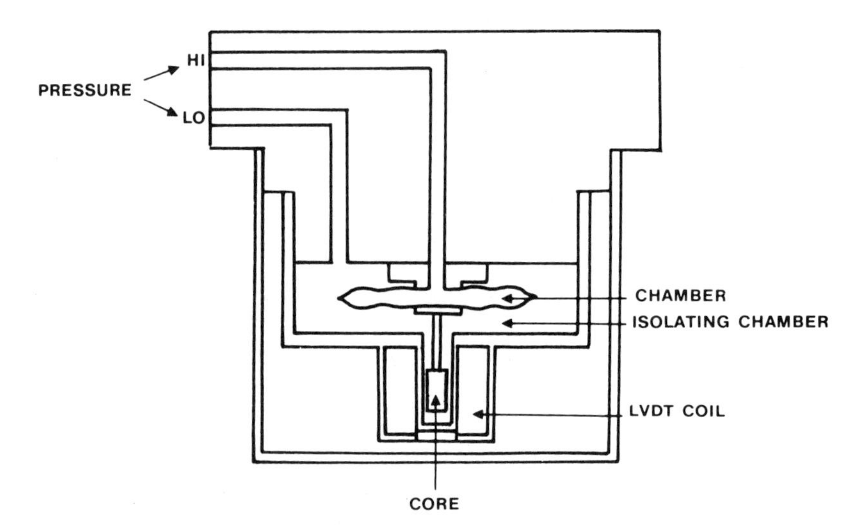

Figure 8-33 LVDT Pressure Transducer

The resistance of the potentiometer can provide specialized output linearities, such as linear, sine, cosine, and exponential. As a voltage ratio device, close regulation of the excitation voltage is not required. High-level output is inherent in the potentiometer concept. Output load impedance must be kept high to limit loading effects.

Ultrasonic Transducers

Ultrasonic waves are vibrational waves of electromagnetic frequencies that are above the hearing range of the normal ear. The term includes waves of a frequency

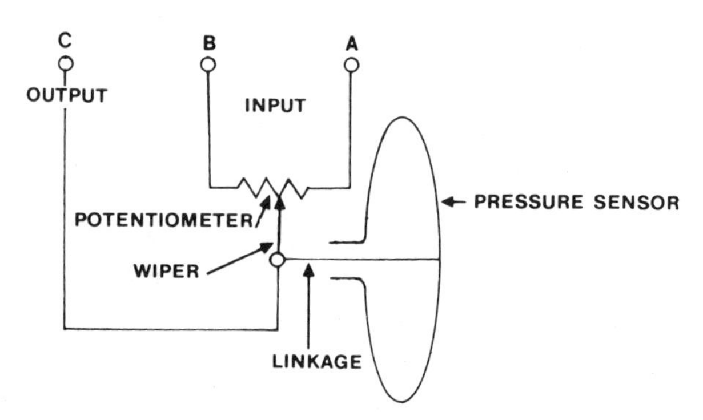

Figure 8-34 Potentiometer Transducer

of more than 20,000 Hz. The presence of a medium is essential to the transmission of ultrasonic waves, and almost any material that has elasticity can propagate ultrasonic waves. This propagation takes the form of a displacement of successive elements of the medium.

A simple sound wave, as it travels outward from its source, loses strength rapidly as the distance increases. This decrease in the strength of sound waves along a path can be greatly affected by discontinuities within the path (see Figure 8-35). With an ultrasonic control system, a sound path through air is established. The strength of the sound wave at any point along the path is a function of the distance from the point of origin. The introduction into the path of any material capable of absorbing some of the sound energy or reflecting it away from the original path can be measured. This change in the normal weakening or attenuation of sound along a path can be used to operate electronic circuitry.

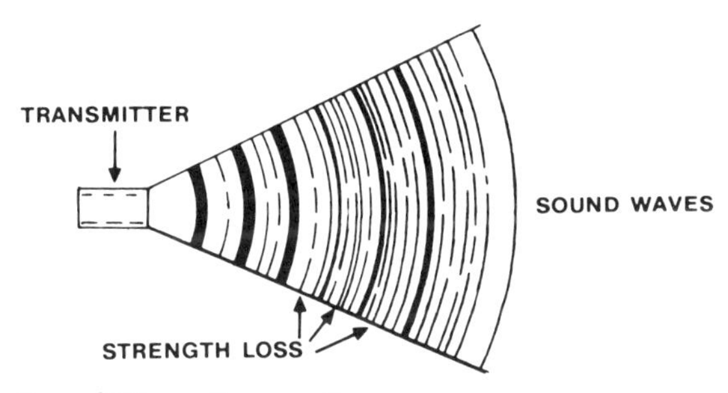

Figure 8-35 Sound-Wave Propagation

A more thorough study of the properties of sound waves in air will demonstrate that factors other than distance cause attenuation of sound waves. Relative humidity effects, temperature effects, and the presence of standing waves are the major stumbling blocks to this type of a control system. Complicated electronic circuitry can be developed to circumvent most of these problems. However, this results in a device that is expensive to manufacture, critical to adjust, and difficult to maintain.

The ultrasonic system can be used in four basic ways. The sensor uses ultrasonic energy to produce a direct path or a reflective path. First, the sensors can be installed to provide a direct path in which an obstruction between the sensors breaks the acoustic path, thereby operating a relay (Figure 8-36a). A relay may be connected to the desired external function (light, buzzer, switch). Although ultrasonic waves are transmitted from one sensor to the other sensor at approximately a 50° angle, an object large enough to cover the active area of the second sensor will break the beam for short path lengths.

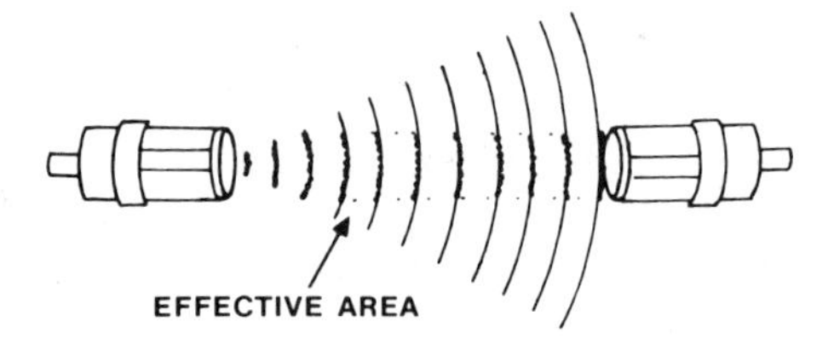

A. DIRECT PATH

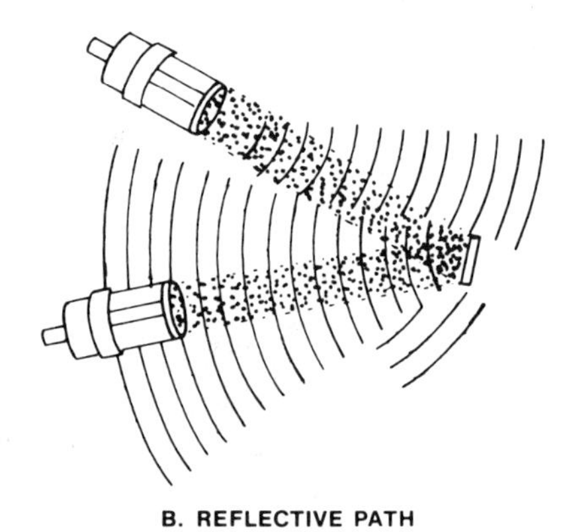

B. REFLECTIVE PATH

Figure 8-36 Ultrasonic Sensing

Second, the direct-path method can also be used to sense the absence of an object from the beam. With the object obstructing the beam, the output relay is closed. When the object is removed, the acoustic path is completed, which opens the relay.

The third and fourth ways in which the sensor can be used are with sensors located to provide a reflective path (Figure 8-36b). As with the direct path, the reflective beam can also be used in two ways: with the relay open to sense objects or with the relay closed to sense the removal of objects from the beam. In this method the sound waves emitted from one sensor are reflected from an object to the second sensor to provide the acoustic path. When the object is removed from the path, the relay is closed. Conversely, when objects establish a path, this opens the relay.

Photodetectors

These devices are described in Chapter 10.

Thermoelectric Transducers

The sensing of temperature is usually accomplished with the aid of a thermoelectric device such as the thermocouple, the resistance temperature detector (RTD), or the thermistor. A thermoelectric device is one that converts heat (temperature) to a corresponding electrical current flow or voltage level.

In all reactions, heat is dissipated when energy is being converted or when some type of work is being done. In some cases, such as the conversion of coal to steam to electric, large losses are caused by heat dissipation. In other uses, such as air conditioning, it is indeed desirable for heat to be removed. Whichever is the case, the temperature transducer plays an important role in monitoring how much heat is present at specific points within the system.

Temperature is monitored by several scales in degrees. The two most used of these scales are Fahrenheit (F) and Celsius (C). The latter was formerly referred to as centigrade. These two scales are the ones the layperson uses in everyday life. The base unit of temperature within the System International (SI) is the kelvin (K). Kelvin is used in scientific work. The SI units represent the metric system. Each one of these scales has desirable features.

Thermocouple The thermocouple consists basically of two dissimilar metals fused together. The measuring junction is the hot junction. As previously discussed, the hot junction is where electrons move from a high-energy state to a lower-energy state and release heat. The measuring junctions are specially constructed in a tube to provide the junction with support while still achieving uninhibited sensing of the environment to be measured. The supporting material is called a *sheath* and is made from metal such as inconel or stainless steel. The sheath is insulated from the junction with ceramic or magnesium oxide. There are three basic junction models: the exposed junction, the ungrounded junction, and the grounded junction (see Figure 8-37).

The thermocouple measurement junction is attached to wires. The wires with junction are called *elements*. The wires are parallel, with their lengths varying depending on the job requirement. The ends of the wire are bent so as to fit neatly into terminal connections.

A typical thermocouple assembly consists of a connection head, extensions, and a thermowell. The *thermowell* encases the thermocouple element in a ceramic-insulated sheath made of stainless steel. The sheath type is dependent upon the local environment. The probe type is dependent on the temperature range required.

The *extensions* are in the center of the probe. It is often desirable to extend

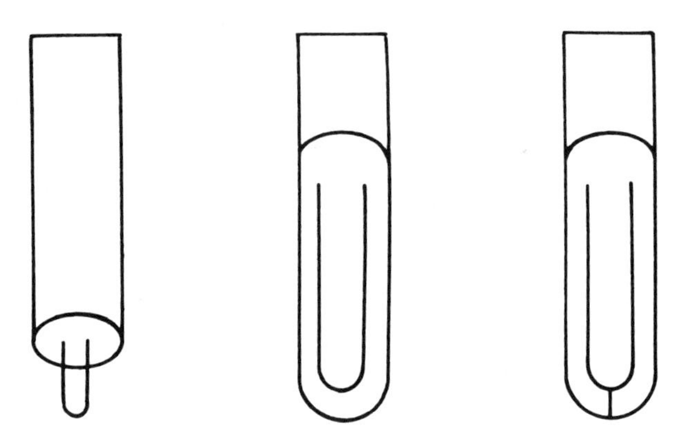

Figure 8-37 Thermocouple Measurement Junctions

the connection between the thermowell and the thermocouple connecting head for operational convenience or to avoid direct contact with hot surfaces.

The top of the thermocouple is called the *connection head*. The head provides protection for the electrical terminations of the thermocouple and connection to the associated instrumentation.

Thermistor The thermistor is a solid-state device that decreases in resistance as temperature increases. The word thermistor is derived from two words: thermal and resistor. In circuit, the decrease in resistance also means an increase in current flow. Thermistors are extremely sensitive. Some may decrease in resistance as much as 5% for each degree Celsius rise in temperature. Thermistors are made from metallic oxide crystals. Thermistors offer extreme sensitivity to temperature differences. For example, a thermistor with practical resistance levels will demonstrate a resistance ratio of 30 between +25°C and +125°C, and a ratio of 9 between 0°C and +50°C. So while thermistors are nonlinear, they offer extreme sensitivity to small temperature changes.

In Figure 8-38a, the thermistor is used as a simple temperature-measuring device. Here the thermistor is placed in series with a meter. The meter is current sensitive and calibrated so that a change in current will cause an equivalent change in degrees of temperature.

In Figure 8-38b, the thermistor is placed in a bridge network to provide a more precise measurement. A meter (galvanometer) is placed across the bridge to monitor minor changes in current, which are reflected as degrees of temperature on the meter.

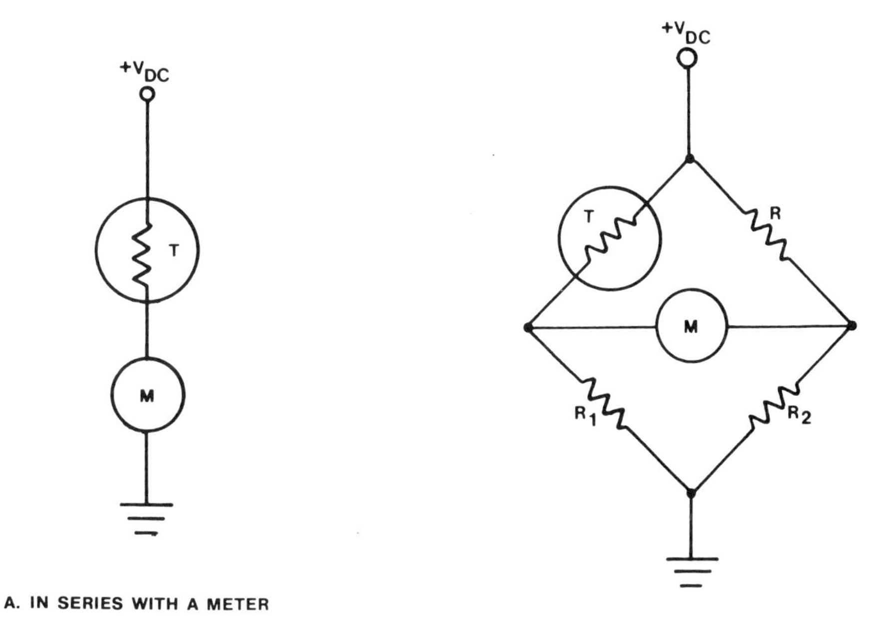

Figure 8-38 Thermistor Modifying Circuit Behavior

Resistance Temperature Detector (RTD) You may recall that the thermistor had a negative temperature coefficient. That is, as temperature increases, the resistance of the thermistor decreases. Most conductors of electricity, such as the copper wire, have a positive temperature coefficient. These conductors increase in resistance as temperature increases. The positive temperature coefficient is termed alpha (α). The thermal component that industry uses is the resistance temperature detector, known as the RTD. As with the thermistor, the RTD is used in applications such as a bridge network to provide precision measurement.

9

Machine-Monitoring Transducers

The purposes for monitoring machine motion are threefold. First is the matter of safety of plant personnel and the potential damage that may be incurred on other plant equipment. The second reason is an attempt to eliminate the expense of repairs, including costs of materials and labor time. The final reason is to preclude the expense incurred by machine downtime or outages due to machine failure.

Two sets of parameters are involved in machine monitoring. The first are the process variables. These variables are measured within the machine piping, exchanges, filters, and the like. They include measurements of temperatures, pressures, and flows of the process material in the machine or in motion through the machine. These process variables have been covered in detail in other chapters in this book. The second major parameter, which we shall concentrate on in this chapter, is measurement that evaluates the mechanical running condition of the machine. The three most common parameters that will be stressed are temperature of machine components, mechanical vibration (dynamic motion) of machine components, and the relative position of moving machine elements with respect to the stationary machine components.

MACHINE-MONITORING PARAMETERS

Machine Temperature Parameters The primary transducers used for temperature measurement are the thermocouple and the resistance temperature detector (RTD). They are used extensively in machine monitoring. For maximum efficiency the monitoring of the temperature of machine components is required, in addition to the temperature of auxiliary systems. Monitoring the temperature of a bearing is more reliable than monitoring the temperature of the lubrication oil in the discharge line. At best, both should be monitored. The monitoring of radial bearings, thrust bearings, and sliding surface babbitt bearings is a necessity. Electric motors and electric generators are monitored at their rotors and stators. Seal and coupling oil system temperatures should be measured along with their pressures. Finally, the temperature of the machine case at various locations and the ambient temperatures of the machine platform should be monitored.

Machine Vibration Parameters The subject of monitoring the mechanical vibrations (dynamic motion) of a machine is more complex than monitoring temperatures. The necessary considerations for selection and application of the proper transducer are more numerous and variable. Rotor vibration can be measured in two ways: shaft motion relative to the bearing (relative motion) and shaft motion relative to free space (absolute motion). Nonrotating components are usually measured with seismic transducers relative to free space. Structural components to be considered for measurement include bearings and/or bearing housing, machine casings, piping, and foundations. It must be noted that rotor and casing vibration can be monitored in all three planes on a machine: axial and two orthogonal radial planes. A parameter auxiliary to shaft vibration is phase-angle measurement. Phase-angle measurement is applicable to all machines. It is used directly for balancing and further for classifying machinery malfunctions. A phase reference should be provided for each different shaft speed or direction of rotation on a machine train.

Shaft Rotation Parameters Shaft rotation parameters begin with rotative speed. Normally, shaft rpm is measured on the prime mover (driver) of the machine. An auxiliary measurement is shaft acceleration or shaft rpm with respect to time. If thermal growth considerations or a particular start-up sequence procedure depends on the rate of increase of shaft speed, then a shaft acceleration monitor should be employed with a readout in rpm per minute. In addition, monitors are available that will indicate zero speed of a rotor to signal machine shutdown function or locked rotor condition for critical electric motor start-ups. Some machines or processes can experience damage due to shaft rotation in the opposite direction from normal. In these cases a reverse shaft rotation monitor should be used. In applications where speed measurement is used for machine or process control applications, a redundant or fail-safe instrument system is mandatory.

Position Measurement Position measurements on a machine are all relative in nature. The position of one machine component is monitored relative to the position of another machine component. Examples of position measurements include shaft position relative to bearings, machine casing position relative to foundation and casing, and piping or foundations relative to other external fixed references.

In general, parameters should be divided into two categories of importance: those that must be continuously monitored and those that must be periodically measured. Extremely expensive or sensitive machinery should have a complete continuous evaluation. Less expensive machinery should be monitored to a lesser extent. Noncritical machinery should be measured periodically or monitored continuously for a specific parameter.

MACHINE-MONITORING TRANSDUCER INSTALLATION

There are two primary methods for mounting transducer probes, external and internal (Figure 9-1). Each has its advantages and disadvantages; however, the external method is generally preferred. With some installations the user has no choice as to which is used.

External mounting is where the probe is physically installed by means of a probe adaptor on the machine case or, in applications where a longer reach is required, attached to a probe sleeve which is installed through an adaptor mounted on the machine case. Integral junction box adaptor assemblies are also used. External mounting allows probes to be adjusted while the machine is running and removed without machine disassembly. It also eliminates potential problem situations that could occur from damage to an improperly installed probe cable in an internal mounting situation or a loose connection if probe and extension cable are connected inside the machine case.

Internal mounting is where the probe is physically mounted inside the machine case close to the shaft. It is used in situations where the physical distance from the machine case to the shaft is too great to allow the probe to be mounted in a probe sleeve inserted through the machine case or where external probe mounting is impractical.

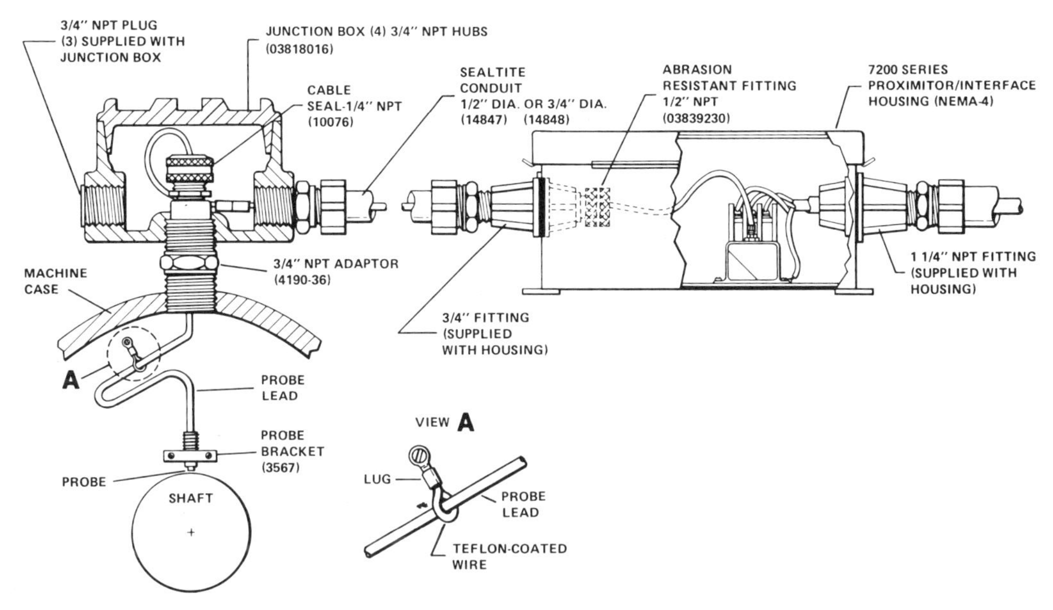

Figure 9-1 Typical Installation of a Machine Monitor *(Courtesy, Bently Nevada)*

The system operates on inductive proximity principles; therefore, the probe does not contact the observed surface and is not affected by nonconducting materials such as air, oil, gas, and plastics in the gap between the probe tip and observed surface.

The selection of probes for specific applications depends on the following requirements:

1. Measuring range
2. Physical shape of probe
3. Probe environment
4. Distance from the probe to the proximitor

Figure 9-1 illustrates a typical machine-monitoring, noncontacting transducer installation. This particular installation is an internal probe mounting.

TRANSDUCER PARTS

The Probe Eddy current proximity probes (Figure 9-2) are typical of machine-monitoring transducers. The proximity eddy current probe can be used

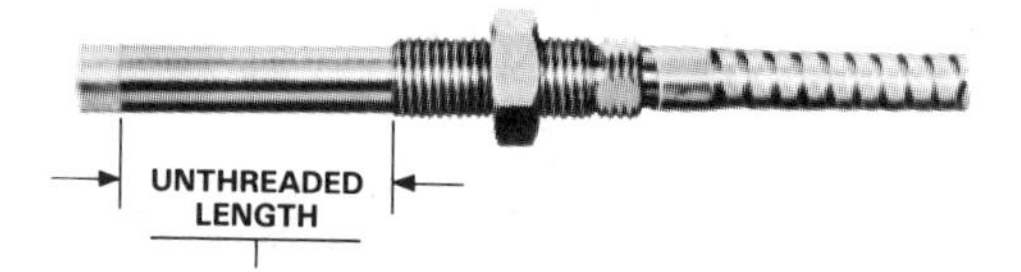

Figure 9-2 Proximity Probe *(Courtesy, Bently Nevada)*

for static gap measurements (axial or radial positions) or for dynamic motion (vibration). The probe is effectively a noncontacting electronic micrometer suitable for slowly changing position measurement or rapidly changing dynamic measurements.

The essential part of the probe is a small coil of wire that is in the tip of a fiberglass body. The fiberglass body can be mounted in several different metal case designs to aid in installation.

The range of the eddy current probe is partly a function of the tip diameter. Generally, the larger the tip is, the longer the linear measuring range. Each probe is equipped with an integral coaxial cable and connector. The cables are available with or without armor sheathing.

Probe Mounting Bracket The purpose of the probe mounting bracket is simply to hold the probe in position. The bracket in Figure 9-1 is used for mounting probes inside a machine case.

A threaded hole holds the probe to allow adjustment of the probe tip in relation to the observed surface. Two other holes attach the probe mounting bracket inside the machine utilizing screws. When the screws are tightened, the slot allows the threaded hole to compress, locking the probe in its preset position.

As indicated in the internal mounting of Figure 9-1, the probe lead should be secured between the probe and the exit point to prevent damage due to whipping caused by turbulent air and/or other fluids inside the machine. Care should be exercised to avoid damage to the coaxial probe lead due to clamping it too tightly and/or bending it more sharply than the recommended minimum bend radius.

Proximitor The function of the proximitor (Figure 9-3) is twofold: first, it generates a small radio-frequency signal that is radiated at the probe coil (tip) and is used to measure the probe tip-to-target dc gap voltage utilizing the eddy current principle. Second, it provides a voltage output that varies as the gap distance varies, providing both a static or average dc gap voltage and a dynamic ac (vibration) voltage. Thus the proximity probe transducer will simultaneously measure average shaft position relative to the bearing clearance and shaft dy-

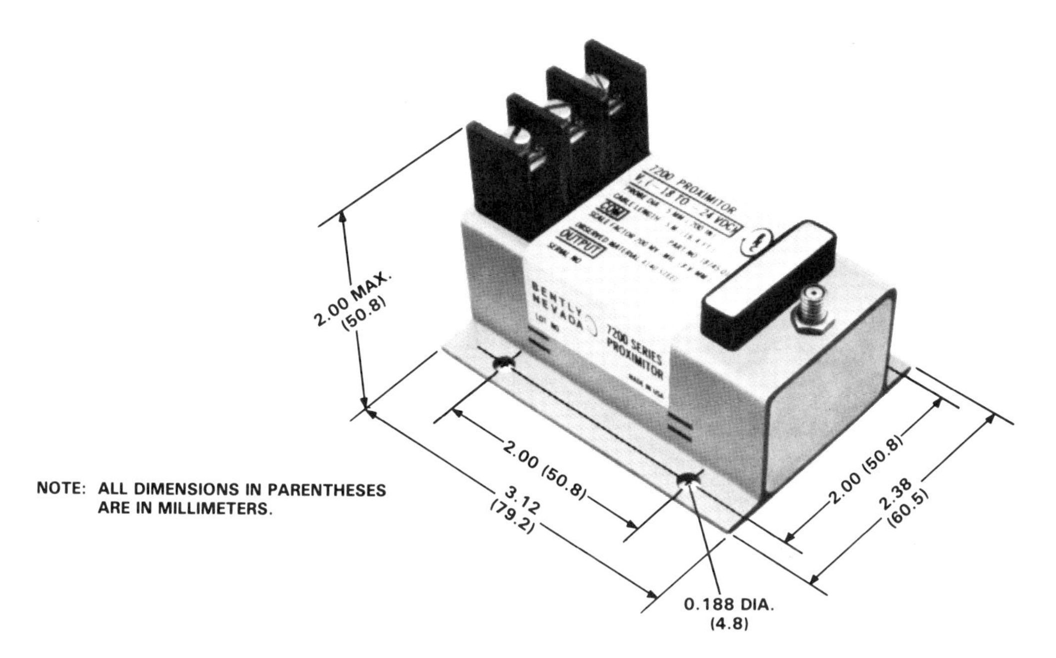

Figure 9-3 Proximitor *(Courtesy, Bently Nevada)*

namic motion (vibration). The signal output and power source input are made to the proximitor through three-conductor shielded signal wire, which can be as much as 1000 feet from monitors without significant degradation of performance. The proximitor is intrinsically safe for hazardous area applications when used with zener barriers.

MEASUREMENT OF ROTATING MACHINERY

The science of the measurement of rotating machinery necessarily includes standards that have been proven through years of experience. To understand the behavior of rotating machinery, it is desirable to communicate standardized and recommended practices defining the polarity and phase referencing of the transducers, data storage, and data presentation devices.

Positioning of a Motion Transducer The motion transducer (Figure 9-4) is installed adjacent to a bearing to monitor a rotating shaft. The probe can be inductive, capacitive, or magnetic. Let us consider for explanation purposes that the probe is an eddy current probe. As the shaft is moved toward the sensitive axis of the probe, the gap decreases and the probe produces a less negative (−) or more positive (+) magnitude of voltage or current. Likewise, movement of the rotating shaft away from the transducer produces a more negative (−) voltage or current. Therefore, the general polarity rule for all transducers is that *motion toward the transducer should produce a positive-going signal.* Figure 9-4 is representative of a typical eddy current probe.

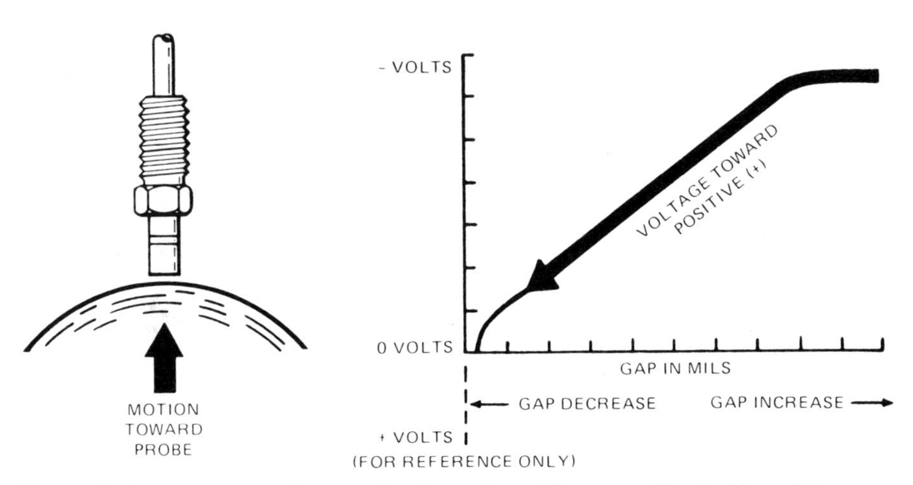

Figure 9-4 Positioning of a Motion Transducer *(Courtesy, Bently Nevada)*

Checking the Polarity of Seismic-Type Displacement, Velocity, and/or Acceleration Transducers The polarity of inertially referenced displacement, velocity, or acceleration transducers (Figure 9-5) is easily tested by means of a tap test. This tap test consists of tapping the transducer lightly in the direction of the sensitive axis. As shown in Figure 9-5, the transducer can be displacement (D), velocity (V), or acceleration (A). The resultant wave form as depicted on a time base graph of an oscilloscope has a sharp peak and then levels out. The initial spike is positive, thereby proving the accuracy of polarity. The response to motion is applicable to displacement, velocity, and acceleration. The response relationship, however, is separate from the phase relationship of the three. In the event all three are observing the same sinusoidal motion, the velocity and acceleration signals will lead displacement signals by 90° and 180°, respectively. This relationship will always be the case except where an inverted polarity is deliberately required.

Testing Rotor (Shaft) Rotation at 90° Transducer Angles In Figure 9-6, a pair of transducers is mounted vertically (at the top of the rotor looking

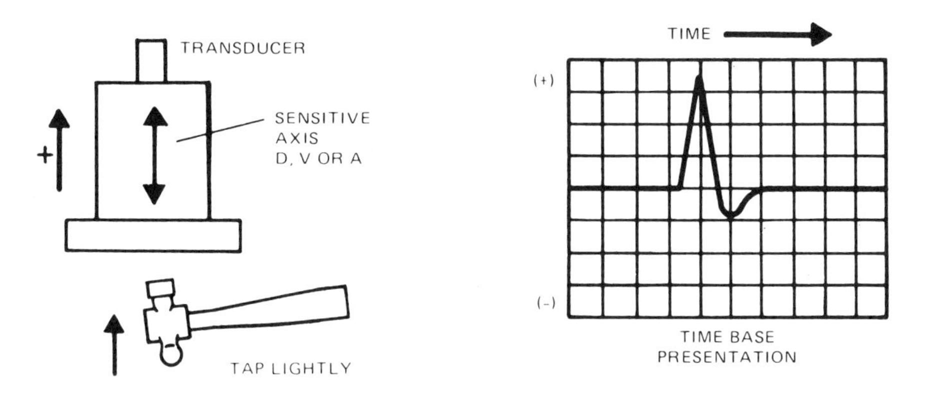

Figure 9-5 Checking the Polarity of Displacement Velocity and/or Acceleration Transducers *(Courtesy, Bently Nevada)*

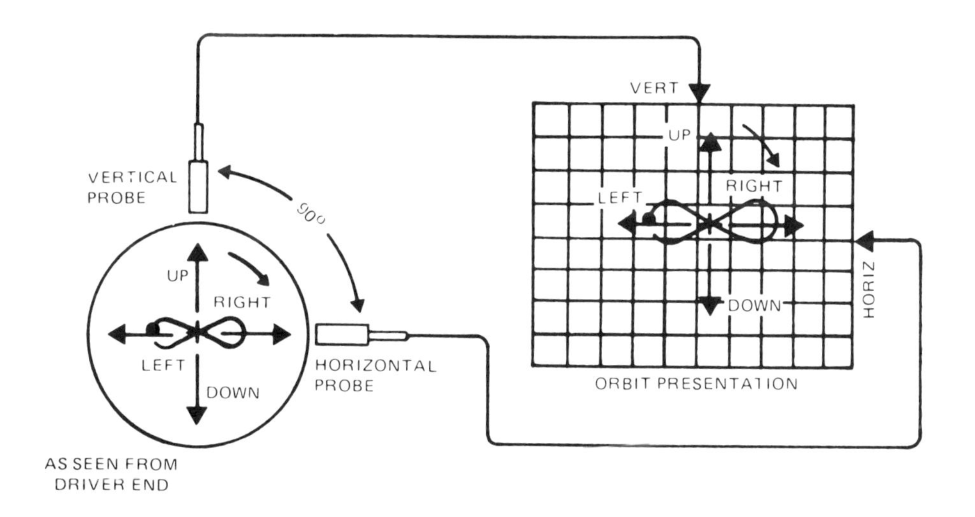

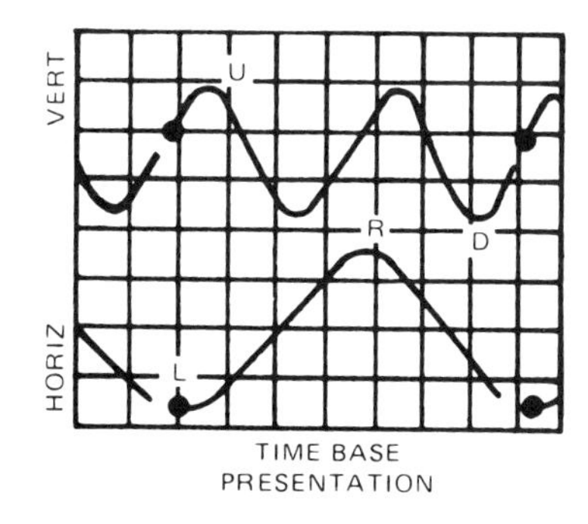

Figure 9-6 Testing Rotor (Shaft) Rotation at 90° Angles *(Courtesy, Bently Nevada)*

downward) and horizontally (at the right as seen from the driver end of the rotor), respectively. The angular relation between the two transducers is 90°. The vertical transducer is connected to the vertical jack of an oscilloscope and the horizontal transducer to the horizontal jack. In this example the resultant display is a perfect bow tie lying horizontally in orbit presentation. The display is a Lissajous pattern. The lower graph is the time base presentation. The upper display represents the vertical presentation of the shaft (rotor) dynamic motion. Note that in the up position on the machine the sine wave is maximum positive, while in the down position, the sine wave is maximum negative (180° shift in angular rotation). The lower display represents the horizontal presentation of the shaft (rotor) dynamic motion. Note that in the right position on the machine the sine wave is maximum positive, while in the left position, the sine wave is maximum negative (180° shift in angular rotation).

Testing Rotor (Shaft) Rotation at 45° Transducer Angles In Figure 9-7, a pair of transducers are mounted 90° from each other, but 45° respectively from

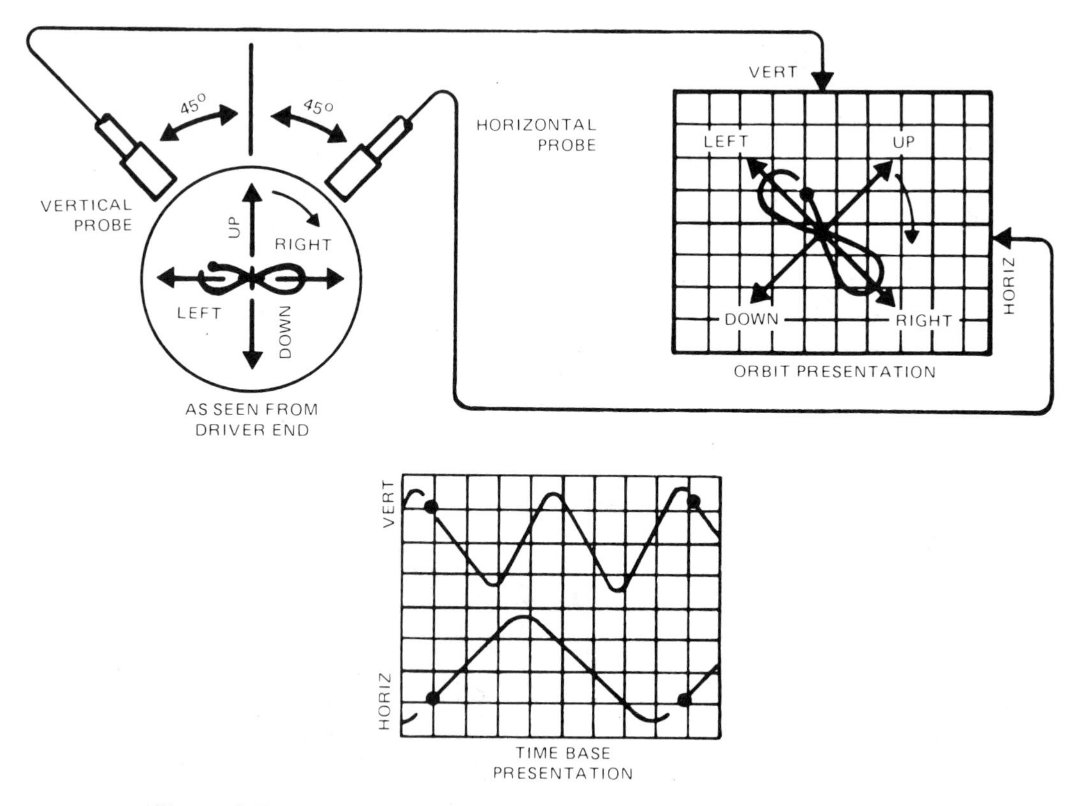

Figure 9-7 Testing Rotor (Shaft) Rotation at 45° Angles *(Courtesy, Bently Nevada)*

a vertical line (at the top of the rotor as seen from the driver end). The vertical probe is located 45° to the left of center and is attached to the vertical jack of an oscilloscope. The horizontal probe is located 45° to right of center and is attached to the horizontal jack of the oscilloscope. The resultant display is a perfect bow tie lying at a 45° angle, with its high side toward the vertical transducer. The lower graph is the time-base presentation. As in the 90° transducer setup in the preceding paragraph, maximum positive sine-wave display is toward the transducer.

THE KEYPHASOR

The Keyphasor is a transducer installed on a machine train observing a once-per-turn event (e.g., keyway) on a rotating shaft providing a voltage pulse that occurs once-per-revolution of the shaft (Figure 9-8). The Keyphasor pulse provides a reference for data taken on a machine train. It is a reference mark and timer for speed, phase angle, frequency measurements, and all data acquisition. The Keyphasor is very valuable when trying to diagnose and correct specific

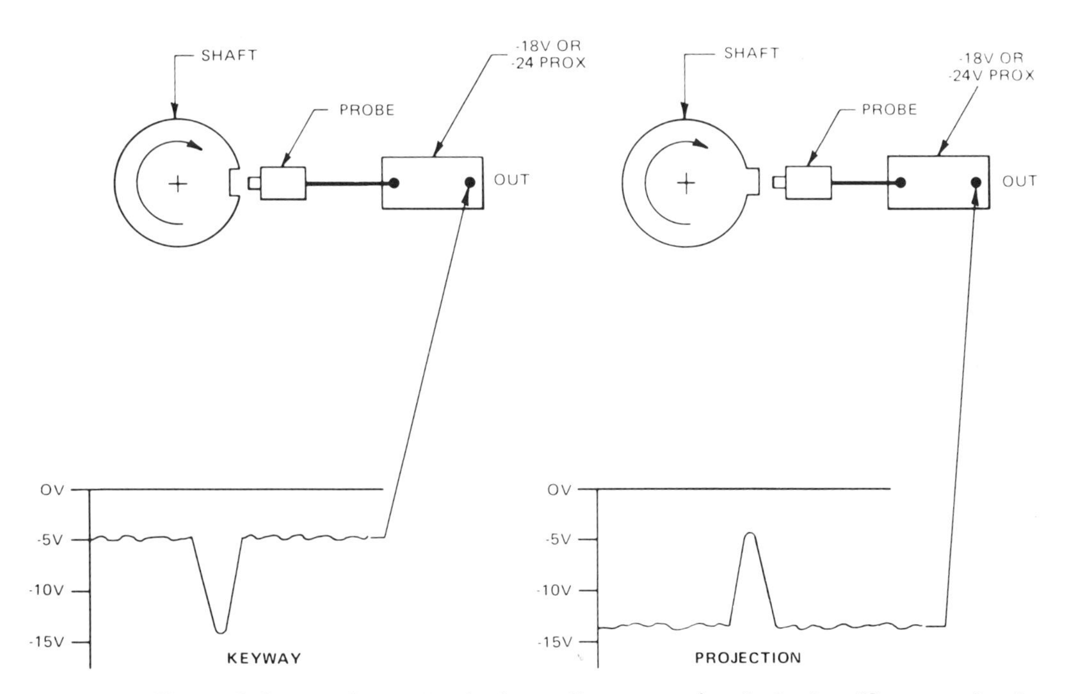

Figure 9-8 Keyphasor Monitoring a Keyway and a Projection *(Courtesy, Bently Nevada)*

malfunctions of rotating machinery. The pulse provided by the Keyphasor transducer is normally input to instruments such as tachometers, vector filter phase meters, or other instruments requiring a timing pulse. A Keyphasor transducer pulse is of special value when it is fed into the *Z* intensity axis of an oscilloscope.

The Keyphasor transducer is usually a proximity probe mounted to observe a keyway or projection, which provides a large change in gap in front of the proximity probe (see Figure 9-9). Since the proximity probe is a gap to the voltage transducer, when the keyway passes the probe a voltage pulse results. Since the keyway passes the probe once-per-revolution of the shaft, the Keyphasor pulse occurs at a frequency equal to the speed of the machine. Therefore, its frequency is the same as the running speed of the machine.

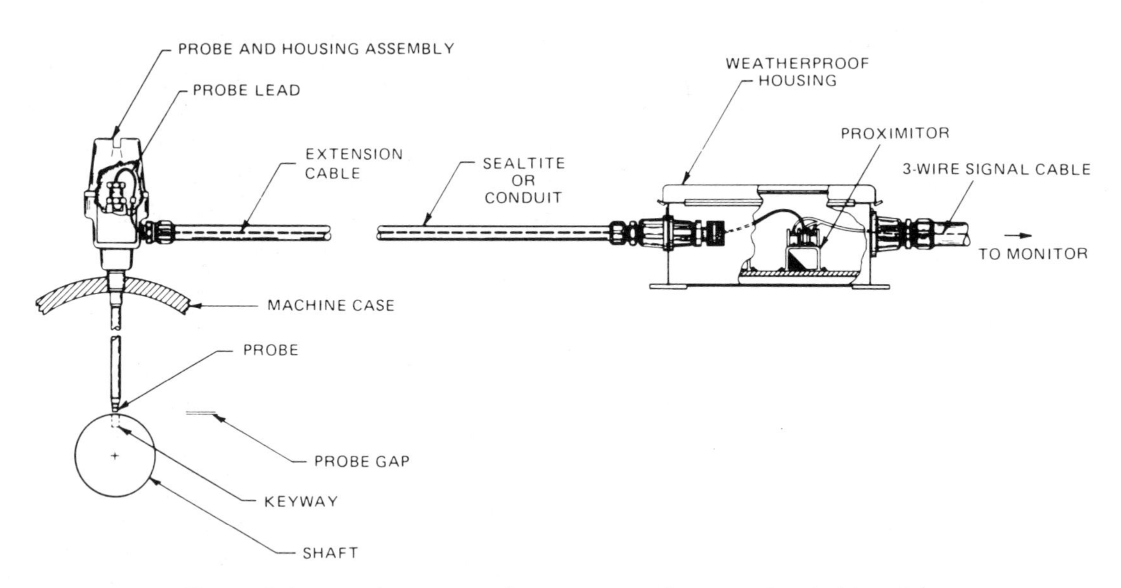

Figure 9-9 Keyphasor Transducer System *(Courtesy, Bently Nevada)*

Wave Forms and Orbits

To properly understand the full operation of a Keyphasor, it is first necessary to understand the wave form or orbital pattern on an oscilloscope (see Figure 9-10). When vibration probe inputs are put into a dual-channel oscilloscope and observed on the cathode ray tube, the dynamic motion (vibration) may be displayed in time-base or orbital form. A wave form or orbit pattern on an oscilloscope is simply a dot moving very rapidly such that to the eye it appears as a continuous line. The rapidly moving dot represents the center-line motion

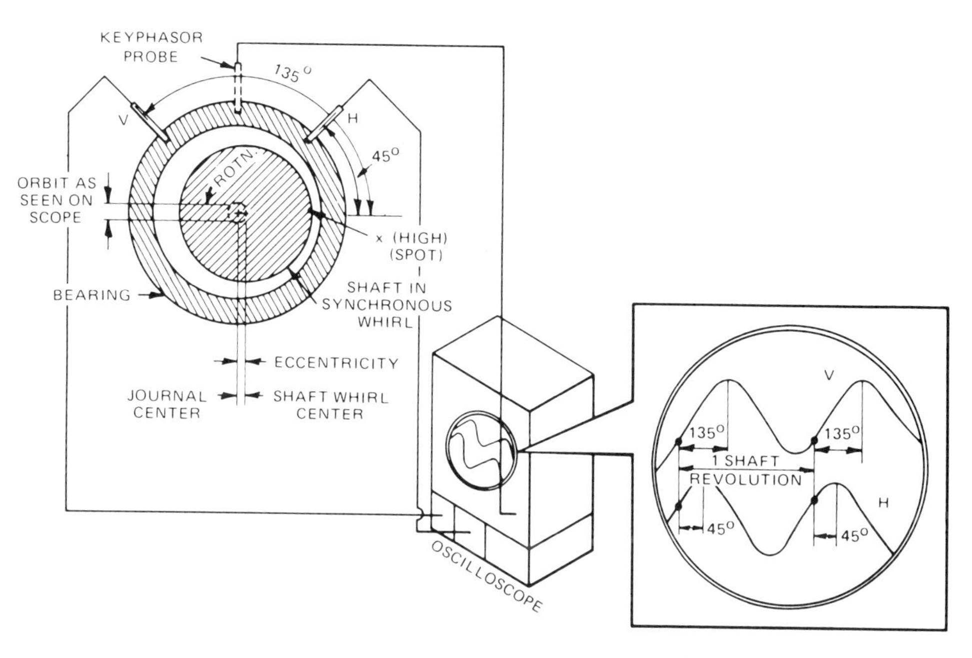

DIAGRAM SHOWING PHASE MARKING OF WAVEFORMS USING BLANKING MARKS ON AN OSCILLOSCOPE DISPLAY.

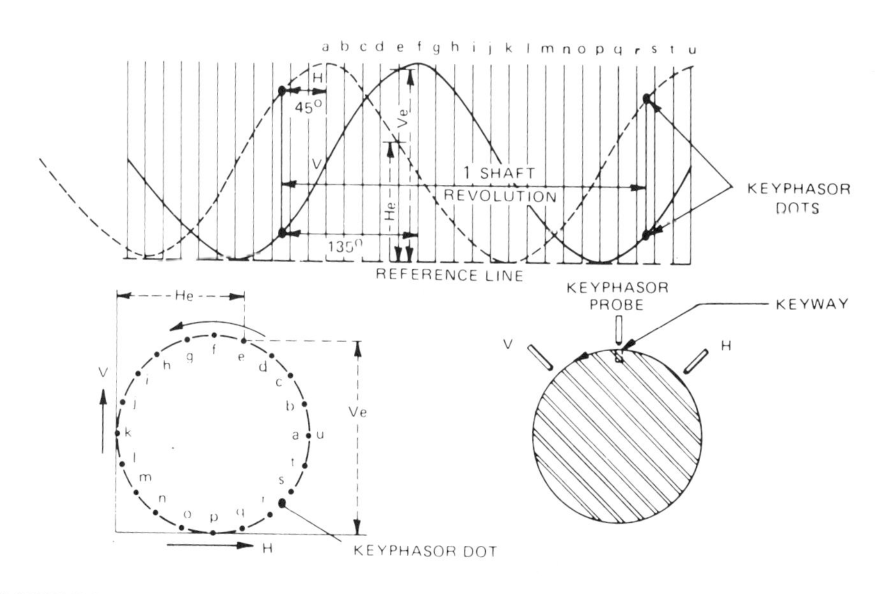

DEVELOPMENT OF AN OBRIT PATTERN FROM SIMULTANEOUS VERTICAL AND HORIZONTAL WAVEFORMS

Figure 9-10 Wave Forms and Orbits *(Courtesy, Charles Jackson, Monsanto Polymers and Petrochemicals Company for Bently Nevada)*

of the shaft. The orbit is a path of the high spot at that lateral position of the probes. The Keyphasor pulse, when fed to the Z intensity input of the oscilloscope, simply intensifies the dot at the instant in time when the keyway (event) is in front of the Keyphasor transducer. Therefore, the Keyphasor dot on the orbit or wave form represents the center-line location of the shaft in its path of travel (or high spot) at the instant that the keyway is in front of the Keyphasor transducer. By utilizing this Keyphasor technique, we get a physical reference to the shaft. The rotor can be stopped with the keyway in front of the Keyphasor transducer, and the exact angular location of a high spot of the shaft that was observed at that particular instant in time can be determined. In addition to Z intensity axis input, the Keyphasor should also be used to trigger the oscilloscope when in the time-base mode.

Phase-Angle Measurements for Balancing

The Keyphasor (shaft reference) technique tells where the rotor high spot is at a particular instant in time when used with a phase-measuring instrument. If we define the phase angle as the number of degrees of shaft rotation from the Keyphasor pulse to the following positive peak of dynamic motion (vibration) filtered at rotative speed as sensed by a vibration probe, the use of the shaft reference technique allows the direct phase-angle readout to define the high spot on the rotor system. The logic of this method of phase measurement is as follows: the Keyphasor pulse occurs the instant the Keyphasor transducer and the shaft reference (notch or projection) are aligned. The positive vibration peak occurs ϕ number of degrees (phase angle) later in time as the shaft high spot passes beneath the vibration input probe. The machine rotor may then be stopped and rotated to align the Keyphasor transducer and its shaft reference. If the rotor is then rotated in the direction of rotor rotation the number of degrees as defined by the phase angle, the high spot of the rotor will lie directly under the vibration probe.

By utilizing the ability of the Keyphasor to define the high spot on a rotor, it is possible to provide a balancing instrument that can read the high spot of the rotor directly with a physical shaft reference. This allows the addition or subtraction of weight at the proper angular location on the rotor to improve rotor balance and helps to minimize the number of machine runs necessary to balance or trim balance a machine. By utilizing it with oscilloscope orbit presentations, it is possible to trim balance machinery in the field as well as on the balance stand.

Two techniques are used to plot rotor unbalance response. These are the polar plot and the Bode plot. They require primary measurements of amplitude, rpm, and phase angle.

The *polar plot* is a plot of the phasor (rotating vector) of the amplitude of rotative speed (or forcing function speed) and phase angle as a function of speed. On rotating machinery, the polar plot is the amplitude and phase reading from a transducer showing the response of the machine to its residual or deliberate unbalance as a function of speed. Mass unbalance manifests itself as a phasor. The magnitude (peak-to-peak mils) and the direction (phase angle) of the resultant motion with respect to a known reference point (Keyphasor) can be described on a polar plot, which displays the in-phase and quadrature components of the phasor. These component values will change as the magnitude and direction of the phasor quantity changes with speed. A classic example of a well-balanced rotor is represented by the polar plot in Figure 9-11.

The Keyphasor pulse also allows a plot of rotative speed amplitude (1×) of a given measurement against rpm, along with the phase lag angle of that amplitude phasor against rpm. The Bode plot provides this relationship and is most useful in showing the rpm of various resonances of a machine (see Figure 9-11).

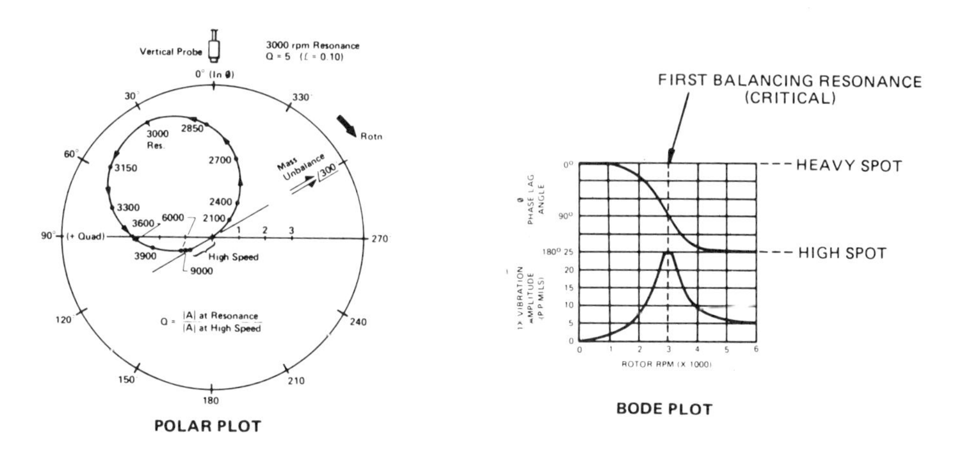

Figure 9-11 Techniques for Plotting Rotor Unbalance Response *(Courtesy, Bently Nevada)*

Optical Pickup Installation as a Temporary Keyphasor

On machine trains where there is no permanently installed Keyphasor proximity probe, a Keyphasor pulse may be obtained by the use of an optical pickup (Figure 9-12). Highly reflective tape is available that, when placed on the shaft,

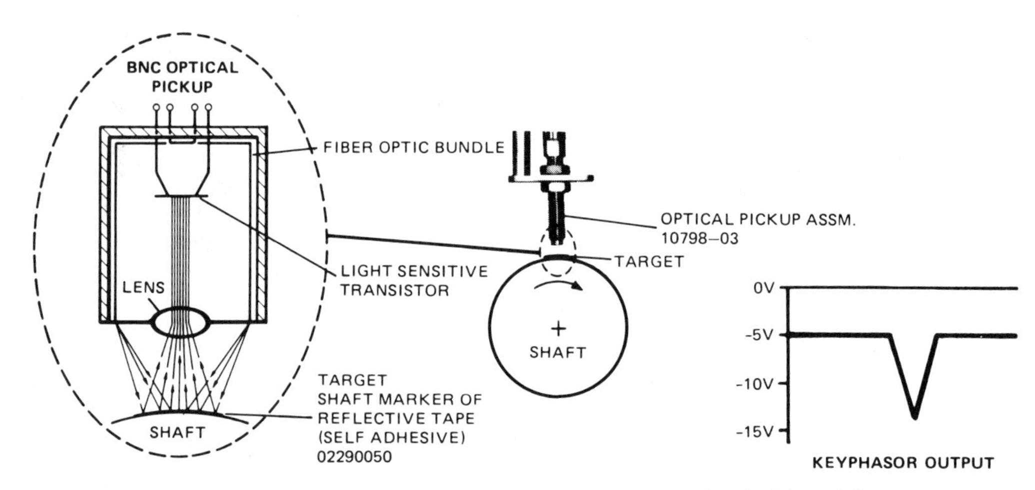

Figure 9-12 Optical Pickup Installation *(Courtesy, Bently Nevada)*

can provide an "event" for an optical pickup to observe. If highly reflective tape is not available, a paint mark on the shaft can serve this same purpose. With the optical pickup, any place where visible shaft is available a Keyphasor pulse can be observed. Stated another way, any place where a strobe light can observe a shaft, an optical pickup can be used.

VIBRATION MEASUREMENT

Preventive maintenance for rotating machinery has been enhanced by the addition of vibration monitoring. It has become recognized that vibration monitoring can help to prevent machinery failures and downtime. The concept called *predictive maintenance* is based on a machine's condition rather than elapsed time.

Generally, the parameters monitored are amplitude of vibration in peak-to-peak mils displacement, peak inches per second velocity, or peak *G*'s acceleration. Periodic checks of vibration must be included in a preventive maintenance program.

This section is dedicated to the discussion of basic dynamic motion (vibration).

Amplitude

Amplitude, whether expressed in displacement, velocity, or acceleration, is generally an indicator of severity. It attempts to answer the question, "Is this machine running smoothly or roughly?" The ability to measure the shaft with proximity probes has helped greatly in providing more accurate information with regard to the amplitude of vibration. In the past, when only casing measurements were available, amplitude of the casing vibration was the only available parameter for severity. Whereas the casing measurements were able to indicate the presence of some machinery malfunction conditions, by and large the casing measurements proved inadequate for proper machinery protection. This was primarily due to the variable transfer impedance between the shaft motion and the casing motion, depending upon the particular machine design, assembly, operating condition, and case pickup location.

Casing measurements have been utilized recently in an attempt to determine the presence of high-frequency vibrations. In this manner blade passage frequencies, blade resonance frequencies, gear mesh frequencies, and the like, are observed with the hope of making early discovery of possible malfunctions. These measurements are generally made on a periodic basis and are not usually continuously monitored. Since these are case measurements, it is somewhat difficult to evaluate the relative amplitudes and corresponding significance of the fact that the vast majority of machine malfunctions manifest themselves at lower frequencies, usually less than 4× running speed. High-frequency measurements are useful only a small percentage of the time in machinery evaluation.

Amplitude of vibration on most machinery is expressed in peak-to-peak mils displacement. With proximity probes mounted at or near the bearings, vibration to tolerances can be established, which provide for the maximum excursion that the shaft makes with respect to the bearing. Today, most continuous monitoring of critical machinery is provided with a peak-to-peak displacement measurement either in mils or micrometers (microns). A normal operating machine will generally have a stable amplitude reading of an acceptable low level. Any change in this amplitude reading indicates a change of the machine condition. Increases or decreases in amplitude should be considered justification for further investigation of the particular machine condition.

Frequency

The frequency of vibration (cycles per minute) is most commonly expressed in multiples of rotative speed of the machine. This is primarily due to the tendency of machine vibration frequencies to occur at direct multiples or submultiples of the rotative speed of the machine. It also provides an easy means to express the frequency of vibration. It is only necessary to refer to the frequency of vibration in such terms as one times rpm, two times rpm, 43% of rpm, and so on, rather than having to express all vibrations in cycles per minute or hertz.

The emphasis on frequency analysis developed primarily from casing measurements, where amplitude and frequency were the only major parameters available to be measured and evaluated. Also, the tendency of certain malfunctions to occur at certain frequencies has helped to segregate certain classes of malfunctions from others. It is extremely important to note, however, that the frequency–malfunction relationship is not mutually exclusive. That is to say, a vibration at one particular frequency often has more than one malfunction associated with it. There is no one-to-one relationship between malfunctions and frequencies of vibration. One must not be easily swayed into attempting to directly correlate certain frequencies with particular specific malfunctions. Frequency is an important piece of information with regard to analyzing rotating machinery, and can help to classify malfunctions, but it is only one piece of data. It is necessary to evaluate all data before arriving at a conclusion.

The typical means of expressing frequency are as follows:

- $1 \times$ rpm: The frequency of the vibration is the same as the rotative speed of the machine.
- $2 \times$ rpm: The frequency of the vibration is twice the rotative speed of the machine.
- $\frac{1}{2} \times$ rpm: The frequency of the vibration is one-half the rotative speed of the machine.
- $0.43 \times$ rpm: The frequency of the vibration is 43% of the rotative speed of the machine. And so on.

It is important to note a means differentiating two types of vibration: synchronous and nonsynchronous. *Synchronous vibration* occurs at a frequency that is some direct multiple or integer fraction of the rotative speed of the machine. For example, $1 \times$ rpm, $2 \times$ rpm, $\frac{1}{2} \times$ rpm, $\frac{1}{3} \times$ rpm. In these examples the vibration frequency is "*locked in*" with the rotative speed of the machine. Nonsynchronous vibration occurs at some frequency other than a locked-in frequency with running speed. When viewed on an oscilloscope in time-base presentation, a nonsynchronous wave form will appear to be continuously moving across the screen. If a once-per-turn event (Keyphasor) is superimposed on the wave form, it will appear to move along the wave form instead of remaining constant in one fixed position as it would with a synchronous vibration. Basic frequency measurements can be made with the use of a Keyphasor and an oscilloscope. It is possible, with some minimal practice, to be able to pick out the major components of vibration present in a vibration wave form. For more discrete frequency analysis, it is necessary to employ some additional instrumentation, such as a tunable filter, swept frequency display, or spectrum display.

Phase Angle

The phase angle of vibration has long been ignored as an important criterion for analysis of rotating machinery by people in many areas of rotating machine use. The power-generation people in the utility industry have, however, long recognized its value.

Phase-angle measurement is a means of describing the location of the rotor at a particular instant in time. A good phase-angle measuring system will define the location of the high spot of the rotor at each transducer location relative to some fixed point on the machine train. By determining these high spot locations on the rotor, it is possible to determine the balance condition and locations of the residual unbalances on a rotor. Changes in the balance condition of a rotor changing this high spot will be shown as a change in phase angle. Accurate phase-angle measurements are essential in the balancing of rotors and can be extremely important in the analysis of particular machine malfunctions. The phase angles of the rotor as determined by various transducers along a machine train can provide valuable information as to the performance of that machine train. It is this phase angle that provides timing information that helps answer the questions: What is happening, where, when, and how? Phase angle is also valuable in determining the rpm location of the natural rotor balance resonances ("criticals").

The most accurate and reliable means of measuring phase angle is with the use of a Keyphasor (shaft reference). The reader should refer to the heading, "The Keyphasor," in this chapter.

Vibration Form

The form of the vibration is perhaps the most important means of presenting vibration data for analysis. It is through this type of presentation that an understanding of a particular machine's behavior can be realized. The previously discussed three parameters have all been measurable quantities that can be displayed on an indicating meter or digital display, whereas the vibration form is the raw wave form itself, displayed on an oscilloscope. The oscilloscope is one of the most valuable, if not the most valuable, data-presenting instruments for machinery diagnosis.

Vibration form can be separated into two separate categories: (1) time-base presentation and (2) orbital presentation. Time-base presentation is provided by displaying transducer inputs on the oscilloscope in the time-base mode. In this mode the oscilloscope displays the sinusoidal-type wave form representing vibration amplitude with respect to time. When using proximity probes, this mode of the oscilloscope displays the dynamic position of the shaft relative to the input transducer versus time horizontally across the cathode ray tube of the oscilloscope. The orbit presentation is provided by displaying the output from two separate proximity probes at 90° angles to one another in the *XY* mode of the oscilloscope. In this mode the oscilloscope displays the center-line motion of the shaft at that lateral location along the rotor. If the probes are mounted at the bearing, the orbit is a presentation of the motion of the shaft center line with relationship to the bearing clearance.

These two presentations give the maintenance engineer the most data in one presentation. Basic amplitude, frequency, and phase angle can be determined by viewing the vibration form. The vibration form presentations inherently help the individual to understand "what the machine is doing" by observing the actual motion of the observed part. This is an important concept. The vibration form allows for the transition from determining what the amplitudes and frequencies are, to determining what the machine is doing. This is the ultimate parameter that we are attempting to measure in any preventive or predictive maintenance program.

Vibration Mode Shape

A recommended practice for providing information on a rotating machine is to provide an extra set of radio *XY* (90°) probes some lateral distance away from the bearing. This extra-horizontal set of probes would not normally be monitored but would be available for diagnostics. The extra-horizontal probes provide a third dimension to the machinery data and allow for an estimate of the mode shape of the machine rotor for the determination of nodal points. It is important to recognize that any set of *XY* probes along the machine train will provide the motion of the rotor at that lateral location along the machine. By utilizing the extra set of *XY* probes at a different lateral location along the machine train, we can attempt to determine the basic mode shape of the rotor itself. This mode shape can help to give closer estimates to the internal clearances between the rotor and stator elements and to give an estimate of the nodal points along the rotor shaft.

VIBRATION PARAMETERS

The parameters are all a means of analyzing the vibration (dynamic motion) of a particular machine. They are all means of looking at what the machine is doing on a dynamic basis. The presentation of the amplitude, frequency, phase angle, form, and mode shape are all applicable to casing measurements as well as to shaft or rotor measurements. It is important from an overall system analysis to know what the casing is doing dynamically as well as what the rotor is doing. Such things as structural or piping resonance, loose or cracked foundations, and external vibration input sources, can be determined from measurements on the nonrotating machine parts. In the overall system analysis of a machine's mechanical performance, casing measurements can indeed be important.

The comparison of shaft or rotor vibrations with casing vibrations also can be an important parameter in determining the overall condition of a machine. The transfer impedance between the shaft and the casing can vary widely due to various machine parameters. A comparison of both the amplitude and phase relationships of the casing and shaft vibrations may indeed prove valuable in solving a particular machine malfunction problem.

Relative Versus Absolute Measurement

No discussion of dynamic motion (vibration) would be complete without the discussion of *relative* versus *absolute* measurements. A proximity probe, by its very nature, when mounted rigidly to the bearing cap or casing of a machine provides a vibration measurement of the relative motion between the shaft and the mounting of the proximity probe. This relative measurement has proved a satisfactory parameter for continuous monitoring of most machines. However, on some machines the absolute motion of the rotor (motion relative to free space) becomes an important parameter for continuous monitoring. This absolute motion can be provided with a *dual probe* that utilizes a relative proximity probe providing shaft motion relative to the casing and an absolute seismic-type transducer mounted on the casing of the machine in the same radial and angular orientation plane as the relative proximity probe. A vector summation of these two transducer inputs provided in the monitoring circuitry for this dual probe gives the *absolute* shaft motion. In this manner, four separate pieces of information are available.

1. Shaft motion relative to the casing
2. Casing absolute (seismic) motion relative to free space
3. Shaft absolute (seismic) motion relative to free space
4. Shaft position measurement as supplied by the dc output of the proximity probe transducer

The absolute measurement is most important on machines with flexible support structures or machines subject to high casing vibrations, as compared to the relative shaft vibrations.

The importance for *XY* (two-plane) monitoring has been well established for most types of machines. It is entirely possible to have totally different vibrations in the vertical and horizontal directions at one particular bearing. It is entirely possible, for example, to have different amplitudes and different frequencies (and normally corresponding different phase angles) in two different planes at one bearing. This has been documented on many, many machines, and the importance of *XY* mounted probes at radial bearings should not be underestimated.

POSITION MEASUREMENT

Eccentricity Position

Eccentricity position is the measurement of the steady-state radial position of the shaft in the journal bearings. Under normal operation with no internal or external preloads on the shaft, the shaft of most machine designs will ride where the oil pressure dam places it. However, as soon as the machine gets some external or internal type of preload (unidirectional steady-state force), the eccentricity position of the shaft in the journal bearing can be anywhere. This eccentricity position measurement can be an excellent indicator of bearing wear and heavy preload conditions such as misalignment. In installations where only single-plane monitoring is present, it is imperative that eccentricity position be measured on a periodic basis. This is due primarily to the possibility of heavy preload condition precluding an increase in amplitude of vibration in the plane of sensitivity of the probe. This could possibly allow for a machine malfunction to occur without significant amplitude warning. By observation of the eccentricity position, an early warning is possible.

Eccentricity position should also be closely watched during machine start-up. During a machine start-up, one would normally expect the shaft to rise from the bottom of the bearing to some place toward the center of the bearing. This is due fundamentally to the oil flowing under the shaft making the shaft rise in the bearing. It is generally believed that the oil film is about 1 mil in thickness. Observations on many bearings show that it is more often about one-third of the bearing clearance in the preloaded direction of the shaft. The measurement of eccentricity position is accomplished by monitoring the dc output of the proximitor associated with the radial vibration probe at the bearings.

Because of the ability of the eccentricity position to change under varying conditions of machinery load, alignment, and the like, it is important that the proximity probe transducer system have a long linear range sufficient to allow for these eccentricity position changes to occur without having the shaft move outside the linear range of the proximity probe. This is especially true in large machines where large bearing clearances are normally present. Whereas eccentricity position is monitored continuously on some machines, on most machines a periodic check of eccentricity position appears to be acceptable for predictive maintenance. Especially where misalignment or other preload conditions may be considered as a possible malfunction condition, eccentricity position should be monitored very closely. It is important to document the cold eccentricity position and the hot eccentricity position so that a frame of reference is established for comparisons of eccentricity position at later dates.

Axial Thrust Position

Axial thrust position is the measurement of the relative position of the thrust collar to the thrust bearing. This measurement is perhaps one of the single most important monitored parameters on a centrifugal compressor and/or steam turbine. The primary purpose of an axial thrust position monitor is to ensure against an axial rub between the rotor and the stator. Axial thrust bearing failures can occur very quickly and can be catastrophic, and every attempt should be used to protect against this possible machine failure mode.

At least one and preferably two (in a voting logic configuration) axial thrust position probes should be mounted to provide axial thrust position protection. Care must be taken in the selection of probe mounting locations to ensure minimum effect of thermal growth of the rotor and minimum effect of springiness of the thrust bearing assembly in the accuracy of the reading. In early applications of proximity probes for axial thrust position measurements, it was very often found that the alarm and danger set points were established too close to the initial cold float zone of the machine. It was found that deflections occur in the thrust bearing assemblies, and thermal growth of the rotor occurs such that under normal operating conditions the position of the rotor in the hot float zone can indeed appear to be wider than the normal position of the rotor in its cold float zone. It is important to note that most machines have sufficient axial clearance between the rotor and the stator that wide set points can be established allowing the thrust collar to wipe the babbitt of the thrust bearing shoes to some degree without having rotor to stator rubs. Under normal operating conditions of a centrifugal compressor or steam turbine, thrust position can vary with load of the machine, so varying thrust position measurements under differential loads and conditions of a machine are not uncommon. The thrust position measurement may also be important in the determination of surge or incipient surge conditions.

An added benefit received when applying thrust position probes for axial thrust position measurements is that axial vibration measurements can be read from the same proximity probe. Whereas axial vibration is not normally monitored continuously on most centrifugal equipment, it has proved valuable in diagnosing some particular machinery malfunctioning conditions. If axial vibration is to be monitored or used for diagnosis of a particular machine, it is necessary for the observed surface to be smooth and to be perpendicular to the center line of the rotor. This will minimize any affect of mechanical runout on the dynamic output of the probe, thus providing accurate axial vibration readings.

Eccentricity Slow-Roll (Peak-to-Peak Eccentricity)

In large steam turbines for power-generation service and in some industrial gas turbines and large fans, it is very desirable to provide an indication of eccentricity slow-roll, also called peak-to-peak eccentricity. Eccentricity slow-roll is the amount of bow the rotor takes while it is at rest. This bow can be indicated by the slowly changing dc peak-to-peak measurement from the proximitor as the rotor turns on turning gear or at very low (slow roll) rotative speed. When the peak-to-peak amplitude is at an acceptable low level, the machine can be started without fear of damage to seals and/or rotor rubs caused by the residual bow and its corresponding unbalance. Eccentricity slow-roll is best measured with a probe mounted as far away from the bearing as possible so that maximum bow deflections can be measured.

In large steam turbine applications, eccentricity slow-roll is often recorded on a strip-chart recorder where the operators can visually see the amount of bow in the shaft and its rate of decreasing amplitude while the machine is on turning gear and is allowed to warm up. Peak-to-peak eccentricity can also be displayed on a multipoint-type recorder or fed to a computer through sample-and-hold techniques, thus providing a low-frequency peak-to-peak measurement with an update of once-per-turn of the shaft.

Differential Expansion

In very large machines, such as large steam turbines in power-generation service, it is extremely important that during start-up the casing and the rotor both grow thermally at the same rate. If the rotor or the casing grow at different rates, there exists the possibility of damage to the machine caused by axial rubs. To measure this differential expansion, a long-range proximity probe is mounted at the end of the machine opposite the thrust bearing where the maximum relative growth between the case and the rotor can be observed. The typical range for this proximity probe is 1 in. In very large machines, the range required of the proximity probe system can be as great as 2 or 3 in.

Case Expansion

On very large machines it is also very common to provide, in addition to differential expansion, a case expansion measurement. The case expansion measurement is usually provided by a contacting linear variable differential transformer (LVDT) mounted externally to the machine case and referenced to the foundation.

Case expansion measurement helps to provide information with regard to the relative growth of the case to the foundation. By knowing the amount of case growth and the amount of differential growth, it is possible to determine which is growing at a more rapid rate, the rotor or the case. If the case is not growing properly, the "sliding feet" of the case may be stuck.

Alignment

The need for proper alignment between different machine cases of a machine train has been well established. Misalignment has been determined to be one of the more frequently occurring malfunction conditions, especially in installation of process compressor trains and gas-turbine-driven pump machine trains. The normal alignment procedures generally allow for the determination of an alignment drawing through calculations, using the growth of the various different machines involved. This has often been shown to be a "guesstimate" at best, and some means of determining the final, hot running condition of the machine has proved to be desirable and necessary.

Considerable work has been done in the area of hot alignment measurements. The more popular techniques include optical measuring techniques, utilizing optical instruments similar to those used for surveying land, and the proximity reference system. The proximity probe has been used to determine the relative position change between shafts of different machine cases and is usually a more accurate representation of machine alignment than casing measurements.

SPEED (RPM)

The measurement of speed (rpm) of the rotor has long been standard procedure. Most major centrifugal machine trains have a continuous indication of the rpm of the machine. General practice in the past has been to provide this indication by a means of some analog-type meter indication. Most of these analog meter indications have simply provided a general indication of rpm of the machine.

With the advent of new, reliable digital circuitry, digital tachometers have become more popular for speed indication. Besides providing greater resolution and being more accurate and easier to read, the digital tachometer lends itself very well to providing redundant overspeed trip protection. The digital indication of rpm eliminates many of the older problems associated with rpm measurement. Accurate, easily readable indication of rpm is provided generally in the control room for operators to observe. Many users also prefer a remote digital indicator at the machine train for use during start-up procedures.

For machines where digital tachometers are not normally supplied, portable digital indiators have become popular for use during start-up for monitoring of rpm and for checking of mechanical overspeed trip mechanisms.

Transducer inputs for the digital tachometers can be from a variety of transducers. Among the more popular are the Keyphasor proximity probe input, the photoelectrical pickup, and the magnetic pickup. These transducers are all designed to observe from one to any number of events per revolution of the shaft. This basically digital input is then translated into a direct readout of rpm by means of the digital tachometer circuitry.

The correlation of vibration measurement with rpm can be important for the final analysis of the mechanical performance of a particular machine. Centrifugal equipment is designed to operate in a speed range that will not coincide with the balance resonances of that particular machine and at speeds that will not excite these particular resonances. A start-up piece of information that is important for determining the balance resonances is an *XY* plot of amplitude and phase angle of the synchronous or overall vibration versus rpm of the machine. In plotting and correlating these parameters, it is possible to easily determine the machine balance resonances (criticals).

TEMPERATURE MEASUREMENT

In the final analysis of the condition of a particular piece of rotating machinery, other parameters also become important. One of the most popular and important parameters not yet discussed is that of temperature measurements. The temperature of the bearing babbitt in both radial and axial bearings is becoming a more widely used parameter. A correlation of this temperature information with vibration and/or position measurements helps to give a better indication of possible machinery malfunctions.

CORRELATION OF ROTATING MACHINE OPERATIONAL MEASUREMENTS

Correlation of temperatures, pressures, flows, and other process parameters that could affect the operation of a piece of machinery is extremely important for the overall system analysis of the machine in service. It is through this correlation that a good predictive maintenance program can be established. The ability of the engineer to utilize all the available information in determining the mechanical running condition of the piece of machinery is extremely important in the overall objective of maintaining proper operation and continuous on-line service of any piece of equipment.

The reader who has a thorough understanding of the parameters discussed in this chapter will have a good beginning toward understanding of the mechanical performance of centrifugal equipment. It is through such understanding that the reader will ultimately be able to determine what a particular piece of machinery is doing.

CASE EXPANSION TRANSDUCER

On large steam turbines the machine normally expands as it warms up to operating speed and temperature. Large turbines are mounted so as to allow this thermal growth of the machine. The linear variable differential transformer (LVDT) is mounted in such a manner as to measure the thermal growth of the machine case relative to the machine foundation. The LVDT is normally mounted on the opposite end of the turbine from where the case is tied to the foundation from the thrust bearing. The LVDT is mounted to either the machine case or the supporting foundation; the rod is then attached to the other. When this procedure is followed, the thermal growth of the machine case or shell is then shown on the indicator readout. Failure of the case to grow is an indication of potential trouble.

DIFFERENTIAL EXPANSION TRANSDUCER SYSTEM

The term differential expansion is the net difference in the expansion of two or more components due to an increase in temperature. In rotating machinery, it can be more appropriately described in terms of temperature variations affecting the expansion, as well as contraction, in the axial plane of rotating and stationary parts. Differential expansion then represents the variation in length of a turbine rotor relative to its casing.

Differential expansion is a natural occurrence in all turbomachinery. In large steam turbine–generator sets, it can be quite significant and should be monitored all the time. During the loading and unloading phases of machine operation, internal temperature will rise and fall, causing the rotor and machine case to "grow" and "shrink" at different rates. If the loading and unloading are not properly controlled, this differential expansion could approach and exceed a critical limit and (since axial clearances between rotating and stationary parts are purposely kept to a minimum), internal axial rubbing could result. By monitoring differential expansion, an operator can control the steam flow in response to changing load requirements and thus reduce the possibility of an axial rub due to excessive differential expansion.

Differential Expansion Measurement Consideration

Axial movement of a machine rotor and case due to thermal expansion is directed away from the corresponding mass centers. The machine case or shell is secured to the foundation (earth) at one end and allowed to grow in a direction away from that secured point while the rotor grows away from the thrust bearing. If the secured end of the case and the thrust bearing are at the same end of the machine, the rotor and case will grow in the same direction. If the casing is secured at the end of the turbine opposite the thrust bearing, rotor and case growth will be in opposite directions.

Most steam turbine casings are secured at the end opposite the thrust bearing, and it is at this point that the rotor expansion measurement is taken (Figure 9-13). A turbine rotor is known to respond to temperature cycling faster than the turbine case for the following reasons:

1. Rotor has smaller mass.
2. Rotor has higher heat transfer coefficients.
3. Rotor is closer to the steam path.

So it seems natural that the terms *long rotor* and *short rotor* were adopted as designations for the amount of differential expansion.

A *long rotor condition* occurs when rotor expansion is greater than case expansion during machine loading with increasing temperature. Also, during machine unloading with decreasing temperature, a long rotor condition occurs when the rotor contraction is less than case contraction.

A *short rotor condition* occurs when the maximum case expansion is greater than rotor contraction.

A *cold rotor condition* exists when the rotor and case are at ambient temperature "prestart-up."

Figure 9-13 shows the rotor axial growth in relation to the differential expansion transducer system. Rotor position nomenclature is referenced accordingly. For the purpose of illustration, the system is mounted on the turbine side of the rotor collar. It could just as well be mounted on the other side of the collar. Long rotor would then be decreasing gap. Long, short, and cold rotor positions relative to the probe face are illustrated in Figure 9-14.

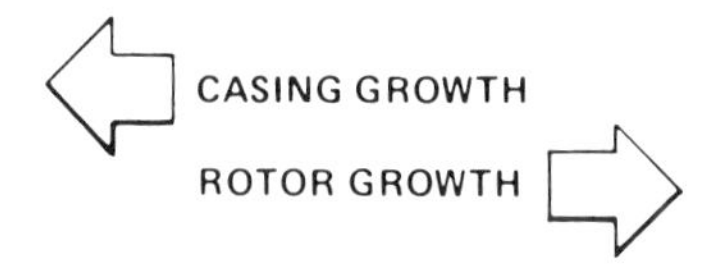

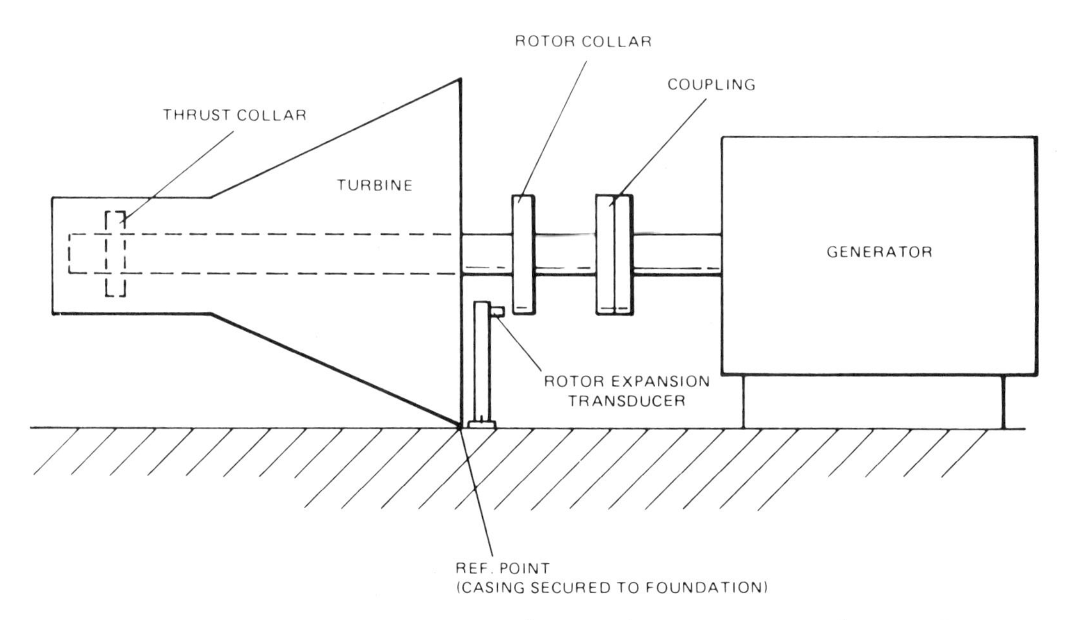

Figure 9-13 Noncontacting Transducer Machine Monitoring Applications *(Courtesy, Bently Nevada)*

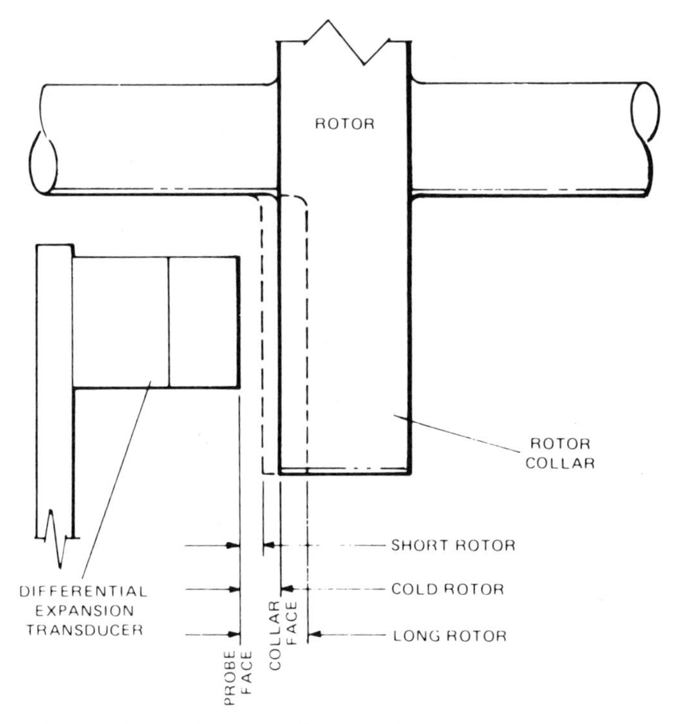

Figure 9-14 Rotor Thermal Growth in Relation to the Transducer *(Courtesy, Bently Nevada)*

ACCELERATION TRANSDUCER SYSTEM

The purpose of an acceleration transducer system (Figure 9-15) is to measure structural motion in terms of accelerations (g's or m/s^2). Its output is a voltage proportional to the acceleration along its sensitive axis of the surface on which it is mounted.

The accelerometer portion of the acceleration transducer system is a "contacting" transducer that is physically attached to the vibrating machine part. The accelerometer uses a piezoelectric crystal situated between the accelerometer base and an inertial reference mass. When the crystal is strained (compression or tension force), a displaced electric charge is accumulated on the opposing major surfaces of the crystal. The crystal element performs a dual function. It acts as a precision spring to oppose the compression or tension force and it supplies an electric signal proportional to the applied force.

A 95-Ω coaxial extension cable (with an interface connector on both ends) is used to connect the accelerometer to the interface module for the standard

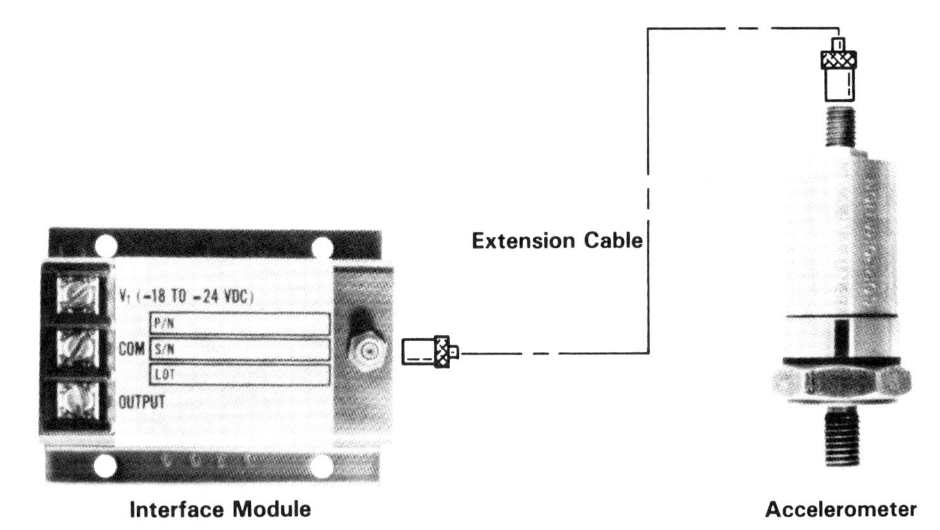

Figure 9-15 Acceleration Transducer System *(Courtesy, Bently Nevada)*

system. A 50-Ω high-strength cable with interfacing connectors on both ends is used to connect the accelerometer to the interface module for the special high-frequency system. The acceleration transducer interconnecting cable length is not critical for proper system operation. Typical maximum length is set at 30 ft.

The accelerometer interface module performs five separate functions. It (1) provides the constant current drive for the accelerometer amplifier, (2) provides amplification for the ac signal, (3) generates the proper dc bias level

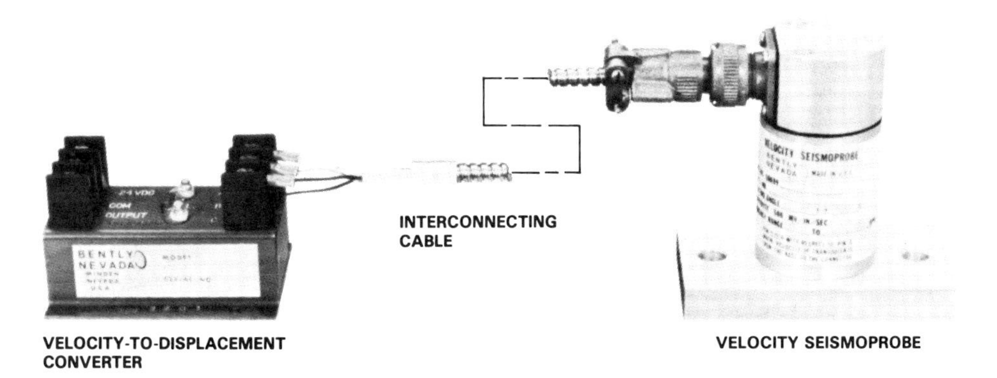

Figure 9-16 Velocity Transducer System *(Courtesy, Bently Nevada)*

for the signal output, (4) provides the proper OK function comparator circuitry, which allows detection of an open- or short-circuit fault condition, and (5) provides a low-output-impedance amplifier that buffers the accelerometer signal and enables capacitive cable loads to be driven. The output can then be delivered through suitable cable to suitable readout devices.

VELOCITY TRANSDUCER SYSTEM

A velocity transducer system (Figure 9-16) consists of a seismic pickup and interconnecting cable and may include a velocity-to-displacement converter

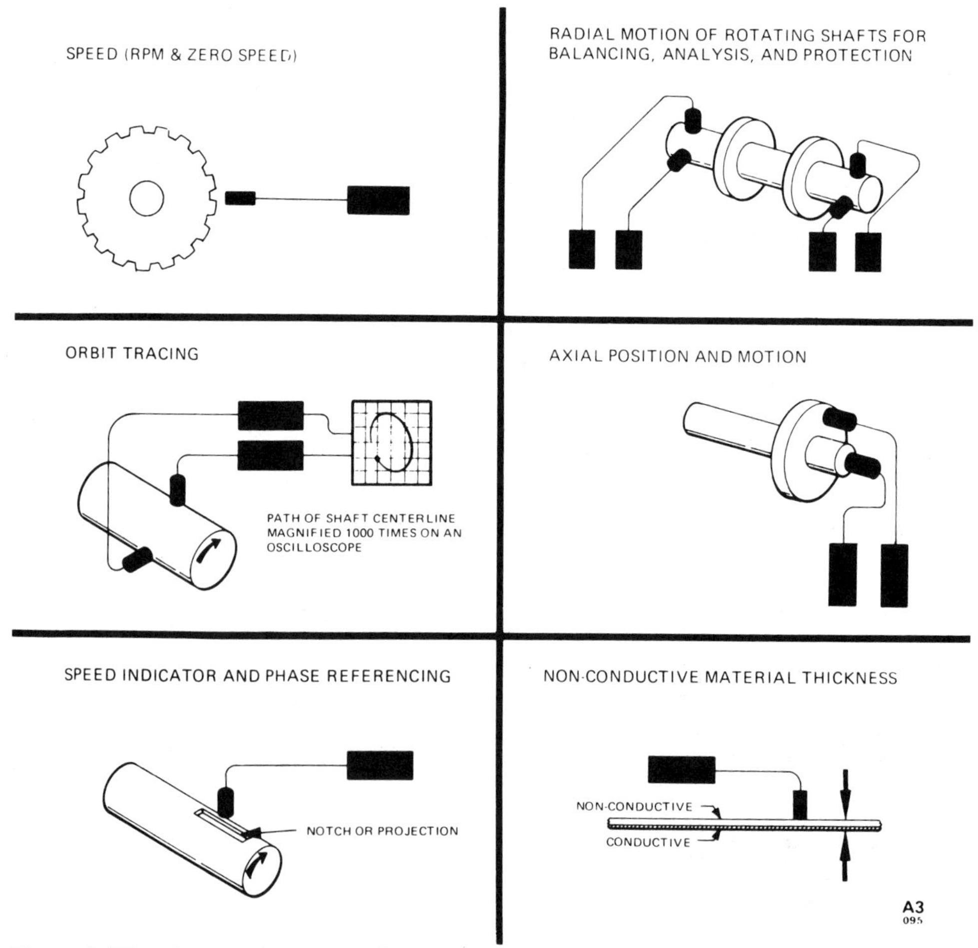

Figure 9-17 Noncontacting Transducer Machine Monitoring Applications *(Courtesy, Bently Nevada)*

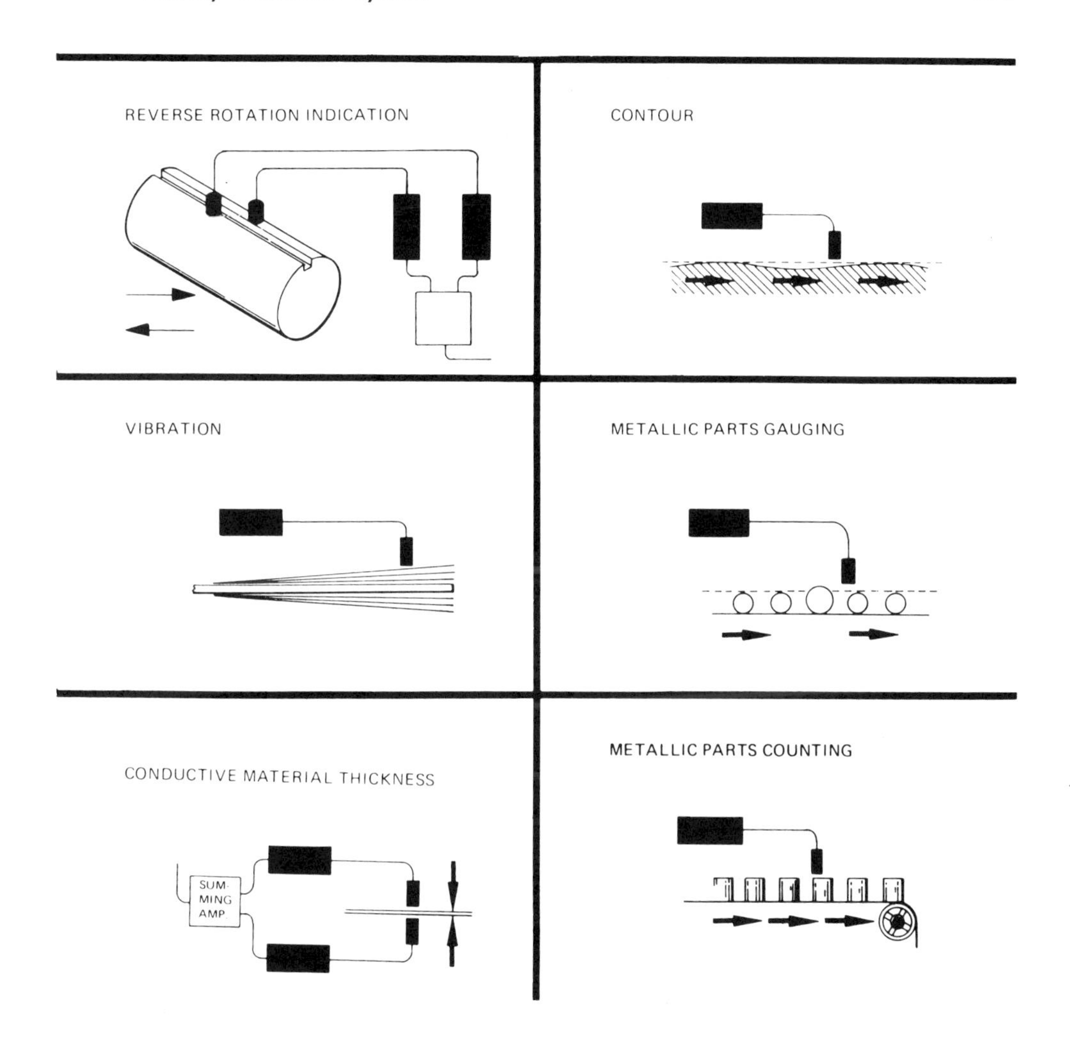

Figure 9-17 *continued*

(VDC). The purpose of the velocity transducer system is to measure machine case or structural vibration velocity, and using the VDC it can convert the vibration velocity into an electrical signal that represents the displacement of the machine case. Without the VDC, the output signal represents the vibration velocity of the machine case.

The seismic pickup portion of the system operates on the inertial mass–moving case principle. The inertial mass is a cylindrical copper wire coil wound

CASE MOTION

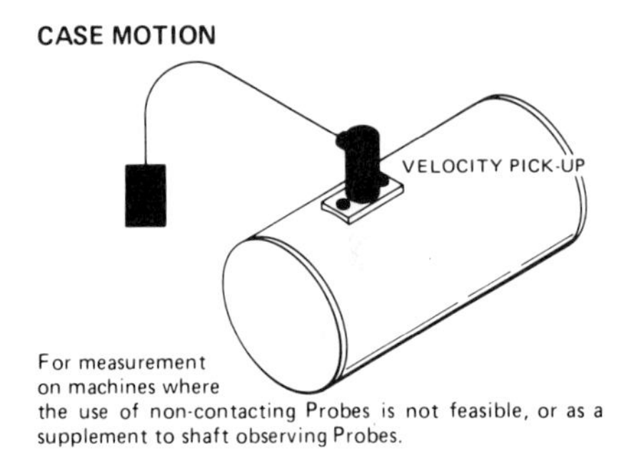

For measurement on machines where the use of non-contacting Probes is not feasible, or as a supplement to shaft observing Probes.

CASE ACCELERATION

For gear mesh and other high frequency information. A supplement to shaft measurements.

CASE EXPANSION

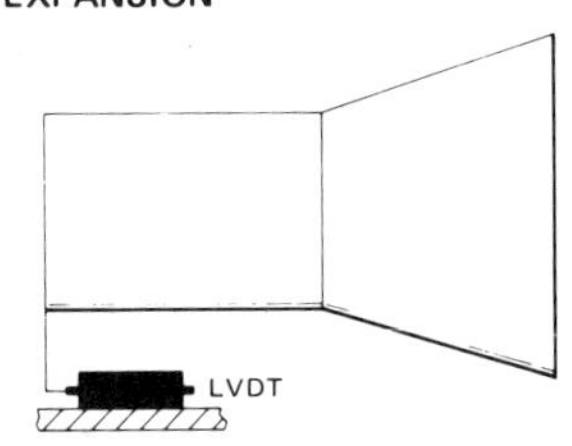

For large steam turbine applications.

VALVE POSITION

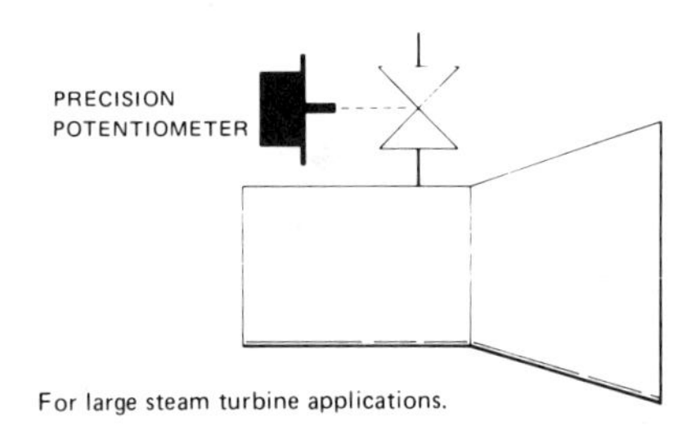

For large steam turbine applications.

ABSOLUTE SHAFT MOTION

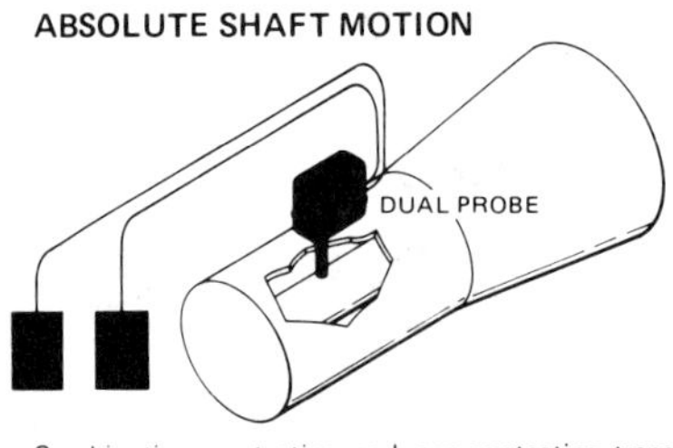

Combination contacting and non-contacting transducers for machinery applications where both shaft and casing motion are significant.

STRUCTURAL MOTION

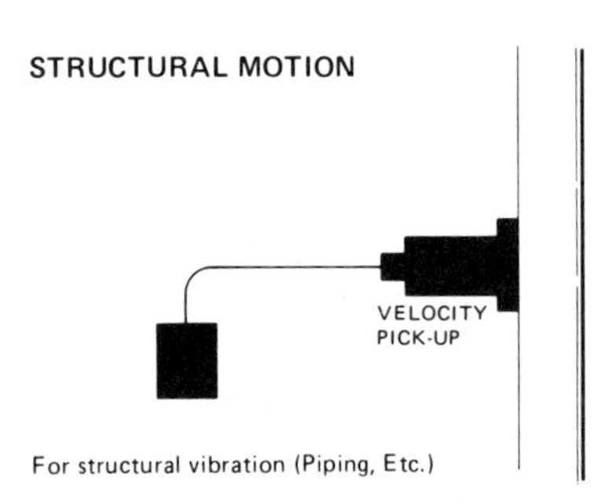

For structural vibration (Piping, Etc.)

Figure 9-18 Contacting Transducer Machine Applications *(Courtesy, Bently Nevada)*

on a thin brass bushing and suspended by sensitive springs inside the pickup case. Rigidly attached to the pickup case and physically located inside the coil is a permanent magnet. The seismic pickup case and the magnet vibrate in direct accordance with the machine case, causing the magnet to move with respect to the relatively stationary coil. When properly installed, this movement induces a voltage in the coil that is proportional to the velocity of vibration, provided that sufficient frequency of motion is present. This current may be measured directly or may be carried by the interconnecting cable to the velocity-to-displacement converter. The VDC electronically converts (integrates) the vibration velocity signal from the seismic pickup to a proportional displacement signal superimposed on a constant dc voltage.

TYPICAL NONCONTACTING TRANSDUCER SETUPS FOR MOVING OR ROTATING MACHINE-MONITORING APPLICATIONS

Figure 9-17 provides typical applications for machine monitoring of moving or rotating parts.

TYPICAL CONTACTING TRANSDUCER SETUPS FOR MACHINE MONITORING OF VIBRATION, ACCELERATION, AND OTHER MOVEMENT

Figure 9-18 shows typical applications for case motion, acceleration, and expansion. It further provides application setups for value position and shaft and structural motions.

REFERENCES

Technical data and illustrations in this chapter were provided by Bently Nevada Corporation, Minden, Nevada.

10

Optoelectronics

Optoelectronics is the branch of electronics that deals with light. Optoelectronics has suddenly become the dominant growth area in the electronics field. From light emitting and receiving devices, through solar cells and fiber optics to lasers, optoelectronics must now be dealt with as an emerging giant of the industry.

Devices such as light-emitting diodes (LEDs), injection laser diodes (ILDs), optoisolators, and detectors are being built stronger, in more and different sizes, shapes, and colors, and with more current-carrying capacity. Displays are now manufactured with built-in intelligence and computer capabilities.

Fiber optics is no longer an infant in the electronics field. It has grown from a laboratory science into a practical tool for transmission of information. Fiber-optic electronics is spreading rapidly to aerospace, construction, medicine, mining, oceanography, and most other high-technology fields.

Today, lasers are no longer fictional aspirations, such as Buck Rogers' ray gun. The broadest possible applications for lasers are now being realized. Some of these applications are in surveying, medicine, manufacturing, direction finding, and communications.

LIGHT SOURCING, TRANSMITTING, AND RECEIVING

Light is either leaving a source or arriving at a detector or sensor. At all other times it is in motion between the two. The sun and the stars are the natural sources of light. Reflective sources of light include planets and moons. The amount of light energy received from the sun (solar irradiance) outside the earth's atmosphere is called the solar constant and is equal to 140 milliwatts per square centimeter. The atmosphere absorbs much of this. If we consider a clear day with no fog, smog, or cloud cover, the amount of solar irradiance reaching the earth is 100 mW/cm^2. The amount of light energy received from the sun is an inverse curve that is dependent primarily on the distance from the sun.

Light is the form of electromagnetic radiation that acts upon the retina of the eye and the optic nerve. This makes sight possible. The term light has been altered by technical advances to include electromagnetic wavelengths not visible to the human eye. The entire frequency spectrum that deals with visible and adjacent wavelengths is called the *optical frequency spectrum.*

There are three basic levels of the optical frequency spectrum:

1. *Infrared band* of light wavelengths that are too long for response by the human eye.
2. *Visible band* of light wavelengths that the human eye responds to.
3. *Ultraviolet band* of light wavelengths that are too short for response by the human eye.

Optoelectronics deals with this entire spectrum. Wavelength ranges for these areas of the spectrum are 0.005 to 0.3900 μm for the ultraviolet, 0.3900 to 0.7500 μm for the visible, and 0.7500 to 4000 μm for the infrared.

Since light is a measureable source, two scientific fields have been created to measure it. Photometry is involved with the visible electromagnetic wavelengths. Radiometry deals with those wavelengths that are too short or too long for the human eye to perceive.

The theory of light called the *quantum theory* defines the photon. The *photon* is an uncharged particle and has an energy level of its own that is dependent on its frequency or wavelength. The higher the frequency of the energy is, the greater the strength of the photons.

Light Sourcing

A *point source* is theoretically a dimensionless (no length, width, or height) point in space from which light is propagated in all directions away from the source (see Figure 10-1a). In reality this cannot be, for any point or light source must have a finite size. Point sources can, however, appear to be a defined point such

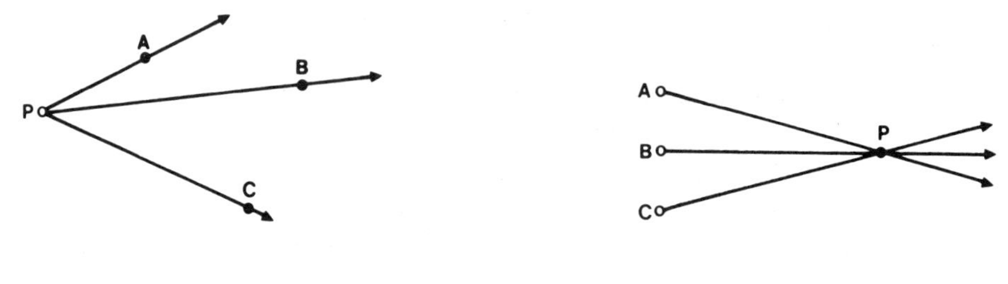

Figure 10-1 Light Sourcing

as a star that is light-years away and a laser beam. In the illustration, rays that leave point *P* arrive at point *A* in parallel lines of the direction *P* to *A*. Likewise, rays leave point *P* to points *B* and *C* in the same manner.

In the Figure 10-1b, the *extended source* of light is shown. This source is depicted as a point in space that is illuminated by light being radiated from several different directions. Points *A, B,* and *C* (the extended source) are illuminating point *P* from several directions. Point *P* then is illuminated by each source point. The farther you move from point source (*A, B, C*) the light lines become more nearly parallel. From an extreme distance the extended source appears to be a point source. For example, a flashlight viewed from a distance of 5 ft would appear to be an extended source that radiated from many points on its face. That same flashlight when viewed from 500 ft would appear to be a point source.

Light Transmitting

Light is fairly predictable, tending to repeat itself in the same manner. Light energy travels in waves such as the electromagnetic waves of electricity (see Figure 10-2). The waves in electricity are called *sinusoidal* or *sine waves.* This name comes from a trigonometric ratio. The sine wave has an angular rotation of 360°. It is maximum at 90° and 270°, and minimum at 0°, 180°, and 360°.

The sine-wave phenomenon is the same in light waves as in electrical waves. Each wave is characterized by five basic quantities: velocity, amplitude, frequency, polarization, and wavelength. The characteristics of waves are also the same. Each wave has a positive and a negative alternation. These are identical to each other. Each light wave has a time period in which it goes through its complete cycle or alternation. Some waves take a shorter time to complete the cycle. For instance, from a home electrical socket a current of 60 cycles per

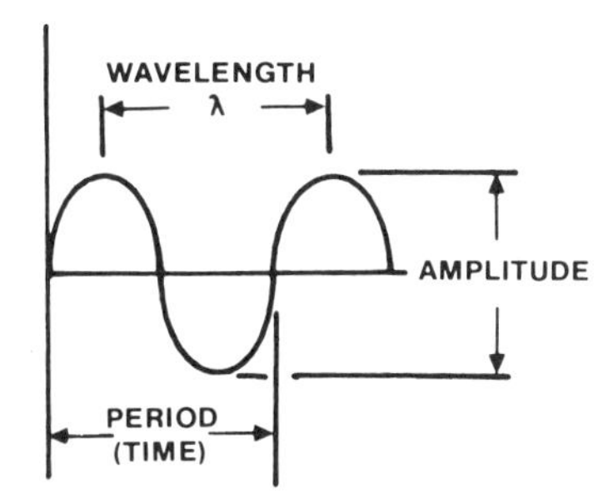

Figure 10-2 Electromagnetic Waves

second is taken. This means that during 1 second the cycle takes place 60 times. The time for one cycle to take place is 1/60 of a second or 16.6 milliseconds (one period).

The time for one cycle to take place is called the *period* and is measured in seconds. The number of cycles that takes place within a time period is called cycles per second or *hertz*.

The physical length of the wave is called its *wavelength*. It is this entity that concerns us in optoelectronics. All electromagnetic waves travel at the same velocity, that is 300,000,000 m/s or 186,000 miles/s. The *amplitude* of the wave is the magnitude of the wave vibration. The *frequency* of the wave is the number of waves passing a point in a second. The wavelength is measured in meters per second and is called lambda (λ). The wavelength is the distance between two consecutive wave peaks. Calculations of wavelength are as follows:

$$\lambda = \frac{300{,}000{,}000 \text{ meters}}{\text{frequency}}$$

The optical spectrum does not deal only with frequencies and wavelengths. It deals with color. Each wavelength is a color. The human eye sees red on one side of the spectrum. Center frequencies are blue, green, yellow, and orange. White light such as that from the sun or a flashbulb can be separated or dispersed into all the colors of the frequency spectrum. Rays of light are propagated from one of three basic sources: *panchromatic light* is radiation that includes all the wavelengths within the spectrum; *heterochromatic light* is light that has several very distinct wavelengths; *monochromatic light* has one wavelength or an extremely narrow band of very distinct wavelengths, such as monochromatic laser beams.

Polarization of Waves Light waves are either polarized or unpolarized. If the waves oscillate randomly through space or material and have no general direction, the waves are said to be *unpolarized.* If the waves oscillate at the same amplitude and trace out a circle in a plane perpendicular to the direction of propagation, the waves are said to be *circularly polarized* If the waves oscillate at the same amplitude along a line in a plane perpendicular to the direction of propagation, the waves are said to be *linearly* or *plane polarized* (see Figure 10-3).

A beam of unpolarized light may be the result of two beams combined in two different planes at two different amplitudes with no phase relationship. The directional variations of the two waves are not related nor are their magnitudes.

It is extremely important that waves be polarized in optoelectronics. Within optoelectronic systems such as the laser, linear polarizers are used to separate out all rays that are not parallel to the propagation axis. The beam can be tested for polarization by rotating the polarizer by 90°. This should cancel the beam by orienting the polarizer so that its axis is perpendicular to the plane of the transmitted beam.

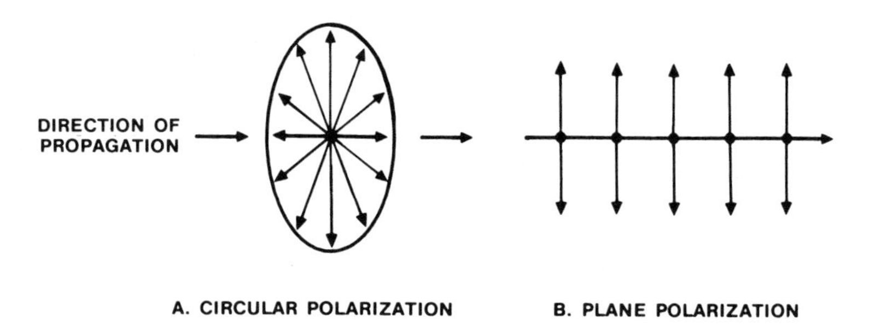

Figure 10-3 Polarization of Waves

Coherence Light sources such as fluorescent lamps do not radiate coherently. The light from these lamps is radiated in many frequencies and just as many phases. As a result, the light beams do not have direction, certainly are not polarized, and act in an extremely unpredictable manner. This incoherency can be compared with the light of a laser beam, which is directional and polarized. Its wave fronts are defined and its frequency along with phase relationships is constant. In effect, the beam is monochromatic (one color) and coherent (see Figure 10-4).

The coherency of the light waves is imperative to good laser operation. To have coherency, the waves must have a consistent relationship between their

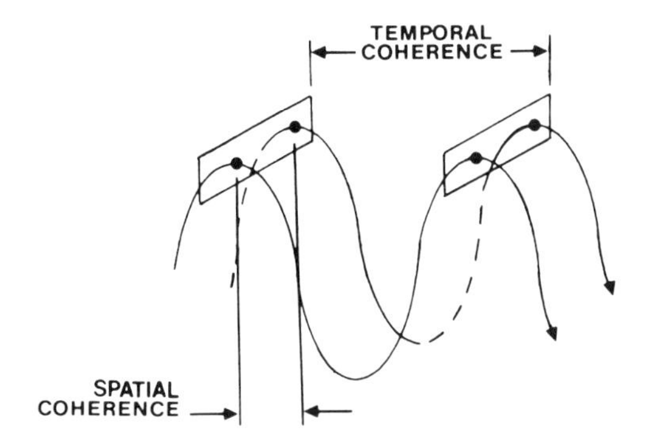

Figure 10-4 Coherence

wave troughs and wave crests. *Wave troughs* are points of minimum vibration, and *wave crests* are points of maximum vibration. Coherency is divided into two major relationships, spatial and temporal. To be spatially coherent, waves must have phase correlation across a wave front at any point in time. The spatially coherent waves also maintain their shape in relation to time. To be temporally coherent, waves must be frequency constant so that phase correlation is in the direction of propagation. The phenomenon of coherence allows the laser beam to be focused to a minute unit area. A conventional source cannot be focused without loss of power. Furthermore, coherent light can be made to produce constructive interference of waves such as those used in holography.

Beam Divergence The light beam in a laser is coherent. It can therefore be focused to a minute beam. The laser beam will tend to fan out or diverge as it travels away from its power source. The beam may have many different cross-sectional shapes. However, for purposes of explanation, we shall assume that the beam is round.

Figure 10-5 illustrates a typical beam. *D*1 represents the diameter of the beam at its aperture (output from the laser). *D*2 represents the diameter of the

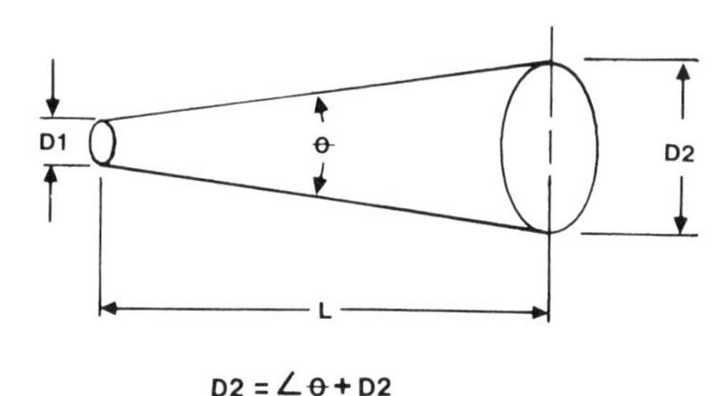

Figure 10-5 Beam Divergence

D2 = ∠θ + D2

beam at the detector, receiver, or somewhere in space. L is the length of the beam at that point. Angle θ is the angle of divergence or fanning out. The angle in the illustration is extremely exaggerated, for the divergence of a laser beam is a very narrow cone.

Light Reception

The reception of light is completely dependent on what use or application is intended. If the desire is to light a room, the light should reflect diffusely from most objects in the room, with absorption dependent on the materials and colors in the room. There is little use for refractive light. If the light were to be used for fiber-optic applications, fiber ends would reflect at 0% if possible, and the light refracted into the fiber would be maximum. The amount of absorption within the fiber would be held to a minimum. If the light were to be used for photodetection, the amount of absorption would be extremely high (100% if possible). The amount of reflection would be held to a minimum. Thus the application decides the use of materials that are low or high absorbers, reflect or do not reflect, and refract or do not refract.

Interference Effects The study of interference and its peculiarities is an extremely broad field. The field has applications in optoelectronics, primarily in laser technology and holography. Interference effects may be produced by coherent light. Broadly defined, *interference* is the interaction between two beams of light. The two beams should be coherent for the interaction cannot take place with incoherent light. The interaction produces interference fringes. The fringes are variations of resultant amplitudes of the interaction as the waves vary from point to point in a field of view. The variations are due to path differences where the waves reinforce each other at certain points and oppose each other at other points. These interference patterns allow the relative coherency of a light source. The measurement is made by analysis of the interference fringe pattern.

If the interference patterns projected on the screen are dark fringes, the two waves are out of phase. Out-of-phase interference patterns are called *destructive interference.* In Figure 10-6a, interference patterns 180° out of phase are illustrated. The wave pattern is a standing wave such as is found in laser cavities. However, if the interference patterns projected on the screen are bright fringes, the two waves are in phase. In-phase interference patterns are called *constructive interference.* In Figure 10-6b, these in-phase waves are summed to a higher-amplitude resultant.

Incoherent light does not produce spatial or temporal coherence. The wave front is not clearly defined and the phase relationships are constantly shifting;

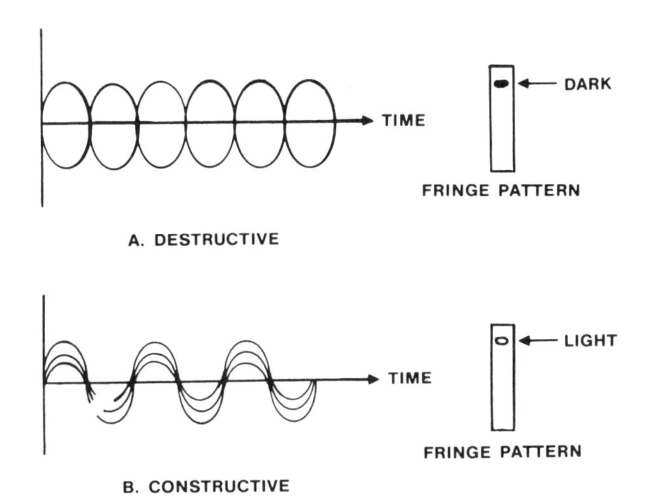

Figure 10-6 Interference Patterns

therefore, incoherent light sources cannot be used for the formation of interference fringes. However, if the incoherent light source is projected through a narrow-band filter, it may be utilized. The filter tends to isolate monochromatic waves of the light. That light may contain some degree of temporal coherence.

Incidence Incidence is the amount of flux per unit area that is normal or perpendicular to a surface or detector. If the flux is not normal (not perpendicular), the normal component of the angular flux is the incidence. In radiometric terms, incidence is called *radiant incidence* or *irradiance.* Irradiance (E_e) is measured in watts per square meter ($E_e = W/m^2$). Incidence in photometric terms is called *luminous incidence* or *illuminance* (also called illumination). Illuminance is measured in lux (lx) or lumens per square meter (lm/m^2).

When light falls on a surface, it is partly reflected from the surface and partly absorbed. It may also be converted by the surface into some other form of energy. In this case it becomes part of an active surface such as a photodiode or phototransistor. In any event, incidence is how much flux falls on the surface area. From a point source, the law of illumination of light falling on a given area is called the inverse square law (see Figure 10-7). You will note that 1 lumen of flux illuminates 1 square foot of area from 1 foot away at 1 footcandle (1 lumen/square foot). At 2 ft from the point source, an area of 4 ft^2 is illuminated to $\frac{1}{4}$ footcandle. Finally, at 3 ft from the point source, an area of 9 ft^2 is

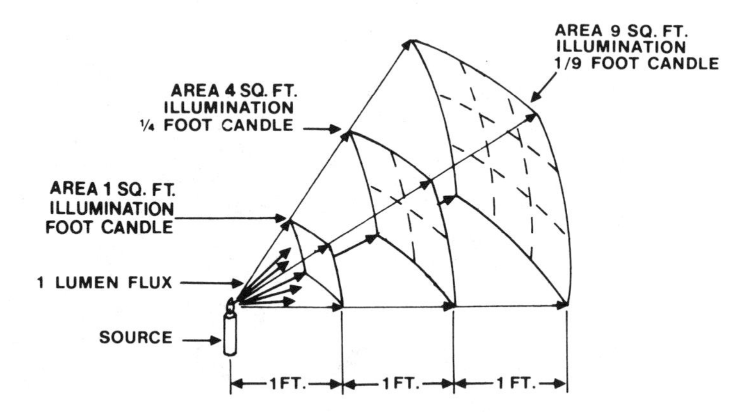

Figure 10-7 Point Source Illumination

illuminated to 1/9 footcandle. The distance away from the source is squared to provide the illuminated area. The illumination of that area is the inverse to the square of the distance.

Illumination from an area source must be looked at from a different perspective. The area source must be assumed to be an extended source. That is, each point or element on the surface contributes to the luminous intensity of the source. The luminous intensity in a given direction, divided by the projected area of the source, is called the *luminance* of the source. In Figure 10-8, area source illumination is illustrated. An angle is made between a normal line (perpendicular to the surface) and the incident flux lines. The angle is used to calculate the effective area presented to incident flux. The effective area is equal

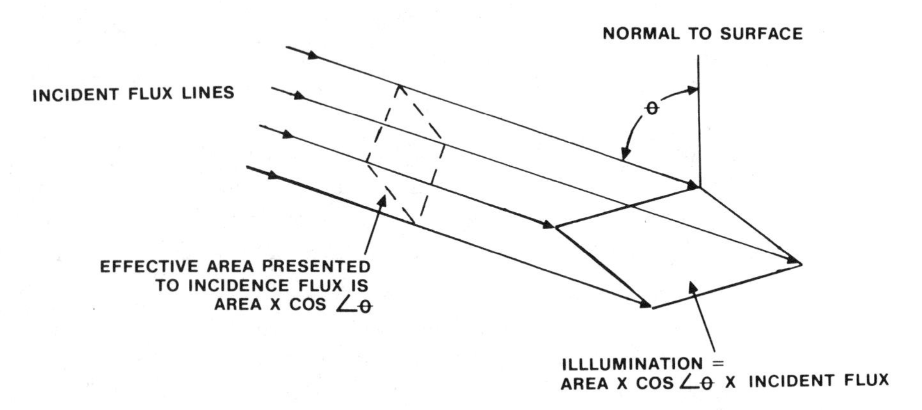

Figure 10-8 Area Source Illumination

to area times cosine $< \theta$. The illumination in lumens per square feet is equal to the area of the receiver times cosine $< \theta$ times the incident flux.

Reflection When light beams or rays strike a surface, they reflect from that surface either in a diffuse manner or a specular manner, or in some cases both. The amount of reflection is largely due to the type of material on the surface of the reflector (see Figure 10-9).

Diffuse Reflection Diffuse reflectors scatter the light outward in several beams, depending on the surface material and the amount of light the material absorbs. Virtually all material absorbs some light. In Figure 10-9a, the principle of diffuse reflection is illustrated. A smooth surface such as a freshly painted house would diffuse light in rather equal and alike reflected rays. Another surface, such as a rock-covered roof, would cause light rays to be reflected in scattered, nondescript lines.

Surface or Specular Reflection Whenever a beam of light from one medium, such as air, strikes a second mirrorlike medium such as glass, part of the beam is reflected and the other part refracted (see Figure 10-9b). The angle at which the incident rays strike the second medium is called the *angle of incidence.* This angle is the angle made by the incident beam and a line normal (perpendicular) to the boundary of the two media. Part of the incident beam is reflected at an angle called the *angle of reflection*. This angle is made by the reflected beam and a line normal (perpendicular) to the boundary of the two media.

Angle of incidence = Angle of reflection

If the angle of incidence varies, so does the angle of reflection by the same amount. The angle of incidence, the angle of reflection, and the angle of refraction all lie in the same plane.

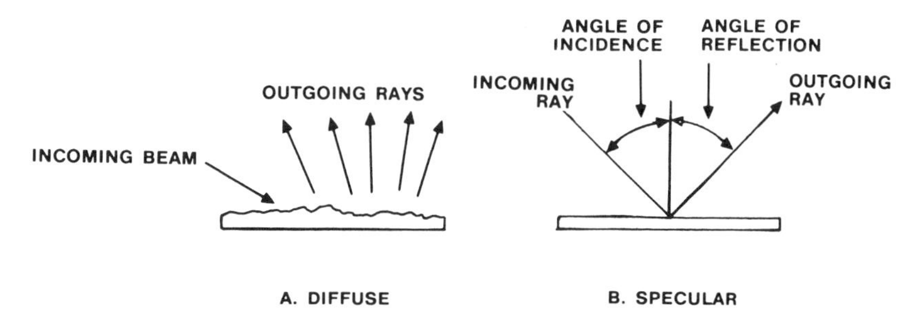

Figure 10-9 Reflection

Fresnel Reflection Fresnel reflections become Fresnel reflection losses when dealing with fiber optics. Fresnel losses are a result of the differences between refractive indexes of the glass fiber core and its cladding. Variations of Fresnel reflections are related to unpolarized light, light polarized perpendicular to the plane of incidence, and light polarized parallel to the plane of incidence.

Surface Refraction Whenever a beam of light passes from one medium such as air to another medium such as water, the beam separates at the intersection. Part of the beam is reflected back into the incident medium (air), and part is refracted into the second medium (water). The angle made by the incident beam and a line normal (perpendicular) to the intersection is called the angle of incidence. The angle made by the reflected beam and the line normal to the intersection is called the angle of reflection. The angle of incidence and the angle of reflection are equal. The angle made by the refracted beam and a line normal (perpendicular) to the intersection is called the angle of refraction (see Figure 10-10).

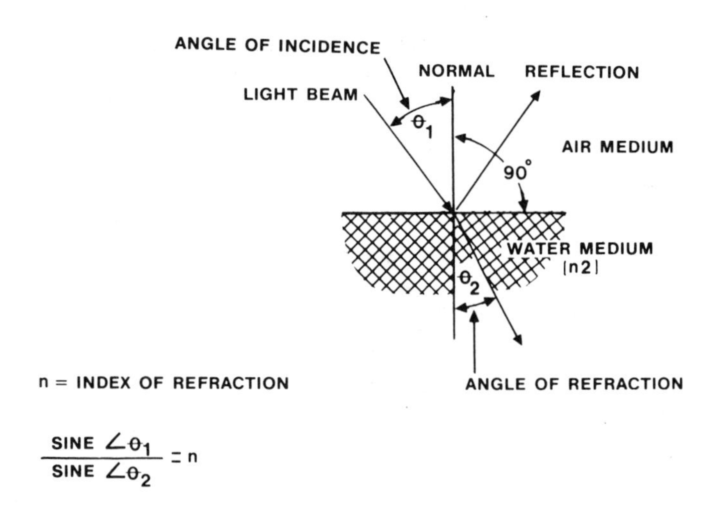

Figure 10-10 Surface Refraction

Dispersion Whenever a beam of light enters matter, its velocity decreases within that medium. The subject of dispersion deals with the speed of light in a medium and the variation in conjunction with wavelength. The speed of light in a medium can be calculated by its refractive index. The *refractive index* (n) of any material is the ratio of the speed of light in a vacuum to its speed in the medium.

The dispersion of each material is different. Curves are supplied by manufacturers that plot wavelength against refractive index. As the wavelength decreases, the refractive index increases and the dispersion ($dn/d\lambda$) decreases. The rate of increase becomes greater at shorter wavelengths. Dispersion is the spreading of light rays. In the propagation of light rays through a medium such as an optical fiber, the problems of dispersion are multiplied. With fiber optics there are two types of dispersion, intermodal and intramodal. *Intermodal dispersion* is caused by the propagation of rays of the same wavelength along different paths through the fiber medium. This results in the wavelengths arriving at the opposite end of the fiber at different times. *Intramodal dispersion* is due to variations of the refractive index of the material that the fiber was made from. Dispersion is measured in nanoseconds per kilometer (ns/km) in fiber-optic applications.

Absorption Whenever a beam of light enters matter, its intensity decreases as it travels farther into the medium. There are two types of absorption, general and selective. *General absorption* is said to reduce all wavelengths of the light by the same amount. There are no substances that absorb all wavelengths equally. Some material such as lampblack absorbs nearly 100% of light wavelengths. With *selective absorption* the material selectively absorbs certain wavelengths and rejects others. Almost all colored things such as flowers and leaves owe their color to selective absorption. Light rays penetrate the surface of the substance. Selective wavelengths (therefore colors) are absorbed, whereas others are reflected or scattered and escape from the surface. These wavelengths appear as color to the human eye. In optoelectronics it is imperative that nearly 100% of incident beams be absorbed in devices such as solar cells and detectors. On the reverse side, 0% absorption is important in the transmission of light waves in an optical fiber cable and lenses.

Scattering Scattering is differentiated from absorption in the following manner. With true absorption, the intensity of the beam is decreased in calculable terms as it penetrates the medium. Light energy absorbed in the material is converted to heat motion of molecules. Consider a long tunnel where you can only see light from one end. As you walk nearer that end, the light gets brighter. The light is absorbed as it travels through the tunnel at the absorption rate of the air medium. If the tunnel is then filled with a light cloud of smoke, the smoke scatters some of the light from the main beams; therefore, the intensity of the light from a fixed distance decreases. You may observe scattering effects by watching dust particles as the sun shines in a window. Parts of the rays are scattered by the dust particles.

A scattering type known as *Rayleigh scattering* is caused by micro irregularities in the medium. A wave passing through the medium strikes these micro irregularities in the mainstream of the waves. The waves reflected from the microparticles are spherical and do not follow the main wave, but scatter. Therefore, the intensity of the beam is diminished. In fiber-optic applications the composition of the glass must be considered to ensure low Rayleigh scattering. Silica has low scattering losses.

Photodetection Photodetection includes three phenomena: photoemission, photoconduction, and photovoltaic action. *Photoemission* is defined as the emission of electrons by light energy. The effect of light on certain materials causes electrons to be displaced from the surface. The emitted (radiated) electrons are called *photoelectrons.* Light is applied to the cathode of a basic phototube. The light causes photoelectrons to be drawn through a load by the positive anode. Thus the light determines the current flow and the voltage drop across the load resistance. Current flow is also controlled by the potential applied to the anode.

Operation through *photoconduction* involves a change in resistance of the photosensitive material. A wafer of photoconductive material is placed underneath a glass window to protect it from exposure. The photoconductive material is tied to a load resistance and a power source. The clear glass window allows light radiation to strike the photoconductive material, freeing valance electrons. The resistance of the photoconductive material decreases, causing current through the load to increase.

The *photovoltaic* cell uses dissimilar metals to generate an electromotive force in response to radiated light. A light-sensitive material is placed beneath a thin layer of transparent metal and next to a dissimilar metal. The light-sensitive material is exposed to radiation through the thin transparent metal that acts as a filter. When exposed, free electrons are removed from the light-sensitive material, causing electrons to flow to the dissimilar metal. This creates current flow and a difference of potential between the two terminals connected to the load.

ELECTROLUMINESCENCE

Probably the most useful of all the optoelectronic effects is electroluminescence. This effect involves the application of electrical energy, in the form of current flow, to a semiconductor material, causing it to emit light. This light, in the form of photons, is used in applications such as light-emitting diodes (LEDs) and in LED combinations to provide digital and alphanumeric displays. Another application of electroluminescence is the injection laser diode (ILD) used to launch light waves in optical fibers.

Electroluminescence is a photoelectric effect in which electrical energy is converted to light energy. This is the opposite phenomenon to photodetection, where light energy is converted to electrical energy.

Figure 10-11 represents the fundamental operation of an electroluminescent device, the light-emitting diode (LED). Voltage is applied to a circuit by a battery or some other direct-current power source. The circuit contains a load resistance and an electroluminescent device (LED). The load resistance limits the amount of current that can be drawn in the circuit. The LED converts the electron flow to radiant energy. The LED is forward biased; that is, the battery polarity is applied positive on the *P* material and negative on the *N* material. Current is caused to flow through a load resistance into negatively (*N*) charged material. Holes (absence of electrons) are created by electrons drawn from the positively (*P*) charged material to the positive pole of the battery. An energy gap is developed between the negatively charged material and the positively charged material at the junction. It is at this junction that electrons combine with holes to create photons. The electron–hole combinations cause the LED device to radiate energy. This radiated energy is seen through a glass or plastic envelope similar to a small light bulb.

The other electroluminescent device is the injection laser diode (ILD). The ILD is a heterostructure device. Electrons are injected into a junction region where there are deficiencies of free electrons in the lattice structure, where electron–hole combination takes place. Photons are emitted. Mirror ends reflect the photons back into the active region, where more electron–hole combinations take place. The mirror ends serve as a feedback mechanism, making the ILD a semiconductor laser. Control of brightness is made by adjusting the current flow.

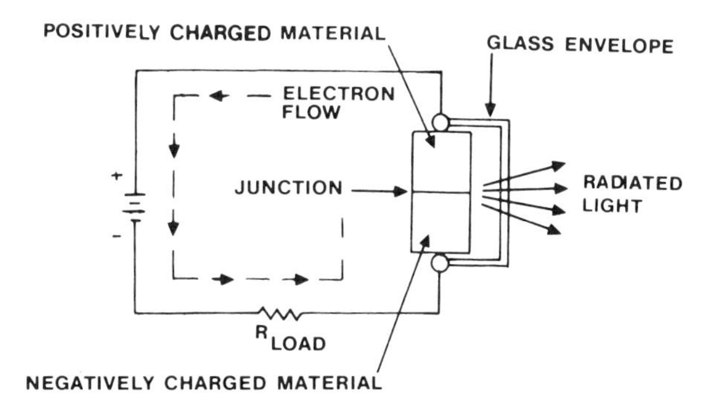

Figure 10-11 Electroluminescence

PHOTODIODE

One of the most basic of all the optoelectronic solid-state devices is the photodiode. The photodiode may be used as a photovoltaic device or a photoconductive device. A photovoltaic device converts radiant energy to a generated voltage. In a photoconductive device, electrical resistance is changed when exposed to radiant energy.

Conventional PN Junction Photodiode Operation

Figure 10-12a is a cross-sectional drawing of a conventional photodiode. Many processes have been used to form *PN* junctions. The technique that has become most popular and most utilized by industry is the *planar process*. This process is utilized by integrated-circuit manufacturers. The process uses silicon as the solid and gases as the dopants. By diffusion, the dopant gases penetrate the solid surface of the silicon. Diffusions may be made on diffusions; therefore, several layers of dopants may be diffused on one device. This serves to provide manufacturing versatility.

The basic *PN* junction used in production of the photodiode is the planar diffused. In Figure 10-12, *N*-type bulk silicon is diffused on one side by $N+$ dopant and on the opposite side by $P+$ dopant. A depletion region between the *N* and *P* exists free of current carriers. It is in this depletion area that the photons should be absorbed. To operate in the photoconductive mode, the device must be reverse biased. To operate photovoltaically, no bias is required. An active area exposes the $P+$ diffusion to light beams. The light beams are absorbed into the semiconductor. When photons of energy are absorbed into the material, electron–hole pairs are formed. Short, medium, and long wavelengths of photon power are absorbed at different depths within the *PN* junction. The depth is dependent on the photon wavelength. Short wavelengths are absorbed near the surface. Long wavelengths may penetrate the entire structure. To be most useful, the wavelengths should be absorbed in the depletion area.

Current is produced by electron–hole pairs being separated and drawn out in directions of more positive or negative sources, whichever is the case. If the electron–hole pairs happen outside the depletion area, they will usually combine and no current will be produced. The active region ($P+$ diffusion) should be extremely thin to ensure maximum penetration. As in other reverse-biased diodes, the depletion area can be made larger by increasing the reverse bias. Figure 10-12b is a typical photodiode. Figure 10-12c is the electrical symbol for the photodiode.

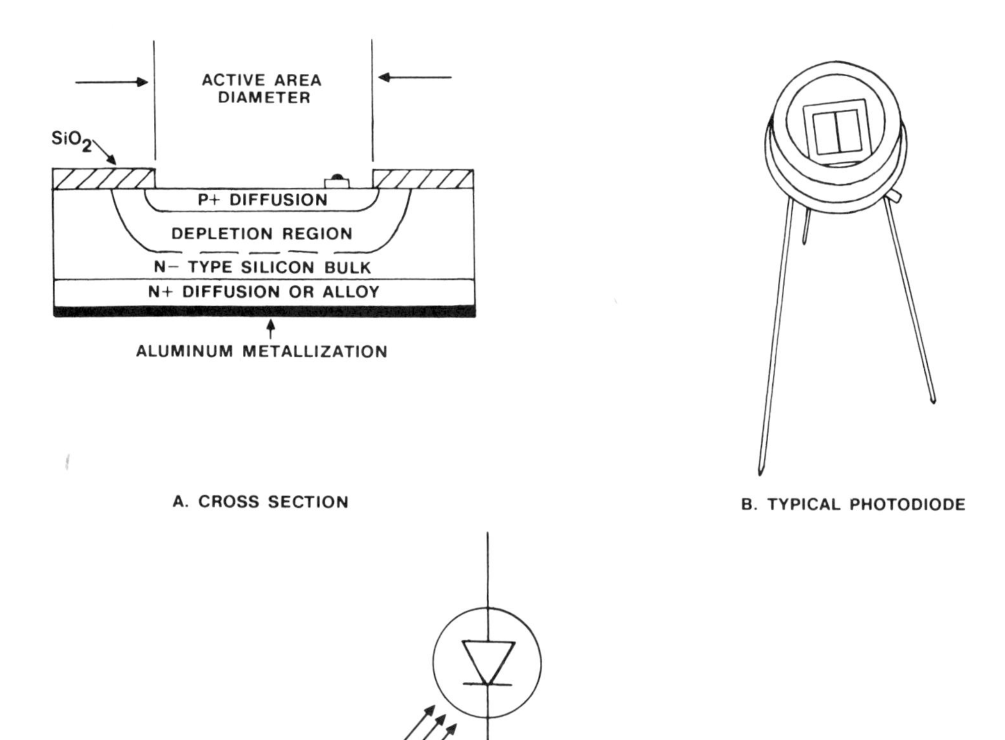

Figure 10-12 Photodiode

Solar Cells

The solar cell is probably the greatest user of the photodiode in the photovoltaic mode. The solar cell converts sunlight energy to electrical energy. In space the solar cell is used to provide power for spacecraft. The efficiency of the cell at high altitudes is very high. Beyond the earth's atmosphere the sun's rays are not inhibited. Solar cells are being used on the earth for many applications, including remote power supplies.

Solar energy has its peak at around 500 nm. Power from this source is nearly 100 mW/cm^2 if it can all be collected. The solar-cell structure is very simple and similar to the photodiode. The depletion area is very narrow. The active area is very large to intercept the maximum radiant flux. The speed of the solar cell is very low. The diode is operated in the photovoltaic mode without bias. Photons of adequate energy create electron–hole pairs. The pairs are drawn from the narrow depletion area with a high probability of recombination. The current created from the recombination is directed through a very low resistance to obtain maximum power transfer.

Solar cells in space may reach 75% efficiency, whereas those on the earth are still struggling to obtain 20%. There have been indications of a breakthrough in efficiency. Even a jump of 5% would be significant. Solar cells are especially useful in remote areas for recharging batteries to operate communication equipment. Solar cells help in operating weather devices.

Cell banks in remote areas must be set to obtain the most sunlight throughout the year. If servos are used to place the solar-cell banks in the right position, the current created by the cell is used up in moving the bank. Therefore, the mechanics of positioning from season to season, along with varying temperatures, cloudy days, and sunlight time, are certainly dominant obstacles against efficiency.

LIGHT-EMITTING DIODE (LED)

The LED represents the best of the electroluminescent devices. Electroluminescence is the emission of light from a solid-state device by application of current flow through the device. The light comes from photons of energy caused by hole–electron combinations. The diode is forward biased from an external source. Electrons are injected into the *N*-type solid-state material. Holes are injected into the *P*-type solid-state material. The injected electrons and holes recombine with majority carriers near the *PN* junction. The result is radiation in the form of photons in all directions and specifically from the top surface.

Cross Section of an LED

Figure 10-13 illustrates a standard red LED. Let's consider some of the materials from which the light-emitting diodes are made. Most are made from compound semiconductors such as gallium arsenide phosphide (GaAsP), gallium phosphide (GaP), and gallium aluminum arsenide (GaAlAs). Gallium arsenide (GaAs) is usually included in this group, but it must be understood that GaAs emits only infrared radiation around the 900-nm wavelength. These cannot be seen and are not generally classified as LEDs. Some manufacturers call them IR LEDs, while others name them infrared emitters.

In the cross section of Figure 10-13a, you will note that the layers of GaAsP are grown on a substrate of GaAs. The substrates are grown in ingots as are other solid-state devices. A lattice structure that is pure and free from imperfections is produced under a controlled environment. The epitaxial layers are gradually changed to maintain a perfect crystalline structure. The outer casing on the red LED in Figure 10-13a is made of aluminum (Al) on the front side with gold–tin (AuSn) on the backside contact. A gold ball bond is used to establish bias contact.

The manufacturing processes involved in the growing of substrate ingots and the further growth of epitaxial layers, impurity diffusions, and thin-film deposition is a complex and highly technical task. Since this is the case, most manufacturers have secret or proprietary procedures because of the high cost of research and development involved.

LED Lamps

Figure 10-13b shows a typical LED lamp. LED lamps are made so as to be as bright as possible within their chemical capabilities. The LED chip is placed within a plastic structure that provides a maximum full-flooded front-radiating area and wide-angle viewing. They are also made to be easily soldered on a PC board or snapped into a mounting clip.

The choice of an LED for a specific use is obvious in many ways. The choice usually boils down to five basic factors.

1. The color of an LED is selected by the user for a particular use. Generally, the color is in various shades of red and often orange. Amber and green are also available, but the difficulty in locating suitable raw material may preclude the choice.
2. The size of the LED depends on where the device is to be installed. Many combinations of chip and casing sizes are available. The larger the size is, the more visible the device radiance.
3. Most LED chips are Lambertian in angular distribution of light. That is, the luminance of the LED can be seen equally well from all directions. With suitable design, the angular light pattern may be changed from a very broad pattern to a very narrow pattern.
4. The luminous intensity of the device will govern the visibility under background contrast conditions when viewed at normal distances.
5. When it is not possible to provide a dark contrasting background, or when the source (LED) is viewed at very close distances, the luminance becomes important.

Figure 10-13c is the electrical schematic symbol for the LED.

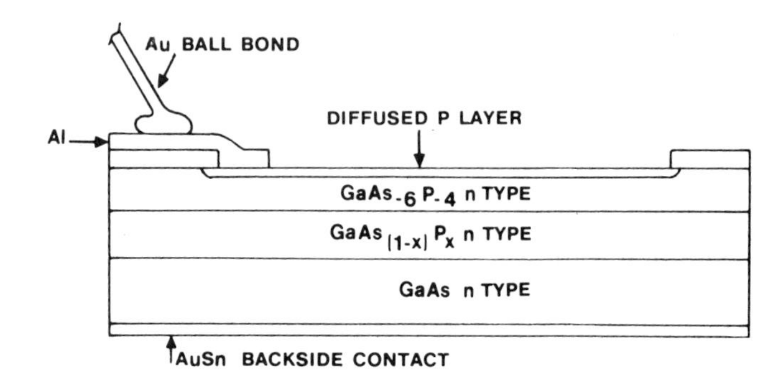

A. CROSS SECTION

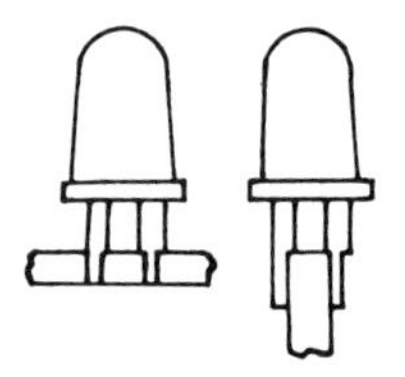

B. TYPICAL LED

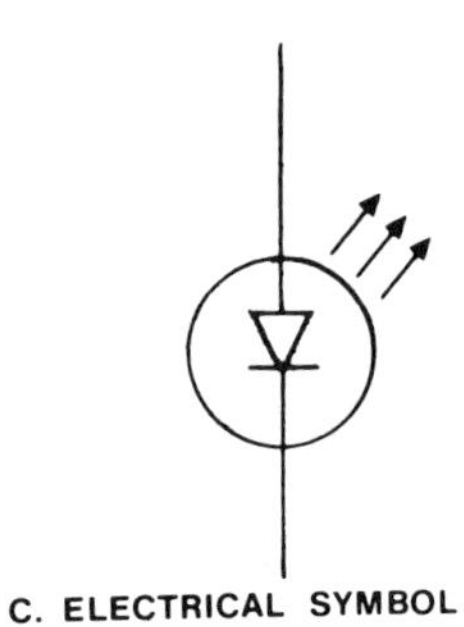

C. ELECTRICAL SYMBOL

Figure 10-13 Light-Emitting Diode

LED Displays

Electronic displays include incandescent, liquid crystals, electrochronic, and the light-emitting diode, among others. This section is dedicated to the LED display. Probably the most useful form of the LED is in digital displays. The LED digital display is reliable and inexpensive. It is usually driven from a low-level dc voltage, but has comparatively high current draw. Some disadvantages are that under high-intensity ambient light they tend to look faded. Large displays are expensive, since they require large quantities of GaAsP material. Manufacturers have developed light-pipe construction to enhance the reflective light. This allows the display to be seen for distances up to 10 ft. The process involves the placement of a reflecting light pipe over the LED chip. The LED emits light that is reflected off the light-pipe cavity walls and emerges as a larger lighting surface. Light pipes are often filled with glass particles to enhance reflection.

Intelligent displays are compact modular components that convert computer data to readable numbers and letters. They combine CMOS ICs, hybrid construction, and plastic immersion optic display technology as a single product to simplify the work of the designer.

PHOTOTRANSISTOR

The importance of the phototransistor has diminished somewhat in the last several years because of the great strides in other optoelectronic devices such as the detector diode. The one important thing that a phototransistor has that other devices lack is the ability to detect light and to amplify within a single device. This is why the phototransistor is still utilized in applications such as position detectors, intrusion alarm sensors, and optical tachometers. In operation, the phototransistor has a large collector–base junction where incident light (photons) can generate electron–hole pairs. The forward-biased base causes light-induced collector current. Operation of the phototransistor is the same as any transistor with photons of light causing the current flow. The phototransistor may not require external bias unless specific transistor operating levels are demanded.

Phototransistors are made from silicon or germanium, much the same as ordinary transistors. They are constructed in the *NPN* or *PNP* configuration. Phototransistors are subject to the typical problem of all transistors, temperature variations. These problems may be resolved by biasing techniques and thermal-stability resistors.

OPTO-ISOLATORS

The opto-isolator is a device with a light source coupled to a light sensor. It transmits while maintaining a high degree of isolation between its input and output. Previously, this job was accomplished by relays, isolation transformers, and blocking capacitors. The opto-isolator replaces these devices and adds better reliability and function. There are several basic opto-isolator types (see Figure 10-14). Two of these are the LED-photodiode and the LED-phototransistor. In each of these cases the LED is the electroluminescent device, whereas the diode and transistors are the detectors. Input leads are supplied from a current source, while output leads may be tied to a variety of electronic circuits.

With no external optic characteristics, opto-isolators are specified as electrical devices. There are functionally two parameters that define the opto-isolator. These are how well they transfer information from input to output and how efficiently they maintain electrical isolation from input to output.

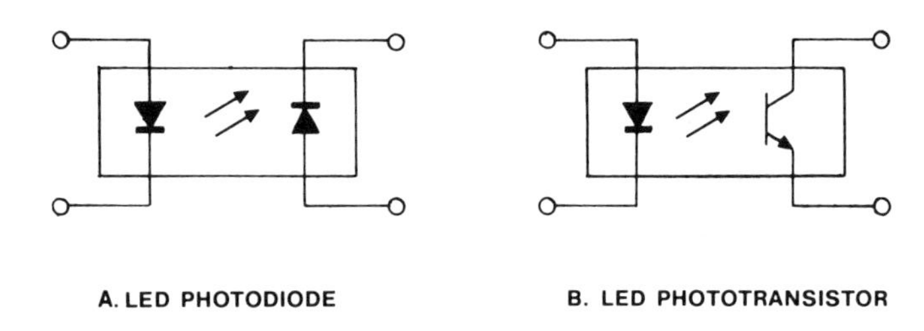

Figure 10-14 Opto-isolators

FIBER OPTICS

Optical fibers are transparent, dielectric cylinders surrounded by a second transparent dielectric cylinder. The fibers are light wave guides used to transmit energy at optical wavelengths. The light is transported by a series of reflections from wall to wall of an interface between a *core* (inner cylinder) and its *cladding* (outer cylinder). The reflections are made possible by a high refractive index of the core material and a lower refractive index of the cladding material. *Refractive index* is a measure of the fiber's optical density. The abrupt differences in the refractive indexes cause the light wave to bounce from the core–cladding interface back through the core to its opposite wall (interface). Thus the light is transported from a light source to a light detector on the opposite end of the fiber.

Fiber Generalities

The discussion of how light is launched into a fiber and arrives on the other end of the fiber is fairly easy to understand. The details involved in transmission and the mechanics of coupling, however, are manifold. Let's first discuss in lay terms how transmission takes place.

Light is propagated through optical fibers in a series of reflections from one side wall to the opposite wall. The acceptance angle *i* must be small enough so that all of the signal is reflected. In Figure 10-15, a fiber core is covered by a layer called the cladding. The core has a refractive index higher than that of the cladding. Index of refraction is determined by the manufacturer when the fiber is produced. When the light enters the optic fiber, it reflects from wall to wall of the cladding back into the core. The fiber illustrated is a step index fiber. That is, there is an abrupt refractive index change between the core and the cladding. In order that the light beam may bounce freely from wall to wall without loss, the cladding was added. The cladding acts as a mirror to reflect without power loss. A second use of the cladding is to prevent light from escaping to nearby fibers in the event of bundling several together. The cladding also contributes to the strength of the fiber.

The acceptance angle *i* is one-half the total cone of acceptance. The *cone of acceptance* is the light-gathering area at the fiber input. Fibers are extemely small. A typical size is around 80 μm.

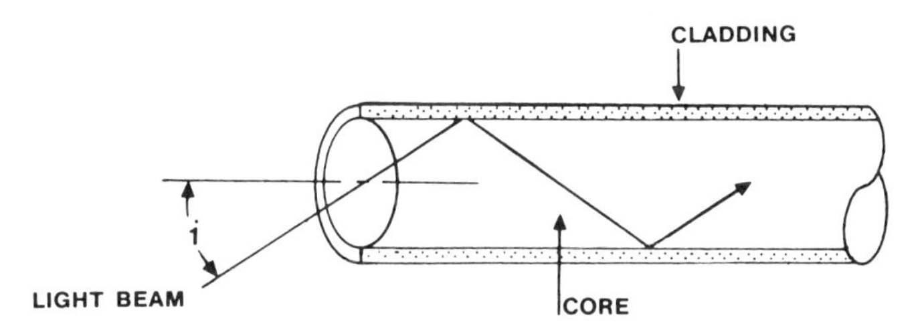

Figure 10-15 Basic Optical Fiber

Manufacturing

Most fibers are made from fused mixtures of metal oxides, sulfides, or selenides. The optically transparent mixtures used in fibers are called oxide glasses. The oxide glasses are common silica (SiO_2); lead silicate, such as crystal; and sodium calcium silicate, such as plate glass. Materials involved in the process are sodium carbonate, calcium carbonate, boric oxide silica (sand), and lead oxide. The chemical mixtures are extremely complex and must satisfy all the requirements of attenuation, strength, and the like, discussed previously.

The manufacturing techniques involved in the production of optic fibers are highly sophisticated processes. Most of these processes are secret or at the least proprietary. Considering the competitive nature of the product, this is understandable. Three general techniques are used in the fabrication process.

The first is to melt glass in one or two containers and pour the glass together to form the fiber. The second is to create a glass core or cladding and then melt a second glass to liquid and deposit the second as a core or cladding on the original. The third is to begin with an exceptional rod of glass and coat it with a plastic cladding.

Fibers are constructed to fill the need of a particular requirement. The major need is in the area of bandwidth, that is, low, medium, high, or ultrahigh. In addition, a fiber may be constructed for its extremely low attenuation per kilometer. In some cases, tensile strength is the most desired characteristic. Two operating classes of fibers are utilized, single mode and multimode. Single-mode fibers are fibers that carry one mode. Single-mode fibers must be designed to accommodate a specific wavelength; otherwise, large attenuation will result. An attractive feature of the single-mode fiber is that it is not sensitive to microbending. These are losses induced by local lateral displacements of the fiber. Multimode fibers allow intermodally dispersed wavelengths. That is, the propagation of rays of the same wavelength follows different paths through the fiber, causing different arrival times.

There are two types of optical fibers, step index and graded index. Cross-sectional views along with the typical light-ray path of these fiber types are illustrated in Figure 10-16. The *step index fiber* has an abrupt refractive index

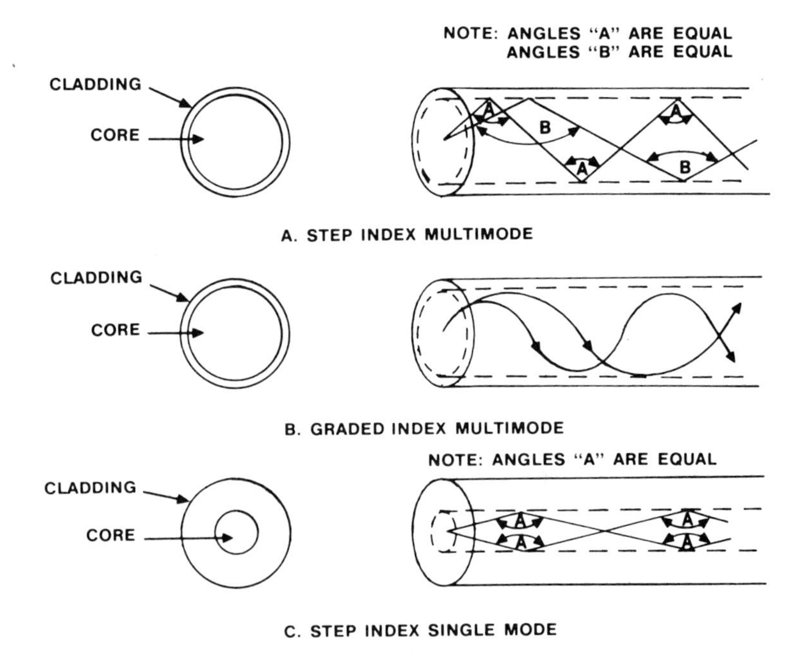

Figure 10-16 Fiber Types

between the core and the cladding. The *graded index fiber* has a variation of refractive index between the core and the cladding. This minimizes dispersion and allows a greater bandwidth of information to be transmitted.

Cable Types

Cables of single or multiple fiber are manufactured usually for protection of the fiber. Fibers are encased within the cable using protective jackets (usually polyurethane). Strength members surround the fibers. Cables are constructed to feature light weight, high flexibility, resistance to kinks, strength, and resistance to crushing. Some tests imposed on cables in the development stage are impact, bend, twist, fatigue under load, high and low temperature, and even storage losses.

As in the choice of fibers, certain trade-offs are often necessary. However, the reasons for cable development are centered on strength and environmental conditions. These should not be traded off in the search for economics.

Power Loss

Power losses are rated in decibels (dB) and are additive regardless of their reason for existence. Some of these losses are as follows:

1. Heat is always a problem in dealing with any engineering effort. Optical fibers must be isolated with heat sinks tied to receptacles, mountings, connectors, and certain splices.
2. Fiber isolation, when in a cable or bundle, may cause some power loss. Sealing off adjacent light sources is necessary. Opaque fiber coverings and sheathing are often added to protect against light sources and the environment.
3. If cable ends are allowed to vibrate, the vibration may actually cause modulation of the light rays.
4. Sand or dust on the critical mirror-finished fiber ends may act as an abrasive. A pitted finish may reflect light rays in a diffuse rather than a specular manner.
5. Water or humidity at the connection may actually improve the coupling. Water has an index of refraction much nearer the fiber core than air.

The most prevalent coupling power loss is due to misalignment of the fiber. There are four major forms of fiber misalignment: core irregularity, lateral (center-line) misalignment, gap loss between fiber ends, and angular loss (non-symmetric cutting of fiber ends). Alignment, then, is a major concern of persons utilizing fiber optics and should not be taken lightly. This is a special science, and the reader should attend to instructions provided by major manufacturers of fibers, cables, and couplings.

Fiber Testing

As with any industry product, testing of optical fiber is accomplished to ensure that the fiber qualifies under the requirements of the specifications. Usually tests fall within the major headings of mechanical, optical, electrical, chemical, and physical. This does not preclude the possibility of a company performing a special test peculiar to its own fiber. Tests provide empirical or historical data that help to improve the product or production cost. Tests are performed under controlled temperature, humidity, and atmospheric conditions so as to obtain the same results. Other conditions such as cleanliness, lighting, accuracy of equipment, and personnel proficiency are considerations to be met.

Basic Fiber-Optic System

A simple fiber-optic system (Figure 10-17) is called a *transmission link*. It consists of a transmitter with a light source, a length of fiber, and a receiver with a light detector. The basic operation of a system is to connect a digital or analog signal to a transmitter. Within the transmitter, the input signals are converted from electrical to optical energy by modulating an optical light source, normally achieved by varying the drive current. The modulated light is launched into a length of fiber where it reflects from wall to wall through the fiber core. At the opposite end of the fiber a detector accepts the light and converts it back to an electrical signal. The electrical signal is converted back to its original form in the receiver. This is a cursory look at fiber optics. The science is new and changing daily.

There are major applications in fiber optics. Perhaps the best known of these is the Lake Placid Winter Olympics Lightwave system. This system was a joint effort of New York Telephone, AT&T, Western Electric, and Bell Laboratories. It was in place for the Olympic Games of 1980. The facility provided an ultrasophisticated communications center, capable of handling a wide range of telecommunications services necessary to support these events.

LASERS

Laser technology is the science that deals with the generation of coherent light in small but powerful beams. The word laser is an acronym originating from

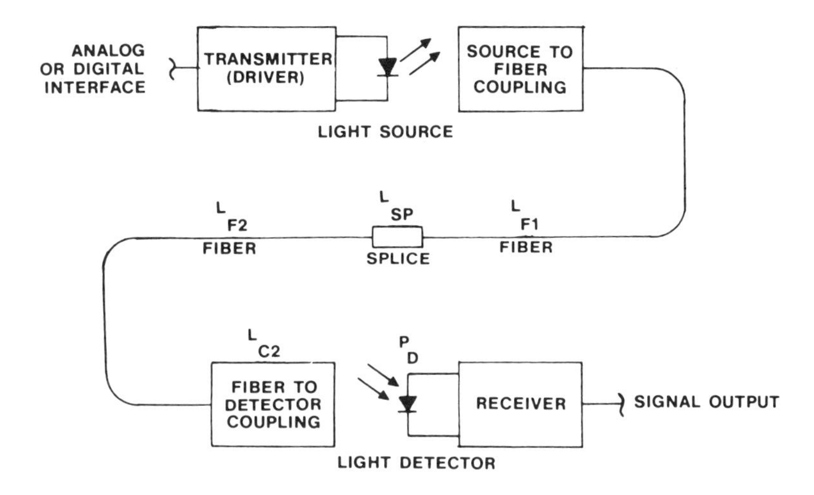

$$P_D = P_S - (\underbrace{L_{C1} + L_{F1} + L_{SP} + L_{F2} + L_{C2}}_{\text{POWER LOSSES}})$$

P_D = DETECTOR POWER

P_S = SOURCE POWER

Figure 10-17 Basic Fiber Optic System

the initials of "light amplification by stimulated emission of radiation." The operating frequency range of the laser is in the optical frequency spectrum from 0.005 to around 4000 μm (a micrometer is equal to 10^{-6} meter).

There are four basic laser types: crystal lasers, gas lasers, liquid lasers, and semiconductor lasers. Crystal lasers use a crystal such as the ruby for an active medium. Gas lasers use gases in combination, such as helium and neon, as an active medium. Liquid lasers use organic dyes such as rhodamine as an active medium. The semiconductor laser is made from semiconductor material such as gallium arsenide.

All lasers have some common characteristics. These are an active material to convert energy into laser light, a pumping source to provide power or energy, optics to direct the beam repeatedly through the active material to become amplified, and optics to direct the beam into a narrow cone of divergence. Other characteristics common to all lasers are a feedback mechanism to provide continuous operation and an output coupler to transmit power out of the laser.

Laser technology is extremely diverse and complex. This section will give a general overview of lasers and is not intended to provide detailed physics and description of the laser field.

Laser Light Properties

Laser light has unique properties. A laser beam is one color or monochromatic. If the light is one color, then it is also one wavelength and therefore coherent. An example of incoherent light is that of an ordinary light bulb, which emits white light. White light is incoherent because it has all visible colors and therefore all visible wavelengths. Figure 10-18 shows some laser light properties. Light such as the light bulb is panchromatic (has all visible wavelengths) and cannot be concentrated or controlled. Laser light can be controlled because it it monochromatic (one color), one frequency. The beams are in step (in-phase) with each other.

The light in a laser beam is extremely directional. It is focused by a lens into a narrow cone and projected in one direction. The direction of a laser beam is the result of optics at the ends or sides of an active medium. The radiance of the beam can be compared with the intensity of the sun. The laser, when focused to a fine, hairlike beam, can concentrate all its power into the beam. Hence, the directionality of the beam has a bearing on its intensity. If the beam were allowed to spread out, it would lose its power. Confinement of large amounts of power into a narrow slit allows the laser to be used to drill holes in and to cut metal, as well as in welding applications.

Laser Classifications

Lasers are classified in three ways:

1. Their active mediums
   a. Solid
   b. Gas
   c. Liquid
2. Their wavelength
3. Their output time characteristics
   a. Pulse
   b. Continuous wave (CW)

Solid Lasers

There are three basic solid lasers: the ruby laser, the Nd:YAG (yttrium aluminum garnet) laser, and the Nd:glass laser. The ruby laser operates in the visible range of the electromagnetic frequency spectrum at 0.6943-μm wavelength. The

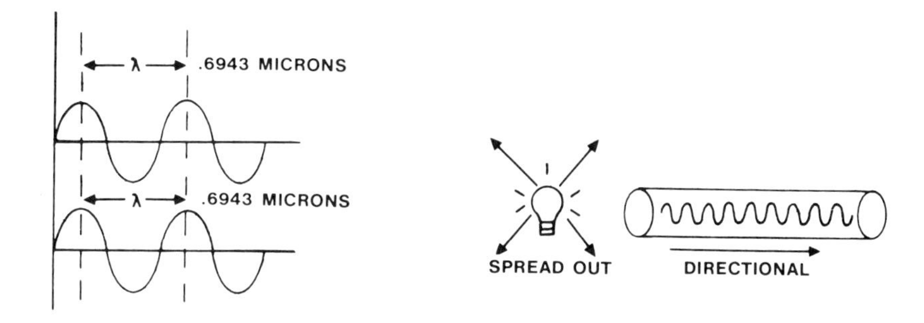

Figure 10-18 Laser Light Properties

Nd:YAG and ND:glass lasers operate in the near-infrared range of electromagnetic frequency spectrum at 1.0640- and 1.0600-μm wavelengths, respectively.

The ruby as an active medium consists of aluminum oxide with small amounts of chromium dissolved in it. Rubies used in lasers are formed in a high-temperature crucible by melting the aluminum and chromium and inserting a seed crystal of ruby. A drawing process is used, with the seed crystal being pulled at a constant speed while rotating. The ruby acts as the host crystal while the chromium acts as the activator. The ND:YAG laser uses a host crystal of yttrium aluminum garnet and an activator of neodymium. The Nd:glass laser uses a host material of glass and an activator of neodymium.

Solid Laser Operation

Let's take a cursory look at solid laser operation. A typical solid laser consists of an active medium, an excitation source, a feedback mechanism, and an output coupler. The active medium in the solid laser illustrated in Figure 10-19 is a ruby. The typical active medium is a tubular-shaped crystal structure. The ends of the crystal are highly polished and are cut parallel to each other. These ends are coated with a reflective coating. One end of the ruby reflects 100% of the light internally. The other end acts as a transmitter, allowing only a small portion of the light to pass outward, while reflecting most of the light back into the ruby for stimulated emission. The excitation source is usually a flashtube. The tube is fitted with a trigger that emits a burst of light to start the process.

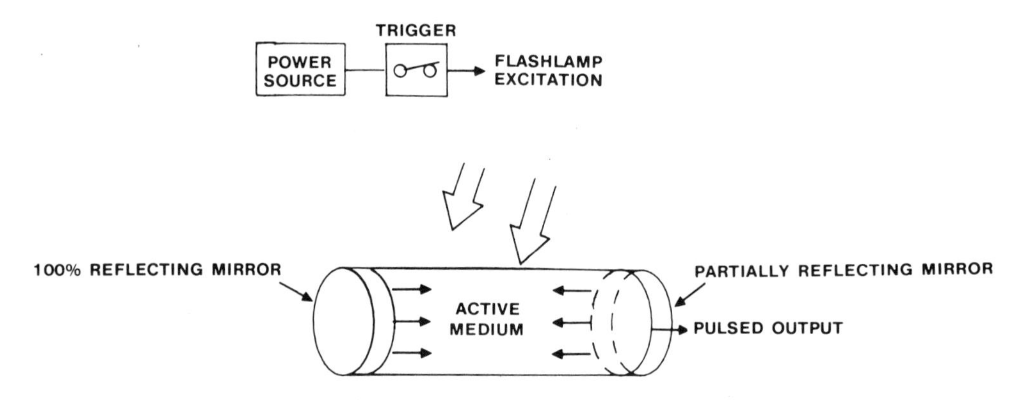

Figure 10-19 Solid Laser Operation

Let's now go through the operation and consider how it all comes together. The flashtube is energized by the trigger. It produces a high-level burst of light similar to a camera flashbulb. The flash, being adjacent to the ruby cylinder, causes the chromium atoms within the crystalline structure to become excited. Population inversion, stimulated emission, and lasing action take place.

When the beam of photons strikes either mirror ends of the ruby, the reflection of the light beam back into the ruby causes an even greater beam intensity because of photon action. The beam is red in color because the laser is amplifying at a wavelength of 6943 angstroms (0.6943 μm). The white light of the flashtube produces many wavelengths; however, only one wavelength can excite the atoms from ground state to excited energy state.

The amplified light is fed partially out of the laser by a transmitting mirror. The transmitting mirror reflects most of the light back into the ruby crystal to attain an ongoing process of stimulated emission and amplification. If too much energy is coupled out of the transmitting mirror, the stimulated emission will decay and die. Therefore, the excitation mechanism must constantly pump energy into the laser.

Gas Lasers

Gas lasers were the first to be built. In a theoretical paper written in 1958 by Arthur Schawlow and Charles Townes, they described an active material of potassium vapor enclosed in a tube. Theodore Maiman's ruby laser was created in mid-1960. Shortly after (2 or 3 months), Ali Javan and associates of the Bell Telephone Laboratories made the first gas laser. Gas lasers include three types: the neutral atom, the ion, and the molecular. Neutral atom lasers are electrically balanced and have an equal number of protons and electrons. An example of the neutral-atom laser is the helium–neon (HeNe) laser. This laser operates in the visible range of the electromagnetic frequency spectrum at a wavelength of 0.6328 μm. An ion gas laser has an active region whose atoms have missing electrons from their valence rings. An example of the ion gas is the singly ionized argon Ar^+. This laser operates in the visible range of the electromagnetic frequency spectrum at wavelengths of 0.4880 and 0.5145 μm. A molecular laser has an active medium whose molecules have two or more atoms either of the same or different atomic species. An example of the molecular laser is the carbon dioxide laser. This laser operates in the infrared range of the electromagnetic frequency spectrum at a wavelength of 10.6 μm. Other prominent gas lasers use krypton, an ion laser (0.725 μm); helium–cadmium, an ion laser (0.4416 μm); and nitrogen, a molecular laser (0.3371 μm).

Gas Laser Operation

Let's take a cursory look at gas laser operation. Each gas laser has several things in common. A typical gas laser consists of an active gas medium within a glass tube, an excitation source, a feedback mechanism, and an output coupler. The active medium in the gas laser illustrated in Figure 10-20 is helium–neon (HeNe). The gas is enclosed in a sealed glass tube called a plasma tube. The glass windows on each end of the plasma tube are made at an angle to transfer maximum light from the reflecting mirrors. The glass ends are called Brewster windows. One end of the plasma tube has a 100% reflective mirror, while the other end is partially reflective and transmits most of the light energy back into the tube for more lasing action.

Now let's go through the operation and consider how it all comes together. Helium is added to neon (90% to 10% respectively) in a plasma tube for

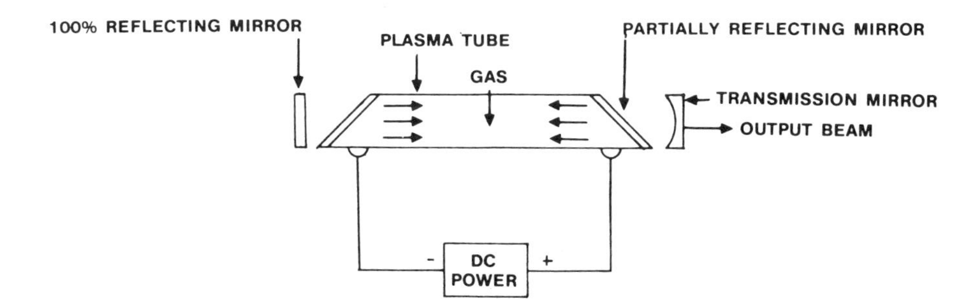

Figure 10-20 Gas Laser Operation

maximum pumping efficiency. Direct current is supplied to a negative and positive end (anode and cathode) of the plasma tube. Population inversion, stimulated emission, and lasing action take place. The beam of photons goes through the Brewster windows on each side of the plasma tube. The beam strikes the 100% reflecting mirror and is reflected back into the active medium. The beam strikes the partially reflecting (transmission) mirror and most of the beam is reflected back into the active medium. A small amount of the beam is released and coupled through an aperture for external use. If too much energy is coupled out of the transmitting mirror, the stimulated emission will decay and die.

The output of most gas lasers is extremely coherent and continuous wave (CW) in operation. Most gas lasers have very low beam divergence and can be easily tuned. They also have low comparative prices and are of fairly rugged construction.

Liquid Lasers

Liquid lasers, also called dye lasers, utilize organic dye as an active medium. The advantage of using a dye laser over others is that the dye laser can be tuned over a wide variety of wavelengths for each dye used. The variety of the dyes literally extends over the entire visible range of the spectrum, with some dyes in the near-infrared ranges. This latter attribute makes the dye laser a highly respected tool in biomedical research.

Some of the dyes utilized in liquid laser operation and their lasing range are listed in the following table:

Dye	*Range (nm)*
Carbostyril	419–485
Coumarin	435–565
Sodium Fluorescein	538–573
Rhodamine	540–690
Cresyle violet–rhodamine	675–708
Nile blue	710–790
Oxazine	695–801

The center of the range provides the user with maximum dye. Optimum dye laser performance is achieved by spectrally matching the dye laser pump with the absorption band (range) of the dye and then tuning the concentration of the dye for maximum absorption. The dye is generally mixed in an alcohol or water solution or some special solution that provides a better mix. Liquid detergent is often added to aid in lasing performance.

Dye lasers can be continuous wave (CW) or pulsed. In operation, dye lasers are pumped by gas lasers or flashlamps. A few are pumped by ruby or Nd:YAG solid lasers.

Liquid Laser Operation

An optical schematic of a dye laser is shown in Figure 10-21. This laser is continuous wave. The CW dye laser operates by rapidly moving dye molecules through a CW laser pump beam, spatially coincident with the optical cavity of the dye laser. Each molecule in effect sees a pulse of light as it passes through the pump beam.

A dye is selected that will provide the spectral range desired. The dye is dissolved in ethylene glycol. The dye is squirted in a flat stream into the laser cavity at a specific angle. The dye is circulated with a gear pump through the cavity, a filter, and a reservoir. The dye solution is excited by a beam from an argon ion laser. The beam is focused to a specific diameter by a focal length mirror *M*4. The focal length mirror *M*4 directs the beam to the cavity and the dye stream. The optical cavity consists of the three mirrors *M*4, *M*2, and *M*1. Mirrors *M*2 and *M*1 are highly reflective. *M*2 is the end mirror, which serves as reflector and a collimator to the output coupler mirror *M*3. Mirror *M*1 is a reflector.

Tuning the CW dye laser wavelength is accomplished with the birefringent filter placed in the laser cavity. The filter consists of three quartz plates of different thicknesses aligned in the cavity and rotated to change the wavelength of the transmission.

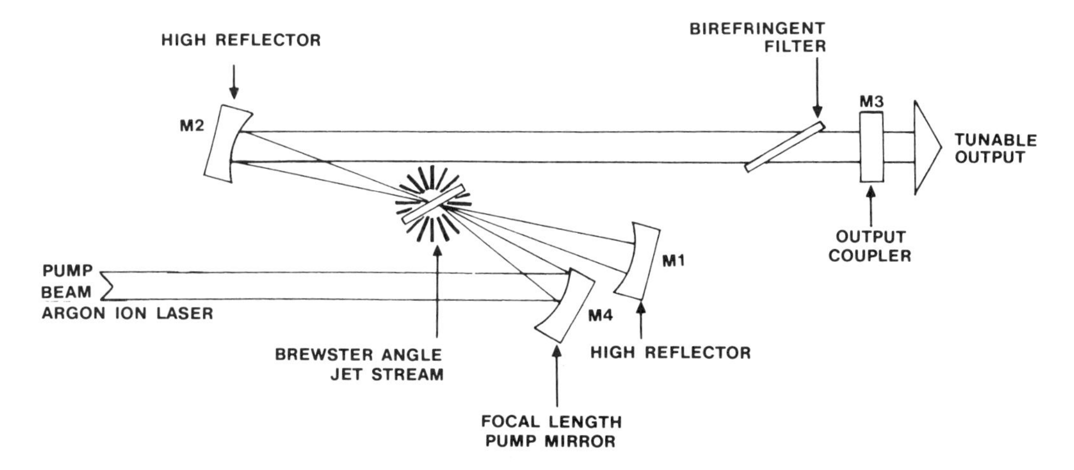

Figure 10-21 Dye Laser Operation

11

Optical Encoders

Modern measurement and control problems and solutions often involve the monitoring of rotary and linear motion. The first stage of either measurement or control is the generation of an electrical signal that represents the motion. When measurement is required, the signal is used to quantify the desired property (i.e., displacement, velocity, etc.), and the data are translated to a format that can be understood by the end user. When control is the objective, the signal is used directly by the associated controller.

Generation of the electrical signal for measurement or control is accomplished with transducers. One major type of tranducer used to monitor motion is the optical encoder, the subject of this chapter.

ENCODERS

Encoders are mechanical-to-electrical transducers whose output is derived by reading a coded pattern on a rotating disk or a moving scale. Encoders are classified into three major categories:

1. The method used to read the code, either contact or noncontact.
2. The type of output, either absolute digital word or series of incremental pulses.

3. The physical phenomenon employed to produce the output, either electrical conduction, magnetic, capacitive, or optical.

The discussion will be confined to the rotary encoder. However, most of the principles will apply to linear encoders as well.

Contact Encoders

Contact encoders employ mechanical contact between a brush or pin sensor and the coded disk. The disk contains a series of concentric rings or tracks, which are thin metallic strips joined at their base as shown in Figure 11-1. The four tracks shown in Figure 11-1 represent a binary code consisting of 2^0, 2^1, 2^2, 2^3. The associated contact sensors are identified at B_0, B_1, B_2, and B_3, and encode the numerals 0 through 15. Disks in presently available contacting encoders may contain up to 30 tracks, one of which is a solid, unsegmented, concentric ring (not shown in Figure 11-1) that functions as an electrical return path. As the disk rotates, the sensors alternately contact conductive strips and adjacent insulators, producing a series of square-wave patterns.

Any pattern that can be produced photographically can be reproduced on an encoder disk and may include uniform or non-uniform tracks, tracks that are out of phase with respect to one another, and other unusual configurations used for special control applications. However, the most common use of encoders is measurement of shaft position, where any non-uniformity is a source of error.

Two common encoder disk problems are non-uniform segment spacing and track eccentricity. Non-uniform segment spacing produces positional error. Eccentricity causes an error that is a sinusoidal function of shaft angle.

In addition to segment placement and eccentricity, the potential performance of contact encoders is limited by the minimum practicable segment and sensor size. The smaller the disk segments, the smaller are the incremental differences in shaft input position that can be detected. It is feasible to manufacture disks with segments and sensors of less than 0.001-in. minimum dimension. However, the life of the resulting encoder would be severely limited by contact wear and bridging of the disk segments. Generally, the finest usable segment width is 0.005 to 0.010 in., and overall resolution is about 2 to 15 seconds of arc.

In addition to the preceding problems, contact encoders are limited to relatively low shaft speeds. Actual limits depend on encoder size, the mechanical interface expected wear, and other factors. In general, 1- to 2-in. diameter encoders are limited to speeds of approximately 200 rpm.

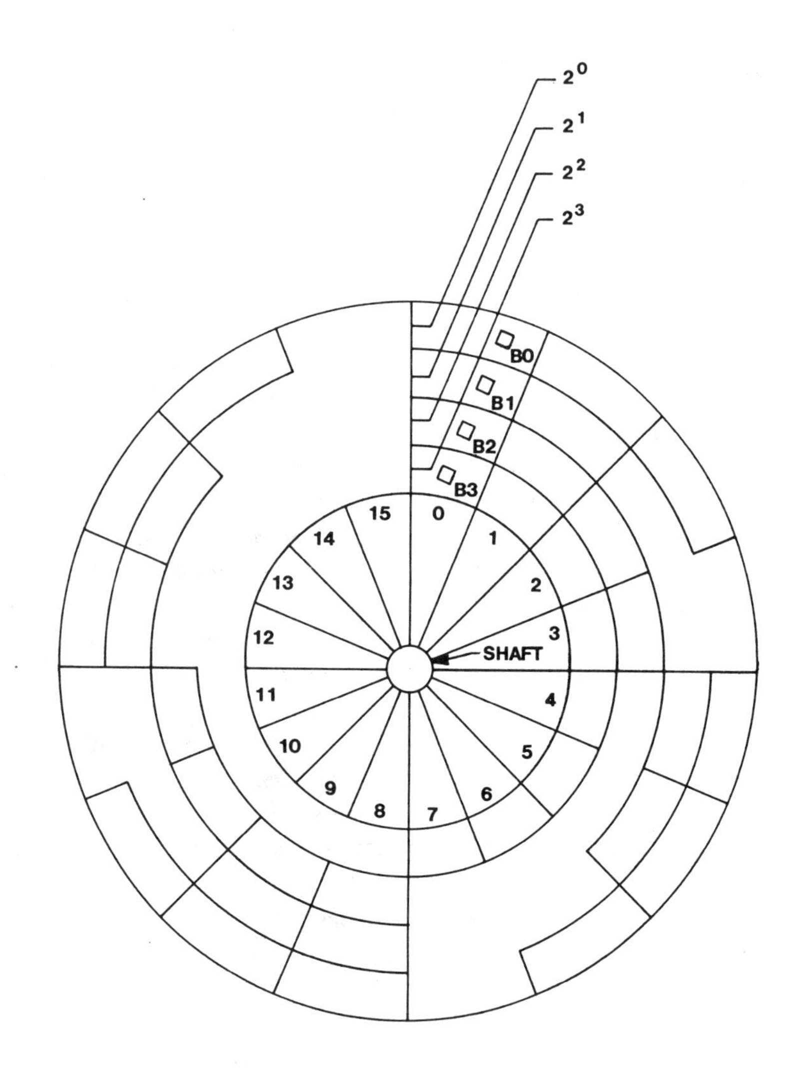

Figure 11-1 Absolute Contact Encoder Disk *(Courtesy, Dynamics Research Corp.)*

Even though contact encoders have been used for many years, occasional improvements are still made. In a recent development, pin contact sensors replaced the conventional cantilevered brush assembly. The pin contact sensor is three spring-loaded pins operated in parallel. This arrangement significantly improves encoder life and reliability. Pin contact encoders have been operated effectively at shock loads up to 150 G's and vibrations of 30 G's at 5 to 2000 Hz.

Non-contact Encoders

Non-contact encoders employ physical phenomena other than electrical conduction to read the coded disk. The most common of these types are magnetic, capacitive, and optical.

Magnetic Encoders Magnetic encoders were developed to replace contact encoders in applications limited by rotational speed. However, magnetic encoders are not significantly more accurate than contact encoders. Magnetic encoders operate by detecting resonant frequency change, a magnetization change, or magnetic saturation in an inductor. With all methods, flux induction by the magnetically coded disk affects the change by aiding or inhibiting an existing state. Thus, for each principle, two normal states exist corresponding to a logical 1 or 0.

The resonant frequency method utilizes a tuned circuit, the normal frequency of which represents one logical state, with the detuning of the circuit representing the opposite logical state.

In the magnetic saturation method (principle of a saturable reactor), the inductor is either magnetically saturated or is nonsaturated. Alternately, the reluctance of the magnetic circuit is effectively translated to logical 1's and 0's.

The inherent accuracy and resolution of a magnetic encoder is no greater than that of contact types. Resolution is limited by the size of the magnetized spot and complicated by interaction between the magnetized spots on adjacent tracks. Magnetic encoders overcome the basic speed limitation of contact encoders and offer somewhat greater longevity by eliminating physical contact between disk and sensor. Also, magnetic encoders have greater resistance to most natural environmental extremes, and all the standard scanning techniques can be employed. However, high ambient fluxes or radiation densities can destroy the disk pattern or inhibit saturated core operation. Therefore, greater precaution against mutual electromagnetic interference is required when magnetic encoders are included in a system. Figure 11-2 illustrates the principal stages of typical magnetic encoding. The requirement for additional shielding to ensure that

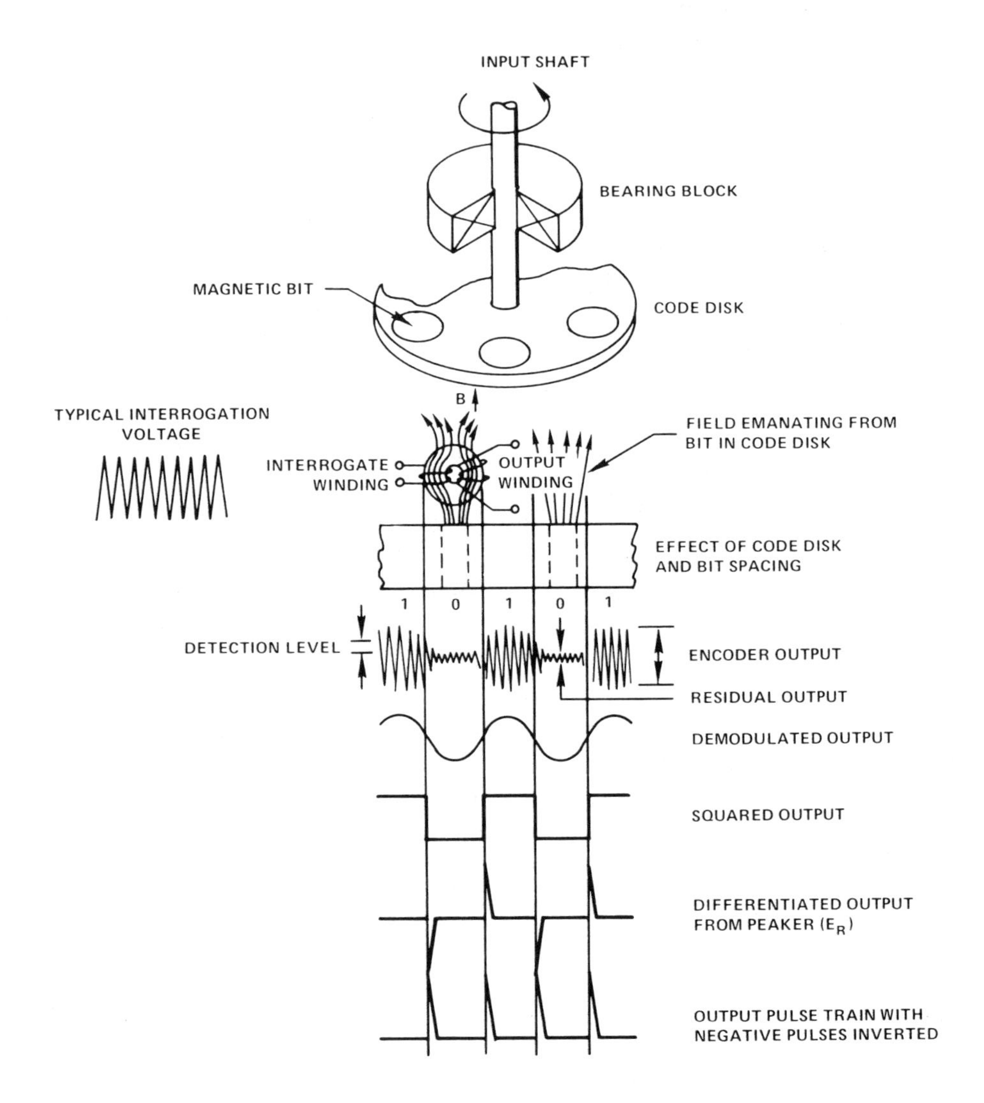

Figure 11-2 Typical Magnetic Encoding *(Courtesy, Dynamics Research Corp.)*

surrounding equipment does not interfere with encoder operation and, conversely, that the magnetic encoder does not disturb the operation of adjacent gear adds to the cost of an already expensive approach.

Capacitive Encoders Capacitive encoders are the least used of the noncontacting types and have been developed in response to unique needs. Readout is effected electrostatically using a phase-shift measuring system or a frequency-control technique to develop the digital output.

Although capacitive devices are not generally available as standard hardware, up to 19-bit, single-turn units have been produced. Theoretically, the capacitive technique can be used to accomplish any of the encoding tasks performed by the contact, optical, or magnetic types. However, practical problems of design, manufacture, and operation have limited the use of capacitive detection, and relatively few of these units are used.

Optical Encoders The optical encoder was the earliest of the noncontact devices developed to eliminate the wear problems inherent with contact encoders. Present-day optical encoders provide the highest-accuracy encoding and can be operated efficiently at high speeds.

Optical encoder disks have opaque and transparent segments. The disks are normally produced by exposing a photographic emulsion to light or, where greater durability is required, by plating a metal on the substrate. The substrate, usually glass, provides mechanical integrity and dimensional stability. Readout is effected by an array of carefully aligned photoelectric sensors positioned in front of the disk. A light source behind the disk provides excitation. As the disk rotates in response to the input variable, the opaque areas on the disk pass between the light source and the sensors, interrupting the light beam and modulating the sensor output in accordance with the selected code. Optical systems focus the light on the coded disk. Mirrors, prisms, lenses, fiber optics, and optical slits or diffraction gratings perform this function.

Fiber optics are light pipes that are used to guide light from the source to the readout station. They are composed of glass fibers made into bundles and operate on the principle of total internal reflection, such that light imposed at the end of one fiber appears at the other end by multiple reflections through the fiber length. By using identical length fibers for any pair of sensors, identical optical paths are formed, stabilizing phase relationships. Imaging fiber-optic techniques may be employed to compensate for rotary encoder disk eccentricity or, as a folded optics system, to double the basic scale resolution of linear encoders.

Light detection can be performed by one of several devices. Materials for all types of light-detecting devices are selected from groups III, IV, and V of the periodic chart and lie halfway in the chemical spectrum between metals and nonmetals. As such they are semiconductors. Each device responds to light in a different manner. Photovoltaic cells, such as selenium or silicon, generate an electric current when exposed to light. The resistance of photoconductive cells varies with light intensity. Such cells are usually made from cadmium sulfide or cadmium selenide, depending on the required response or the portion of the light spectrum for which sensitivity is desired. Current capability varies with the intensity of the incident light. Photodiodes are similar to photoconductive cells. Photodiodes are used because their small areas allow very high frequency response. They are generally run with a back bias, and the reverse leakage current is modulated by the light.

A stiff, lightweight filament is used in the miniature incandescent lamp to achieve superior resistance to mechanical shock and vibration. In this manner, bulb life is lengthened, allowing nonreplaceable bulb assemblies to be permanently mounted within the encoder. Solid-state light sources offer extremely long life, excellent shock recovery in excess of 100 G's, but less total luminescence than the incandescent bulb.

The performance of the contact, magnetic, and capacitive encoders is limited by the minimum practicable dimensions of sensors or of segments on the coded disks. The optical encoder is limited, not by these factors, but by the precision of the mechanical bearing assemblies. Of the encoder types, optical encoders are generally more expensive and provide the highest performance. When first developed, the optical encoder could not withstand extreme environmental conditions. However, there has been much improvement in this area in recent years. Optical encoders are best applied in situations where high speed, accuracy, resolution, and extended service life are required.

Incremental versus Absolute Output

In the preceding discussion, reference was made to a coded disk pattern such as that in Figure 11-1 for which the encoder output is a digital word representing the absolute angular position of the encoder shaft; hence the designation absolute encoder.

If the coded disk pattern is replaced with a uniform pattern such as a series of equally spaced radial lines, encoder output becomes a series of electrical pulses that can be counted to determine shaft position relative to some reference point. This configuration is called an *incremental encoder.*

The incremental encoder is the simpler since the only requirement is generation of a repeated signal for equal increments of shaft angle. The encoder may also contain a zero reference marker. The direction of shaft rotation can be readily detected through one additional code sensor. Circuitry is simple and power requirements modest. High resolutions can be achieved in small units, particularly in the optical types.

There are advantages to such an encoder and weaknesses as well. If, for any reason, power to the encoder is interrupted or a transmission or reception failure occurs, counts will be lost. The error will persist until the zero reference has been sensed. Zero reference is a special signal produced once per revolution of the rotary encoder or at the beginning of linear encoder travel. Some means must be provided for reading the output of either encoder. The incremental encoder normally employs a bi-directional counter with two data lines. The 2^N absolute encoder requires N data lines. In systems with more than one encoder, the counter can be appropriately equipped with memory for time sharing.

In many applications, the incremental encoder represents a nearly ideal solution to the encoding problem. For example, incremental encoders without zero reference are used in many *xy* plotters. In this application, the encoder is used in the floating zero mode because the counter rather than the encoder defines the zero point by being cleared and zero reset on command. The counter is a special register used to accumulate the sum of encoder pulses.

Encoder Error

Encoder error is composed of quantization error and instrument error. Quantization error is fundamental to all digital equipment because of the quantum nature of digital codes. In a perfect encoder with no mechanical, optical, or electronic deviation from the ideal, the mean angular position of the input shaft for a given readout is defined as the angular position midway between the transition from the next lower readout to the transition for the next higher readout. The quantization error is the deviation of the input shaft from the mean position for a given readout, with the maximum error $\pm\frac{1}{2}$ the angular rotation between two successive bits. For example, a rotary incremental encoder that generates 360 pulses per revolution has a quantization error of $\pm\frac{1}{2}$ angular degree (see Figure 11-3).

Instrument error is introduced by the actual manufacturing and assembly process and is defined as the difference between the shaft angle as indicated by the encoder and the actual shaft angle at the midpoint of the given readout. Most manufacturers limit this error to about the same magnitude as the quantization error. The manufacturer's instrument data sheet will normally specify the instrument error. However, in applications involving a large number of readouts, a statistical specification of encoder error is more significant. The

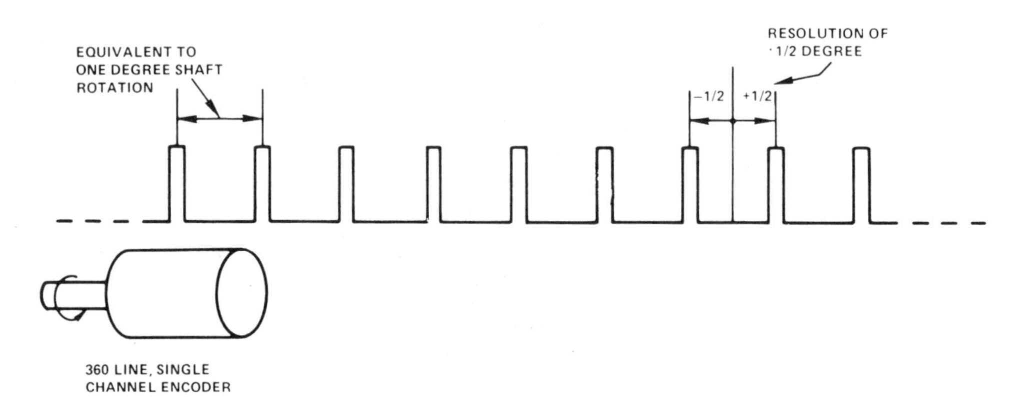

Figure 11-3 Encoder Resolution *(Courtesy, Dynamics Research Corp.)*

standard quantization error combined with the standard deviation of instrument error produces the standard deviation of total encoder error. This figure is generally less than the quantization error alone, but is determined by individual manufacturers for each product.

AN OPTICAL INCREMENTAL ENCODER

Many corporations manufacture the optical incremental encoder. This section deals with the encoder manufactured by the Dynamics Research Corporation of Wilmington, Massachusetts.

Theory of Operation

The principle of incremental encoder operation is the generation of a symmetric, repeating wave form that can be used to monitor the input motion. The basic components of the optical incremental encoder are the light source, light shutter or chopper, light sensor, signal-conditioning electronics, and shaft bearing assembly. The encoder's mechanical input operates the light shutter, which modulates the intensity of the light at the sensor. The sensor's electrical output is a function of the incident light. The encoder's electrical output is produced from the sensor output by the signal-conditioning electronics and can be either (1) a sine wave, (2) a shaped square wave, or (3) a series of equally spaced pulses produced at regular points on the wave form.

In its most fundamental form the light shutter is an optical slit and a glass plate inscribed with alternating lines and spaces of equal width. When the plate is moved relative to the slit, the light transmitted by the slit rises and falls. In practice, a larger reticle that intercepts 20 or more lines is used instead of a single slit (Figure 11-4). The reticle transmits more light and is easier to align. Another advantage of the reticle is that line imperfections such as scratches, pinholes, and dust particles do not significantly affect sensor output since the incident light is averaged over many lines.

The encoder's mechanical input is coupled to the moving plate to operate the light shutter. In the position shown, the sensor will indicate maximum light intensity. After the moving plate has advanced one line width, the opaque lines on the moving plate will cover the transparent lines on the stationary plate, and light transmission will be minimum, theoretically zero. Because in practice the light source cannot be fully collimated, and because some clearance must be maintained between the stationary and moving plates, some light will leak through the shutter in its fully closed position. Consequently, the minimum light transmission will be nonzero.

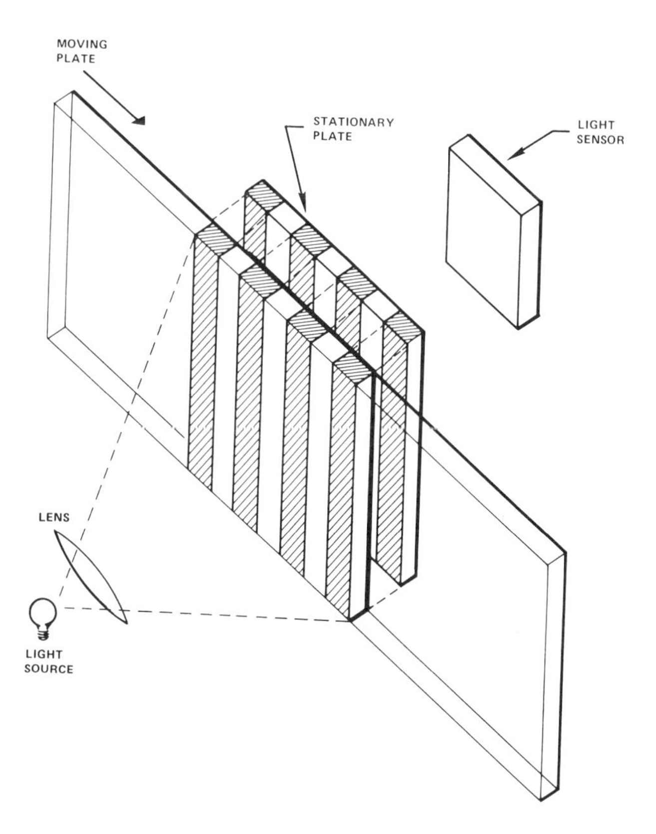

Figure 11-4 Light Shutter *(Courtesy, Dynamics Research Corp.)*

The wave form of the sensor output is theoretically triangular (Figure 11-5) but is, in practice, more nearly sinusoidal. e_{max} depends on the output of the light source, shutter transmission when fully open, and sensitivity of the sensor. e_{min} depends on shutter leakage when fully closed. e_1, the peak-to-peak voltage, is the usable component of sensor output, and, in practice, is limited by the effectiveness of the shuttering mechanism; specifically on the ability to minimize shutter leakage. This is accomplished by collimating the light source as well as possible and by limiting clearance between the glass plates to that necessary to accommodate misalignment of the disk and shaft.

A common method of digitizing encoder output is to produce a pulse each time the wave form of Figure 11-5 passes through its average value, e_{avg}. Ideally, the wave-form crossings are equally spaced and correspond to one line width displacement of the moving plate. The resulting series of equally spaced pulses can be used to precisely monitor the encoder's mechanical input.

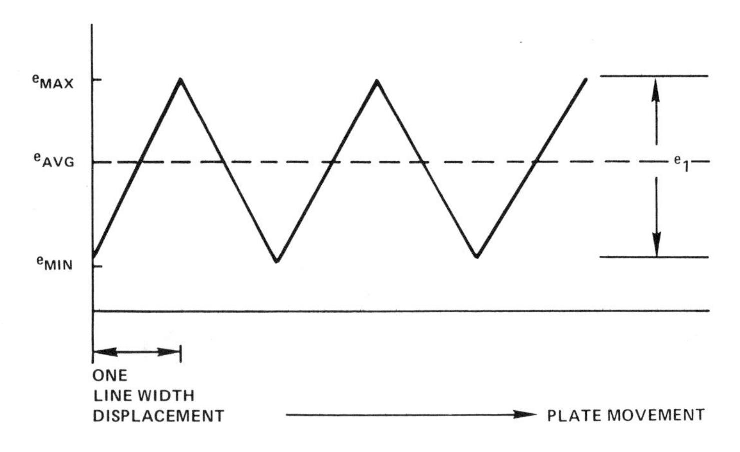

Figure 11-5 Detector Output *(Courtesy, Dynamics Research Corp.)*

The primary problem with this arrangement is illustrated in Figure 11-6. The solid wave form is the same as that of Figure 11-5. The dotted wave form represents signal drift caused by changes of the lamp's excitation voltage or sensitivity of the light sensor. It can be seen that, after drift, the series of pulses generated when the wave form passes through e_{avg} will no longer be equally spaced. Encoder performance is severely degraded since minimum resolution is now twice that if there were no signal drift. (The minimum wave-form dimension that is constant after drift is the spacing between the first and third pulses of any three-pulse series, instead of the spacing between adjacent pulses.)

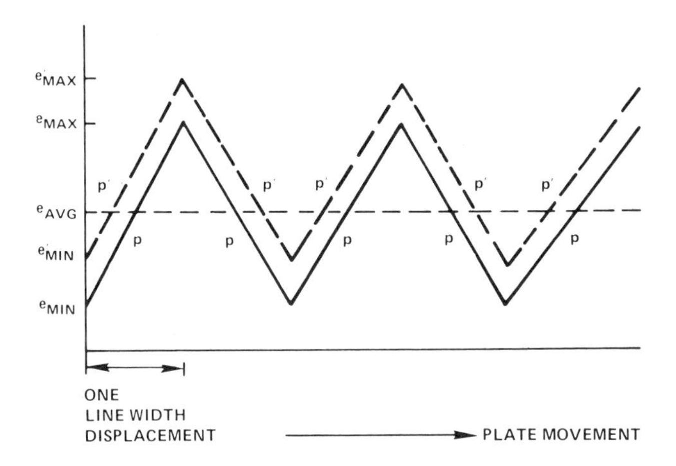

Figure 11-6 Effect of Signal Drift on Spacing of Encoder Pulses *(Courtesy, Dynamics Research Corp.)*

Compensation for signal drift is achieved with another sensor in combination with the same moving plate. The stationary component of the shutter for the second sensor is fixed 180 electrical degrees out of phase with respect to the stationary plate of the first sensor. The same light source illuminates both shutters (see Figure 11-7). When the maximum light is incident on the first sensor, the second will see minimum light. Because the incident light intensity at the two sensors is out of phase, the resulting electrical outputs will be out of phase as in Figure 11-8. The wave form that results when the two sensors are connected in opposition (i.e., "push–pull" or "head-to-tail") is the equivalent of algebraically subtracting the output of sensor 2 from that of sensor 1, as shown in Figure 11-9.

Several characteristics of the resultant wave form are important. First, its average value is theoretically zero. (In fact, the inevitable mismatch between the two wave forms results in some small deviation from zero. This deviation can be suppressed by an external bias voltage.) Second, the peak-to-peak voltage is now twice that of the separate sensors. This means that change of the interpulse spacing for a given shift of the resultant wave form (like that shown in Figure 11-6) is only one-half of what it would be for the same single sensor.

This push–pull arrangement further improves encoder operation by reducing the effects of changes in the light excitation voltage and detector sensitivity. Any light-intensity change that affects both sensors equally will cancel. Likewise, equal detector sensitivity changes will cancel (e.g., temperature-dependent variations).

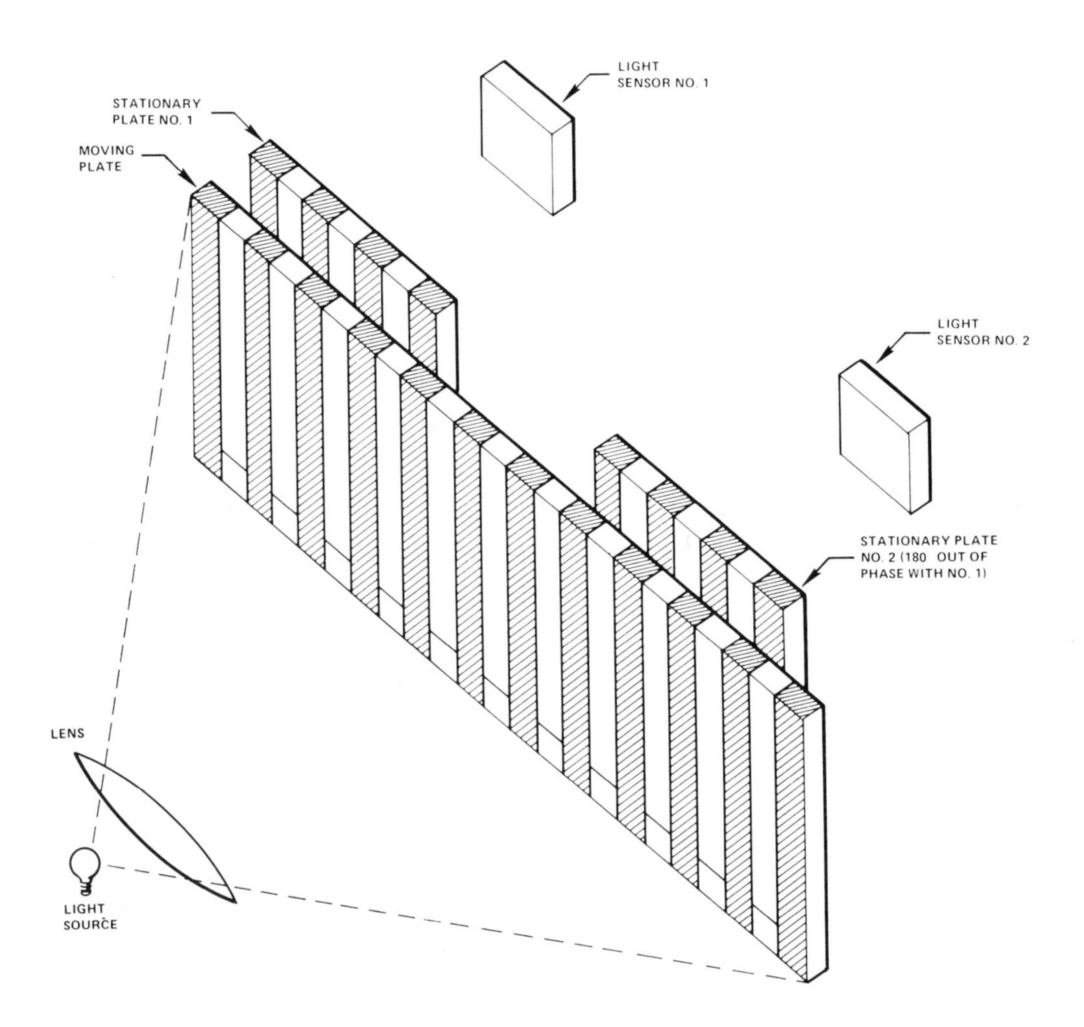

Figure 11-7 Two-Cell Arrangement for Signal Drift Compensation *(Courtesy, Dynamics Research Corp.)*

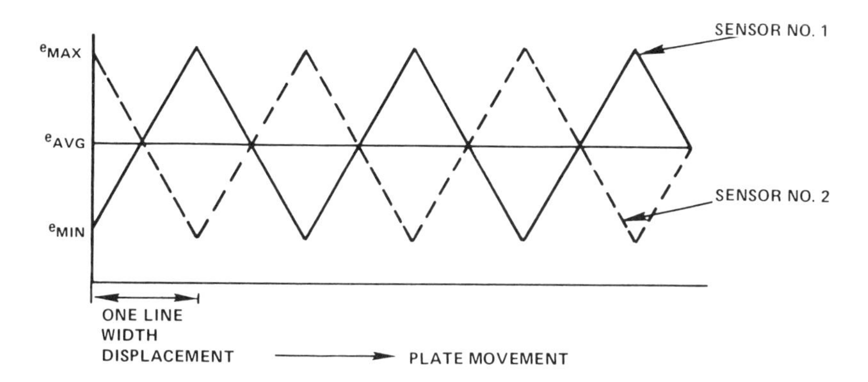

Figure 11-8 Outputs of Two Sensors Fixed 180° out of Phase *(Courtesy, Dynamics Research Corp.)*

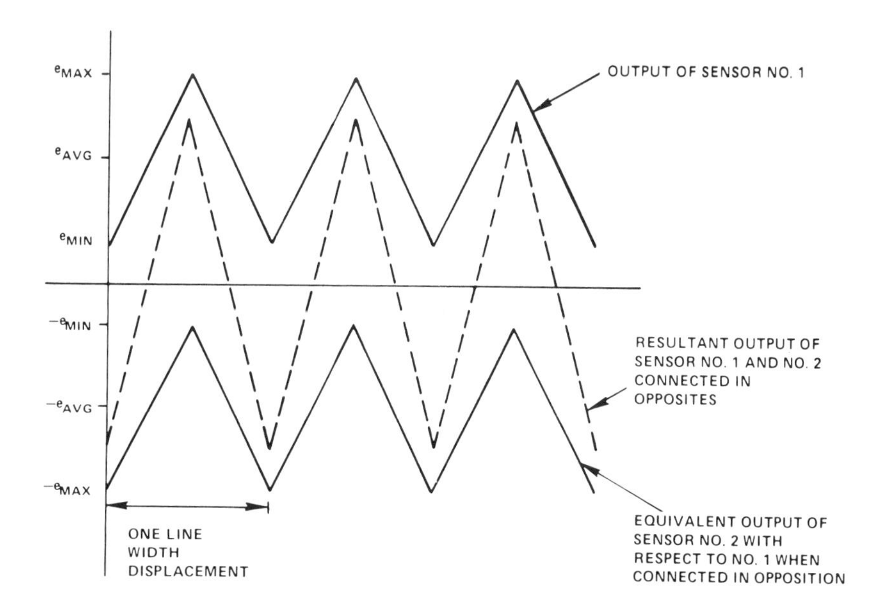

Figure 11-9 Output of Sensors 1 and 2 Connected in Opposition *(Courtesy, Dynamics Research Corp.)*

The output signal just described is called *single-channel output.* In practice, the typical encoder has two channels. The second output is produced with another pair of sensors displaced electrically 180° from each other and 90° from each of the sensors of channel *A.* The resulting wave forms are shown schematically in Figure 11-10. Interpulse spacing for the two-channel encoder represents movement of one-half line width. For example, if the glass plate previously described is a glass disk, the lines become sectors on the disk. If there are 1024 such lines and 1024 spaces, the number of pulses generated per revolution of the disk is $4 \times 1024 = 4096$. The angular movement represented by the interpulse spacing is 5.27 minutes of arc.

$$360 \frac{\text{degrees}}{\text{revolution}} \times 60 \frac{\text{minutes}}{\text{degree}} \times \frac{1}{4096 \dfrac{\text{counts}}{\text{revolution}}}$$

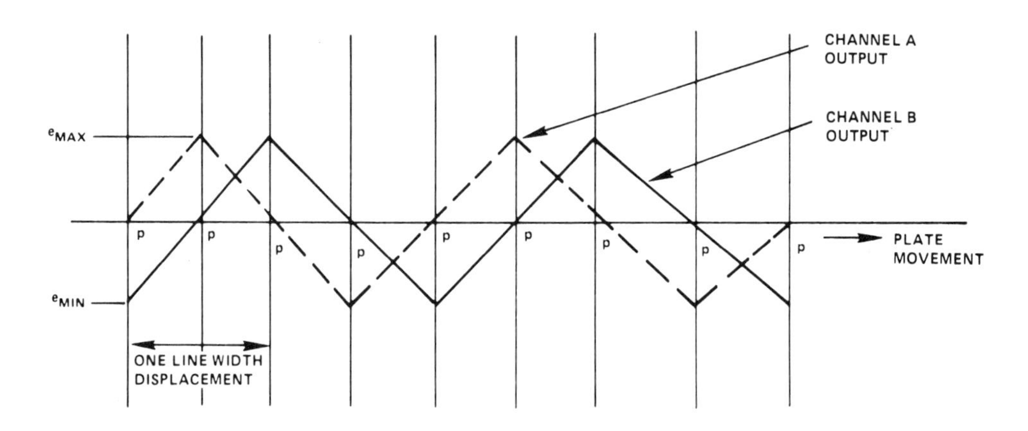

Figure 11-10 Dual-Channel Output Phased in Quadrature *(Courtesy, Dynamics Research Corp.)*

The sensors used in most encoders are large-area silicon junction diodes commonly called cells. To obtain the ideal wave forms shown in Figure 11-11, the cells connected in the push–pull arrangement must track one another. That is, equal changes in input must produce equal output changes. When the components do not track, there will be a symmetry shift in the resultant waveform like that depicted in Figure 11-11.

Symmetry is defined as the condition where the zero dc level of the output wave form divides the wave form into two states of equal duration when the

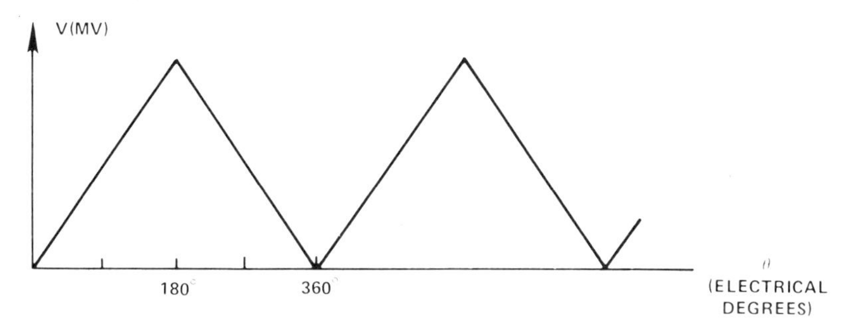

a. CELL NO. 1 OUTPUT VOLTAGE

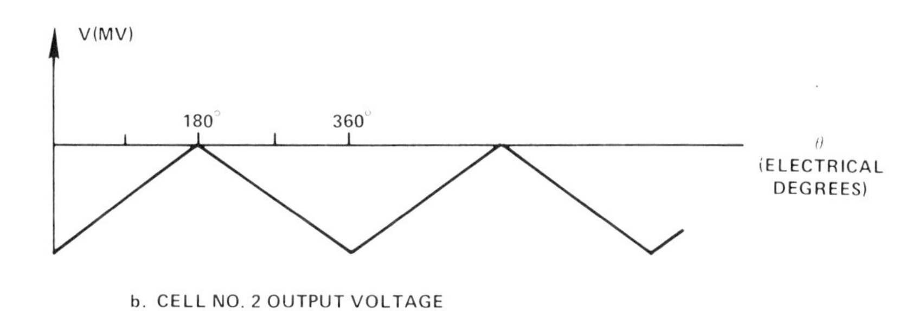

b. CELL NO. 2 OUTPUT VOLTAGE

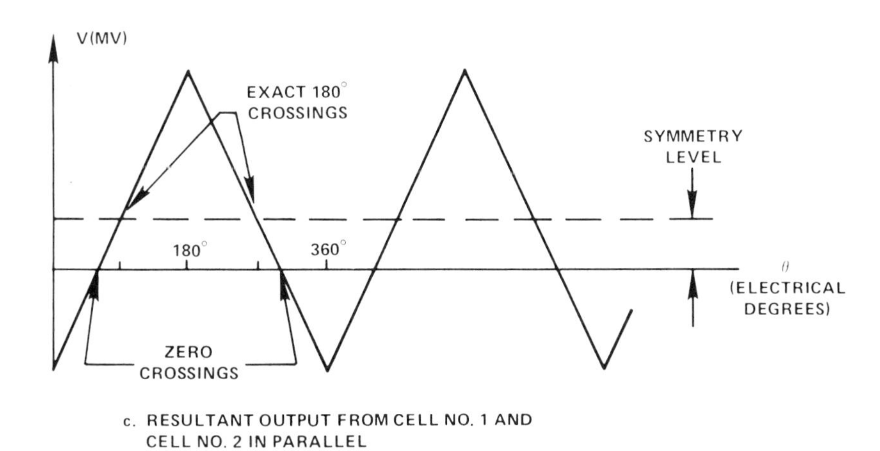

c. RESULTANT OUTPUT FROM CELL NO. 1 AND CELL NO. 2 IN PARALLEL

Figure 11-11 Result of an Unbalanced Pair of Solar Cells *(Courtesy, Dynamics Research Corp.)*

encoder shaft is rotated at constant speed. That is, the wave form is symmetric if it has a zero mean value. If the mean value is not zero, the dc level that divides each cycle of the wave form into equal parts is the symmetry level. This level causes the zero crossing to shift markedly.

Symmetry shift in a given channel is minimized by the following:

1. Using cells of equal temperature and light sensitivity
2. Placing the cells as close together as practical
3. Using one light source to illuminate both cells

Frequency characteristics of the silicon cells are important for high-speed applications. Frequency response depends on load resistance and cell size. The output of a pair of 0.200 in. × 0.050 in. cells can drop 10% to 25% when operated into a 1000-Ω load at 50 kHz. This does not limit the maximum usable cell frequency to 50 kHz, however. What is important is the shift in zero crossing with frequency. With a perfectly balanced channel, the zero crossings are unchanged, but with an unbalanced pair of cells, the shift may be appreciable at high speeds because the outputs of the two cells drop unequally. The average output from each individual cell of a back-to-back pair does not change with frequency. However, as the ac components fall with frequency, any unbalanced dc component becomes more significant and hence has the effect of increasing the phase error. Figure 11-12 illustrates the output of typical cells and the resultant output signal at low- and high-frequency operation. The average value of each cell remains the same regardless of frequency, but the peak-to-peak output changes. The two resultant wave forms show that the symmetry level remains the same. At low frequency the 180° crossing is in error by 10 electrical degrees, but at high frequencies the crossing error increases to 18 electrical degrees. This error can be reduced somewhat by inserting a fixed dc signal of equal magnitude but opposite polarity to the symmetry level.

Optical Incremental Encoder Component Description

The following is a description of the major encoder components and the significant influence of these components on encoder performance.

Light Sensor The sensor used in most applications is a photovoltaic, wide-area, silicon cell. Figure 11-13 shows the construction of the silicon cell. At the heart of the sensor is an area of single-crystal, *N*-type silicon material with a *P*-type layer diffused into it. This crystal generates an output when illuminated.

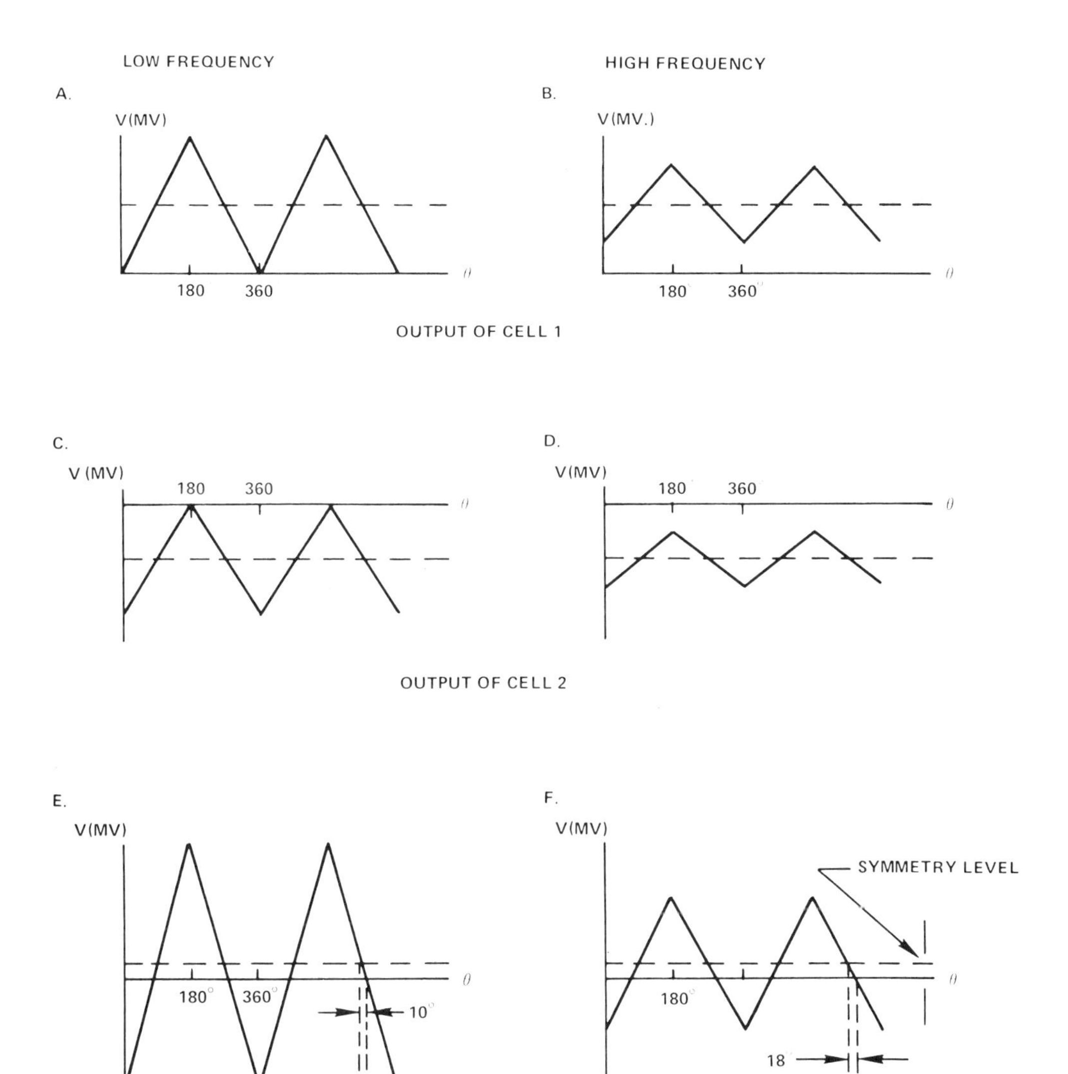

Figure 11-12 Typical Frequency Response of a Pair of Unbalanced Cells *(Courtesy, Dynamics Research Corp.)*

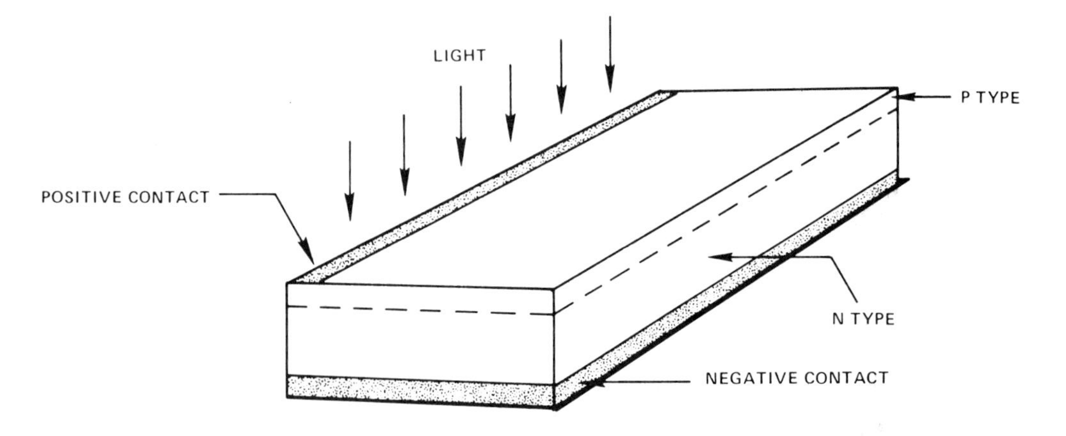

Figure 11-13 Light Sensor Construction *(Courtesy, Dynamics Research Corp.)*

Solder connections are made to the crystal, and the signal is brought out through the leads.

Light energy is comprised of small energy packets called photons. When a photon containing sufficient energy strikes a silicon crystal, a hole–electron pair is created. This pair diffuses and is collected by the *PN* junction. When large quantities of photons are involved, a voltage difference appears between the *P* and *N* regions. When a load is connected across the crystal, the voltage potential causes a current, which is supplied by the light-generated hole–electron pairs.

The silicon cell has several advantages over other photosensitive materials. Unlike the photodiode and phototransistor, operation of the silicon cell requires no external voltage. Furthermore, the silicon cell is a current source and matches the transistor input characteristic. The spectral response of the silicon cell is relatively broad and is well matched to the outputs of both the tungsten filament lamp and the light-emitting diode. The silicon cell's coefficient of output signal versus light intensity in the infrared region is greater than that of any other candidate material. The frequency response of the silicon cell is excellent, and silicon is resistant to most environmental hazards.

Light Source Both lamps and light-emitting diodes (LED) are available in encoders. Lamps offer greater light output than LEDs but have shorter lives. Lamp life can be extended somewhat by dropping the operating voltage. The magnitude of this drop is limited by lamp output and the spectral response of the silicon cell. Lamp output is proportional to the 3.5 power of the operating voltage in the vicinity of the lamp rating. The spectral response of tungsten, illustrated in Figure 11-14, shifts to the right as the operating voltage is lowered. As a consequence, less of the lamps' light output is within the spectrum to which the silicon cell is sensitive. The result is a trade-off between lamp life and signal strength. The lamps currently used are operated at 5 V and have an average life of 40,000 hours at that rating.

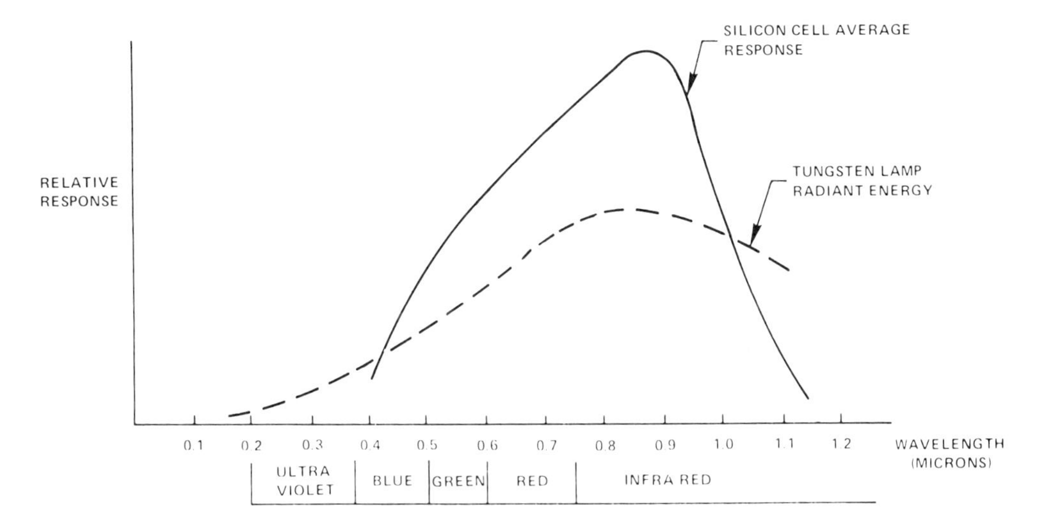

Figure 11-14 Spectral Response of Silicon Cell and Spectral Distribution of Tungsten Lamp *(Courtesy, Dynamics Research Corp.)*

LEDs are used in applications that require improved tolerance for mechanical shock and vibration. Although LED output is less than that of lamps, output is concentrated near the peak of the silicon cell spectral response. The resulting signal is 20% to 50% lower than when a lamp is used. Another penalty of the LED is its high temperature sensitivity. LED output drops about 1% for each 1°C temperature increase. Even so, the LED is an acceptable solution to the problem of mechanical shock and vibration in many applications. Table 11-1 is a comparative summary of the important lamp and LED characteristics.

Shutter A critical encoder component is the shuttering mechanism. The manufacture and assembly of this device is the basis of the encoder's resolution and accuracy. The material of which the scale or disk is made can limit the suitability of the encoder for use in some environments.

Scales can be made by producing (1) slits in an opaque metal, or (2) opaque lines on a transparent material like glass or plastic. Metal scales are normally

TABLE 11-1
LIGHT SOURCES

	Lamp	*LED*
Life	40,000 h max.	40,000 and up
Output	High	½ Lamp
Output versus temperature	None	High
Shock + vibration capabilities	Moderate	Good
Temperature cycle damage	Moderate	Low
Output brightness versus voltage	High	Moderate (linear)
Maximum operating temperature at rated power	200°F	150°F
Source size	Moderate	Small

Courtesy Dynamics Research Corp., Wilmington, Mass.

low-resolution devices because the minimum slit width is approximately the metal thickness, and very thin scales will not support themselves. The solution is to support the metal with a transparent material, the second method of making scales.

Both glass and plastic scales can be used. Glass, being rigid, provides greater accuracy. The disadvantage is breakage. The opaque lines can be metal or photo emulsions. Metal on glass scales are extremely durable and are easily cleaned with acetone and cotton. Photoemulsion scales are more easily damaged and must be carefully handled. Also, light transmission of the photoemulsion scales is reduced about 12% by the clear gelatin that remains in the undeveloped areas.

Linear scales are produced with the process of evaporating inconel on a soda lime plate glass. Master scales for this process are produced with a numerically controlled linear ruling engine equipped with laser interferometer feedback. Dimensional standards for this equipment have been developed by the National Bureau of Standards.

Signal-Conditioning Electronics Encoder electronics perform the following functions:

1. Generation of the output signal waveforms
2. Generation of complementary signals
3. Direction sensing
4. Filtering

The available output wave forms are illustrated in Figure 11-15a.

The square wave form of Figure 11-15b is produced with a shaper circuit called a Schmitt trigger. The shaper has two stable output states, which are controlled by the amplitude of the input voltage.

Shaper output is fed to gates that generate the pulses marking the encoder signal zero crossings (Figure 11-15c). At this stage, circuitry is also provided for sensing direction of movement (depending on whether the channel A wave form leads channel B, or vice versa).

Shaft Bearings Stainless steel, prelubricated, instrument-grade ball bearings are employed to minimize shaft friction in rotary encoders. Bearing assemblies are normally used in preloaded pairs to minimize axial and radial shaft play.

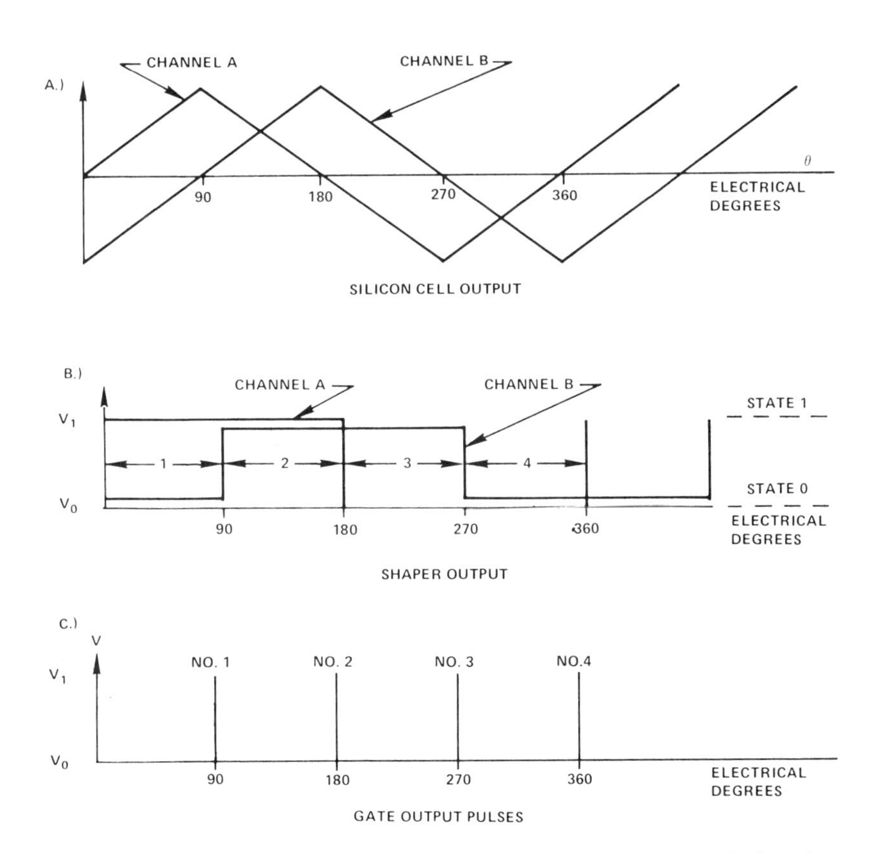

Figure 11-15 Encoder Output Signals *(Courtesy, Dynamics Research Corp.)*

Mechanical Assembly The fundamental mechanical considerations are as follows:

1. Alignment of the disk and reticle patterns concentric with the encoder's axis of rotation
2. Maintenance of a constant, minimum separation of disk and reticle
3. Protection of the functional components from damage during manufacture and installation and from environmental attack during service

Figure 11-16 shows the major rotary encoder components prior to assembly. Figure 11-17 is a cutaway view after assembly.

Accurate concentricity is achieved before the epoxy, with which the disk is attached to the shaft, becomes hard. The disk pattern includes an opaque circle around the outer diameter. The outer edge is accurately round and concentric with the center of the radial sectors of the disk pattern. After the disk is set in the epoxy, the shaft is rotated and the pattern ring is inspected for radial runout with a microscope. If runout is observed, the direction and magnitude of the disk misalignment relative to its center of rotation are determined. The disk is then relocated and clamped in place while the epoxy hardens. This method effectively compensates for bearing eccentricity as well, and total indicated runout can be held within 0.0002 in. When greater accuracy is required, a Lissajous pattern is employed and total indicated runout can be reduced to about 0.00005 in. maximum.

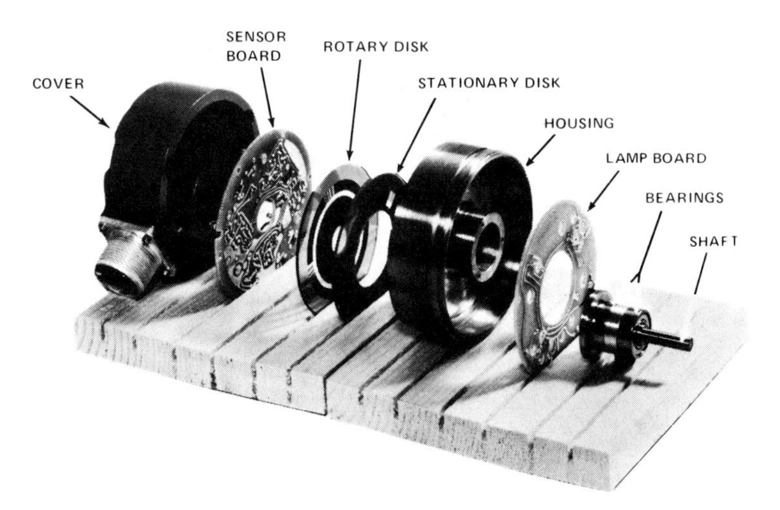

Figure 11-16 Major Rotary Encoder Components *(Courtesy, Dynamics Research Corp.)*

Figure 11-17 Cross Section of the Rotary Encoder *(Courtesy, Dynamics Research Corp.)*

TYPICAL ROTARY ENCODER

Typical of rotary encoders is the Dynamic Research Corporation's Model 77 (Figure 11-18). The unit is a compact optical shaft angle encoder that offers resolution up to 10,000 counts/shaft revolution. It measures 2.3 in. in diameter and is available with English and metric system resolutions of 100/254 or 500/1270.

The Model 77 is offered with any of the following outputs: low-level sine wave (40 mV peak to peak); amplified sine wave (single or differential); square wave (5, 6, 12, or 15 V, with or without complements); pulse type (3 μs $\pm$ 1 μs wide); and dual differential line driver. Typical applications include optical comparators, photogrammetric systems, CAT scanners, computer peripherals, office equipment, and machine tools.

ENCODER INTERFACE CONSIDERATIONS

Lack of general information on encoder applications can frustrate the first-time user. Whatever the encoder application, the following basic decisions must be made:

1. What signal level and wave form will be used
2. How the encoder output will be interfaced with the system

Figure 11-18 Typical Optical Shaft Angle Encoder *(Courtesy, Dynamics Research Corp.)*

3. Whether a rotary or linear encoder will be used
4. How the encoder will be mounted
5. How the mechanical input will be coupled to the encoder

Electrical Interface

The most frequent problems encountered in transmitting the encoder signal to the receiving electronics are signal distortion and electrical noise. Either problem can result in gain or loss of encoder counts. Imprecise monitoring of the mechanical input is the ultimate result.

Signal Distortion Signal distortion is illustrated in Figure 11-19. The receiving electronics will respond to an input signal that is either logical 0 or logical 1. As the leading edge of the wave form is distorted, the transition time increases. At some point the receiver becomes unstable, and encoder counts may be gained or lost.

The primary cause of the distortion is cable length or, more specifically, cable capacitance. To minimize distortion, high-quality cable with capacitance of less than 40 pF/ft should be used. The longer the cable is, the greater the distortion. Beyond some cable length the signal must be "reshaped" before it can be used reliably.

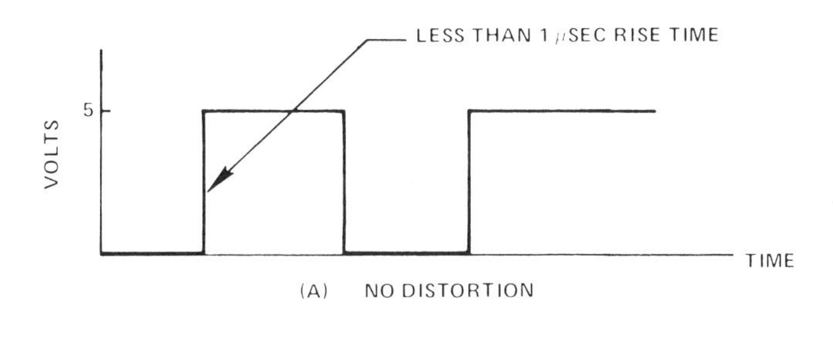

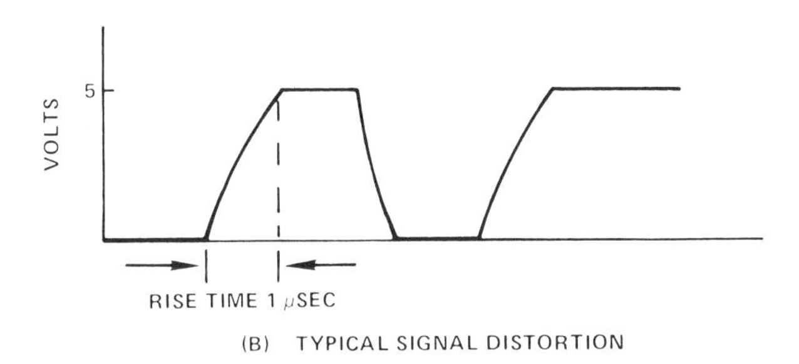

Figure 11-19 Signal Distortion *(Courtesy, Dynamics Research Corp.)*

Electrical Noise The problem of radiated electrical noise, while potentially serious, can generally be overcome with a few simple precautions. Signal cables should always be run in trays isolated from other ac carriers and where possible kept from the vicinity of such noise generators as electric welders and electronic-discharge machining equipment.

When it is known that the cable will be exposed to noise, twisted wire pairs, individually shielded, should be used. The shield should be tied to earth ground through the instrument case of the signal destination (Figure 11-20). In severely noisy environments it may be necessary to also tie the signal ground to the instrument case through a 0.1-μF capacitor.

In addition to radiated noise, encoder operation may be influenced by transients in the encoder's power supply. Encoder power should be regulated to 5 V ± 5%. If load variations are a problem, a line-regulating transformer should be used. Noise spikes on the input power can damage both the light source and the encoder electronics. A diode external to the encoder is recommended for protection. A remote-sense power supply is sometimes required to maintain the proper voltage when long signal cables are used and the encoder power supply is located at the signal destination.

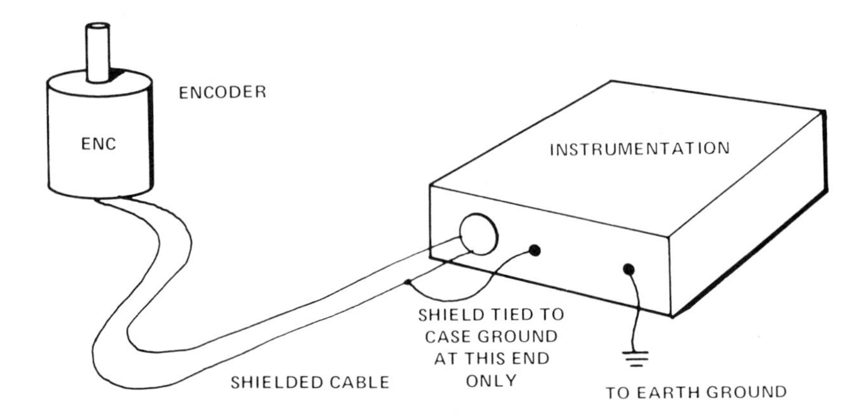

Figure 11-20 Recommended Cable Shield Grounding *(Courtesy, Dynamics Research Corp.)*

Mechanical Interface

Both the encoder mount and the mechanical coupling influence encoder accuracy and life. The rotary encoder can be bolted to its support or attached with the standard synchro mount. The latter allows the encoder to be rotated in its mount for precise orientation of the zero reference with respect to the mount. A pilot circle is provided on each case to ensure concentricity of the mount with the encoder shaft.

The most common methods of coupling the mechanical input to the rotary encoder are as follows:

1. Flexible coupling
2. Rack and pinion gearing
3. Gear to gear
4. Toothed belts

The primary objective of the coupling is to accurately transmit the input motion to the encoder without subjecting the shaft to excessive loads (as specified for each product). Excess loads will damage the shaft bearings and result in premature failure. In addition to static loads, care must be taken to ensure that momentary or shock loads do not exceed specifications.

The flexible coupling is recommended because reasonable misalignment can be tolerated without damaging the encoder bearings. The flexible coupling also normally provides the most accurate transmission of the mechanical input.

When the rack and pinion (Figure 11-21) is used, the encoder should be flexibly mounted to accommodate runout. The rack must also be protected from foreign objects, such as metal chips, that could subject the encoder shaft to a shock load.

Gear-to-gear interfaces (Figure 11-22) must be carefully aligned to avoid bottoming the gears or subjecting the encoder shaft to a moment perpendicular to its axis. The same precautions against foreign objects must be observed as for the rack and pinion.

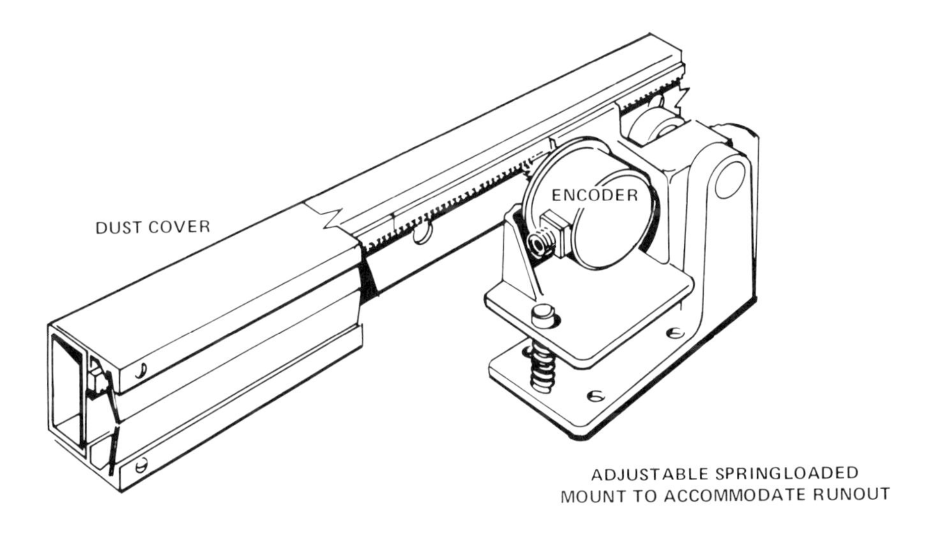

Figure 11-21 Rack and Pinion Spar Assembly *(Courtesy, Dynamics Research Corp.)*

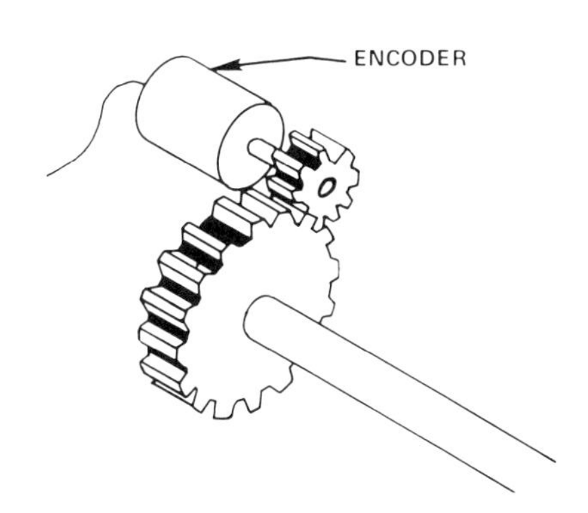

Figure 11-22 Gear-to-Gear Assembly *(Courtesy, Dynamics Research Corp.)*

Toothed belts are sometimes used but are not recommended, because belt whip can subject the encoder shaft to shock loads, and belt drives are relatively inaccurate (Figure 11-23).

Linear encoders may or may not be enclosed, and accuracy depends on system alignment. Mounting the enclosed system is relatively simple. The read head is generally coupled to the mechanical input through a flexure. The unit must be mounted on a reasonably flat surface and the scale aligned parallel with the direction of the input motion.

Figure 11-23 Precision Belt Drive *(Courtesy, Dynamics Research Corp.)*

Greater care is required to mount the unenclosed system. Figure 11-24 illustrates the surfaces that must be aligned. The scale is mounted on surface *C*. The read head travels on surfaces *A* and *B*. To maintain the fixed gap between the reticle and scale, surfaces *B* and *C* must be held parallel, and surface *A* must be perpendicular to *B*. Surface *D* is the top edge of the scale. Parallelism between surfaces *D* and *A* is required to maintain parallelism between the reticle's lines and those on the scale. If the lines are not parallel, the light shutter will not properly modulate the light.

Encoder Life

Encoder life is most often limited by (1) lamp failure, (2) shaft-bearing failure, or (3) an electronics failure. The ultimate of encoder lamp failure is filament evaporation, although premature failure can be caused by excessive mechanical vibration.

Natural failure of encoder electronics is rare and is not a relevant limit to encoder life. However, damage caused by improper troubleshooting or installation practice is far more prevalent. This accounts for the majority of the encoder electronics problems. Most elements of the electronics are current-

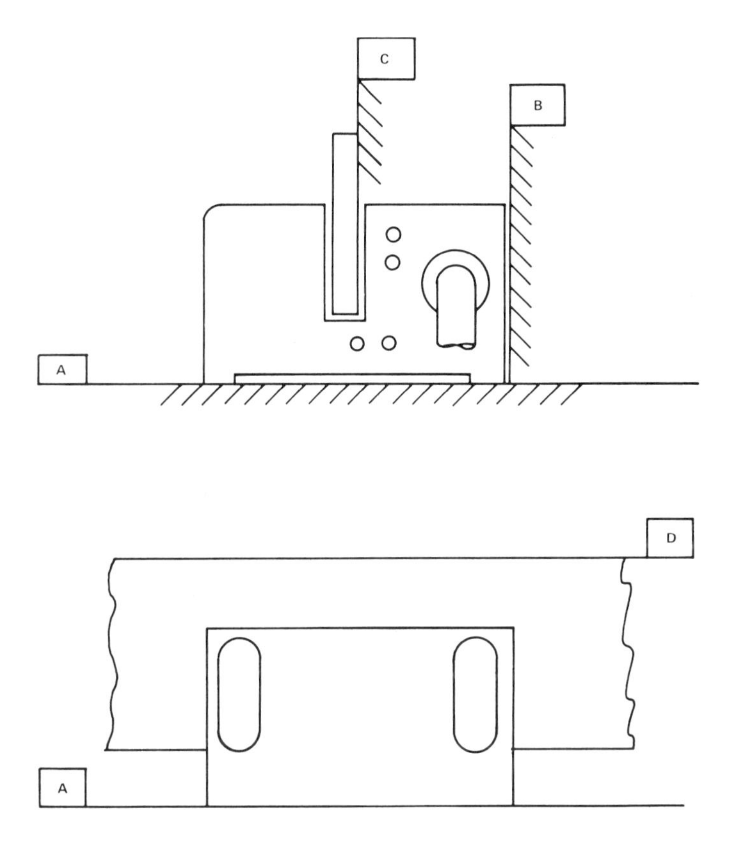

Figure 11-24 Scale and Encoder Mounting Requirements *(Courtesy, Dynamics Research Corp.)*

sensitive devices and are destroyed when an excessive or reverse potential is applied. Installation procedures must be carefully followed and technical assistance should be obtained before troubleshooting an encoder failure.

Error

While the encoder manufacturer is primarily concerned with encoder error, the encoder user must naturally be concerned with total error. Total error consists of encoder error and the errors caused by interaction of the encoder with the rest of the system.

Encoder error consists of quantization error and instrument error. Encoder error is largely determined by manufacturing and assembly tolerances and by how well the characteristics of each silicon cell pair are matched. For example, rotary encoder accuracy is typically limited to disk eccentricity.

Encoder error is relatively easy to quantify and is usually provided by the manufacturer. However, total error depends on the application and is difficult to quantify. When total system error must be determined, it is usually necessary to construct a prototype of the system and conduct tests.

The most frequently encountered sources of error are as follows:

1. Manufacturing and assembly tolerances
2. Silicon cell characteristics
3. Scale or disk alignment
4. Regulation of power supply
5. Electrical noise
6. Temperature variation
7. Mechanical coupling
8. Mechanical vibration

Items 1, 2, and 3 are sources of encoder error. In preassembled encoders these sources are controlled by the manufacturer. In user-assembled encoders, care must be taken to maintain the alignment specified by the manufacturer. Otherwise, the manufacturer's specification of encoder error will be invalid.

The requirements for power-supply regulation and protection from electrical noise were discussed earlier. Failure to properly regulate the power supply may produce intermittent problems that are difficult to trace. Likewise, the error caused by electrical noise is random and usually difficult to pinpoint.

Temperature influences encoder accuracy in several ways. The light source, if an LED, is strongly temperature sensitive. A lamp is not significantly affected by normal operating temperatures (0°F to 120°F) unless operated in a vacuum, which causes higher envelope temperatures and reduces lamp life. Because of the push–pull arrangement, silicon cell output does not vary significantly with temperature.

Dimensional changes with temperature are not compensated. The most significant dimensional change is that of the linear encoder glass scale. Rotary encoder accuracy is not significantly affected by dimensional changes of the glass disk.

The mechanical coupling can cause errors in two ways. The input motion will not be accurately reproduced at the encoder if the coupling flexes (rotary encoder windup). Also, parallel misalignment of the shafts can cause the encoder shaft not to track the input accurately. These characteristics depend on the type of coupling used and are specified by the coupling manufacturer.

Mechanical vibration is a more subtle source of error. Vibration can cause the reticle-to-scale spacing to fluctuate or, more significantly, can cause the relative velocity between reticle and scale to momentarily exceed the encoder frequency response as the vibration passes through its peak velocity. Any error caused by vibration is normally intolerable.

Mechanical wear and decreasing lamp output contribute to a gradual deterioration of encoder accuracy with age. Mechanical wear increases misalignment. Lower light output accentuates silicon cell mismatch. Cell mismatch produces a periodic error in output pulse spacing and is noncumulative. Wear, however, is a source of cumulative error.

ENCODER APPLICATIONS

Inserting Components in Printed Circuit (PC) Boards

Manually positioning electronic components for soldering to printed circuit (PC) boards can severely limit production. Computer-controlled equipment that rapidly and precisely positions the PC board for insertion of each component can increase production rates 10 to 15 times. This equipment is widely used by high-volume PC board manufacturers.

The PC board is typically attached to an *xy* table driven by two low-inertia motors and precision lead screws. Position is provided by rotary transducers attached to the lead screws. The computer, programmed with the coordinates of each component to be inserted, advances the PC board from point to point. The important characteristics of the position transducer used in this application are noise immunity, accuracy, reliability, low cost, and compatibility with digital equipment. One example is a rotary encoder with quasi-sine-wave output coupled through a differential comparator to an up–down counter (Figure 11-25). The differential comparator provides noise protection, and the computer monitors *xy* position by interrogating the counter. Linear resolution of the entire system including the encoder is 0.0005 in.

In applications requiring even greater accuracy, the linear encoder is used. Bypassing the lead screw eliminates one source of error.

The relative simplicity of interfacing the encoder with digital systems is the chief advantage over inductive components. Inductive components require expensive A/D conversion and are also less accurate than encoder-based systems.

Some equipment manufacturers use step motors to simplify stopping the PC board at a precise *xy* coordinate. Although step motors can in theory be operated in an open loop, position feedback is normally employed to compensate for missed steps.

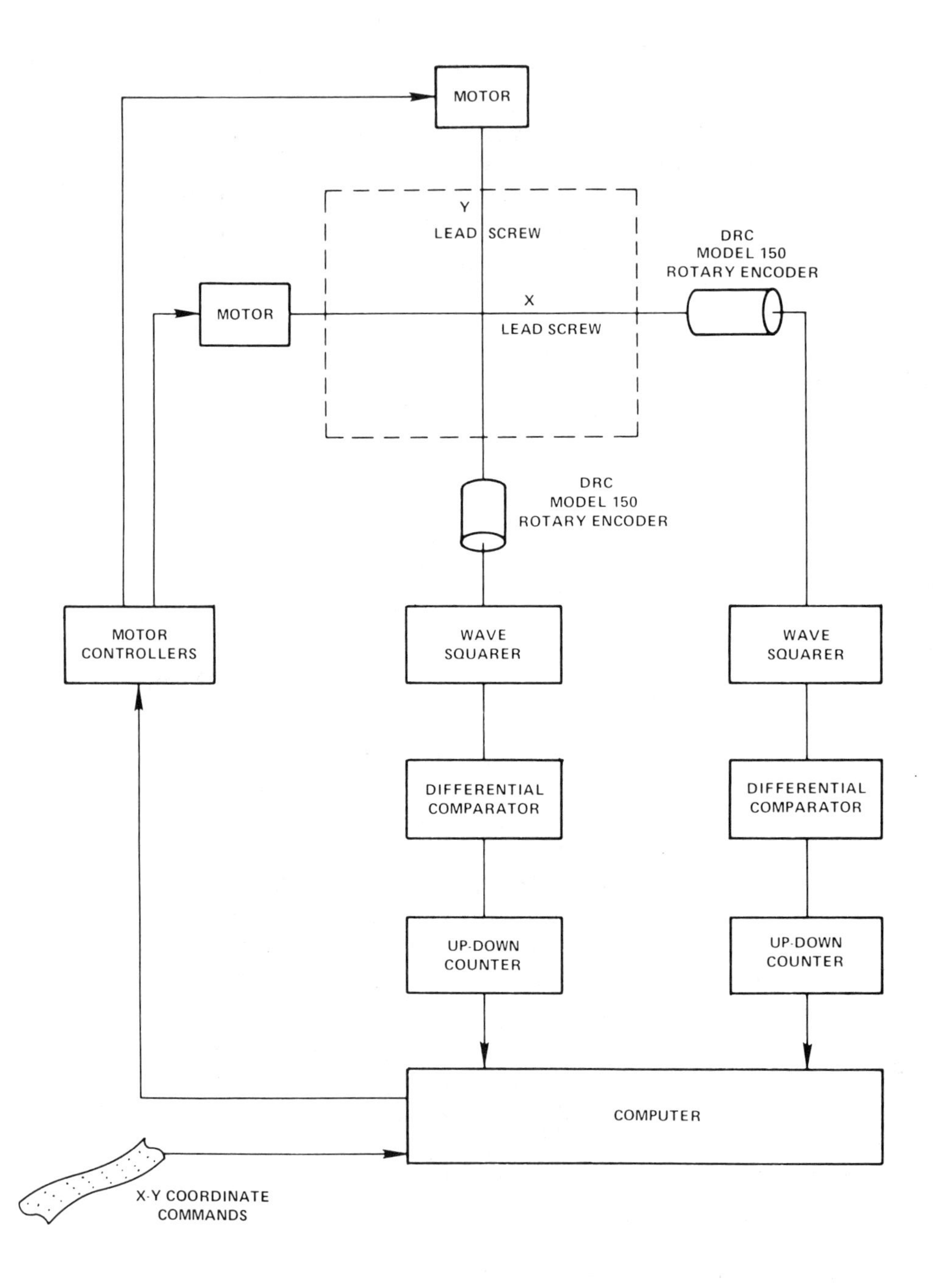

Figure 11-25 *XY* Table for Component Insertion *(Courtesy, Dynamics Research Corp.)*

X-Ray Analysis of Tumors

Tomography, the study of cross sections of the body, is an important medical diagnostic tool. One modern application of tomography is computer-aided x-ray analysis of tumors. The objectives of tumor analysis are determination of shape, size, and precise location. This information is used for surgical removal of the tumor or for other treatment, such as radiation therapy.

Until recently, tumor analysis was performed by fluoroscopy or by a series of x-ray photographs. However, a series of such photographs contains far more information than can be utilized by manual analysis. This problem has been overcome with a computer-aided x-ray scanner. The scanner simulates three-dimensional images of the tumor that can be mathematically rotated for analysis in any orientation.

One manufacturer produces a machine whose scanner consists of an x-ray source and sensor mounted on a doughnut-shaped ring. The source and detector can be moved about the area of interest until sufficient data have been obtained for modeling the tumor. Mathematical analysis requires the precise position of the x-ray source at every point of the scan.

A rotary encoder indicates the position of the x-ray source on the doughnut and a linear encoder indicates the doughnut's position. The square-wave output of each encoder is coupled to an up–down counter. The counters are interrogated by the computer to determine the position of the x-ray source. Figure 11-26 is a block diagram of the system.

In all x-ray diagnostic procedures the patient's radiation dose must be minimized. This requires a short scan cycle, as well as highly reliable equipment so that repeated scans are unnecessary. The relatively noise-free medical environment enhances the basic reliability of the optical incremental encoder. Moreover, the encoder's incremental output is easily interfaced with the computer.

Techniques similar to those described here are being used in many related applications. The common denominator is a need to track the relative position of an object.

Fabric Cutting

Fabric cutting is an important operation in many industries, including the manufacture of furniture, automobile headliners and seatcovers, and garments. Historically, fabric cutting has been a labor-intensive operation, but it is being rapidly automated. Unlike other industries such as machine tools, the driving force for automation of fabric cutting is not reduction of labor, but rather, the material savings to be achieved through higher-accuracy cutting and elimination of errors.

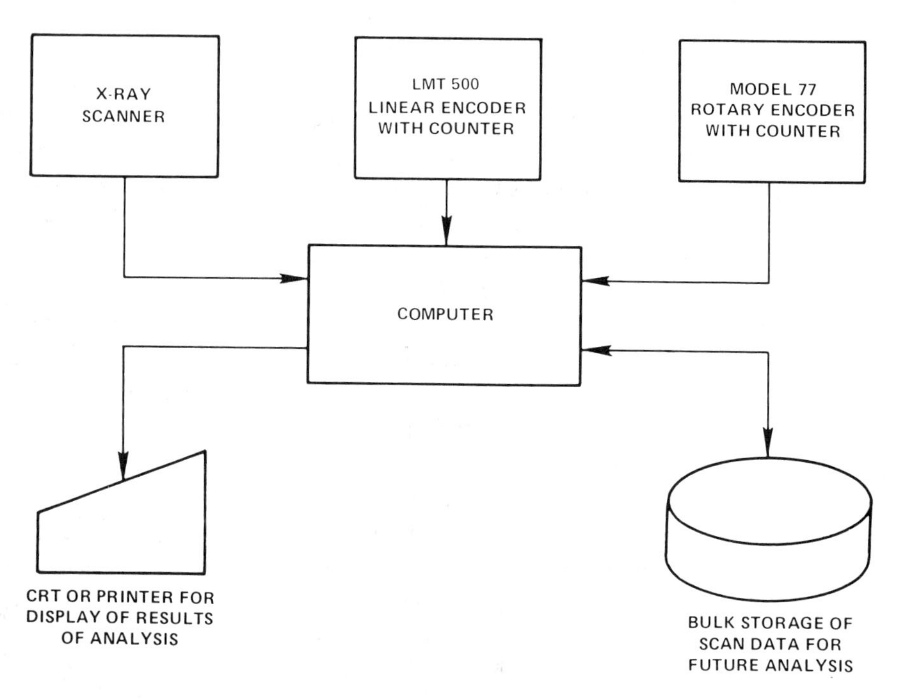

Figure 11-26 Components of System for Computer-Aided X-Ray Analysis of Tumors *(Courtesy, Dynamics Research Corp.)*

Typical fabric-cutting equipment can handle single layers of material or stacks up to 4 in. thick and can traverse areas up to 6 ft wide and 100 ft long. Cutting is performed by a knife blade that is kept aligned with the direction of the cut by rotating the knife about an axis perpendicular to the plane of the fabric (Figure 11-27). A carriage assembly that includes the cutting mechanism, ac drive motors, and control equipment is supported on two sets of round ways (one for x movement and one for y). Depending on the size of the machine, the carriage assembly weighs from 200 pounds to one ton.

Control is provided by a computer programmed with the xy coordinates of the pattern to be cut. Three rotary transducers provide position sensing; two for xy position and one for knife angle. xy input is coupled to the transducer with rack and pinion drives. The knife is coupled directly to a transducer shaft. Low cost, reliability, and resistance to dust or severe mechanical vibration are major requirements of the transducer used in this application.

A major manufacturer of fabric-cutting equipment uses a rotary encoder with 5-V square-wave output coupled to an up–down counter. Tolerance to mechanical vibration is enhanced with an LED light source in place of the more

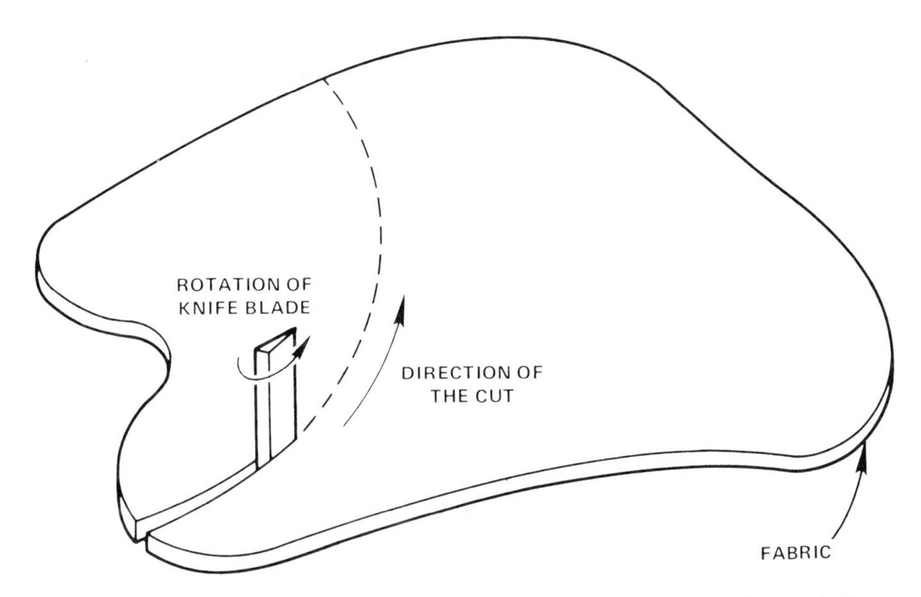

Figure 11-27 Fabric-Cutting Blade Alignment *(Courtesy, Dynamics Research Corp.)*

common tungsten filament lamp. Maximum error of the encoder/rack and pinion assembly is ± 0.001 in. per foot and is far within total system error. Overall error of the typical cutting equipment is about $\pm \frac{3}{32}$ inch. Major sources of error are movement of the fabric during the cut and flexure of the thin knife blade. By using linear encoders or more accurate mechanical coupling, encoder error can be reduced by a factor of 10 where required.

Phototypesetting

Typesetting, when performed manually, requires considerable skill. The two major considerations are as follows:

1. Selecting the proper character of the proper size from the available character set
2. Spacing the characters so that the correct words are formed and so that each line of print is exactly the same length

The large volume of newsprint and other material produced by modern printers provides impetus to automate this process. Phototypesetting is the most widely used of several automatic typesetting methods and illustrates the diversity of encoder applications.

Figure 11-28 shows the basic components of the phototypesetting system. The available characters are negative silhouettes on a continuously rotating disk. The address of each character is a unique angular position of the disk. Type is set by flashing the strobe light when the desired character is positioned over the light source. The projected image of the character is focused on a photosensitive surface by the mirror–lens assembly. Each character's position on the printed line is determined by the position of the mirror–lens arrangement on the lead screws, as illustrated. To print an entire line, the control computer determines, from the input text, the characters to be printed and the spacing required to delineate words and maintain constant margins. The mirror–lens assembly is indexed to the start of the new line. The computer then repetitively executes the following until the line is complete:

1. Address the desired character, on the rotating disk
2. Flash the light at the appropriate time to set the character
3. Index the mirror–lens assembly to position for setting the next character

In some phototypesetting equipment the position of the mirror–lens assembly is monitored with a rotary encoder attached to the drive motor shaft. The chief

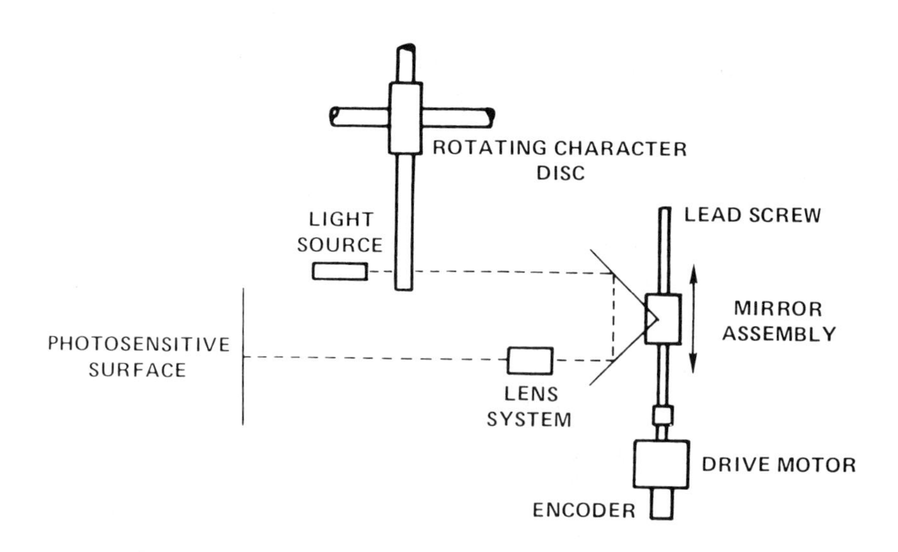

Figure 11-28 Phototypesetter *(Courtesy, Dynamics Research Corp.)*

advantage of the encoder in this application is its low inertia, which facilitates rapid movement of the mirror–lens assembly to each character position.

Synchronization of Spindle and Tool Movement on Numerically Controlled Lathes

Thread machining is a multipass lathe operation. The workpiece is fixed in a chuck and rotated at constant speed. The cutting tool engages the workpiece at the beginning of the threadform and is translated parallel to the axis of rotation. When the full thread length has been cut, the tool is retracted, and the spindle and tool movement is reversed to bring the tool to the beginning of the threadform for cutting the second pass. This process is repeated until the required thread depth has been reached.

To avoid cutting multiple threads, the tool must engage the workpiece at the same circumferential position at the start of each pass. The relation between spindle and tool speeds must also be constant. Otherwise thread pitch will vary along the length of the workpiece.

The conventional lathe uses cams and gears to maintain synchronization and control thread pitch. The desired threadform is cut by selecting the proper combination of cams and gears. Skilled operators are required, and long-term wear of the mechanical parts causes accuracy to deteriorate.

The numerically controlled (NC) lathe achieves synchronization without cams and gears. Having fewer mechanical parts reduces wear and simplifies lathe setup. Less skill is required to operate the NC lathe than the conventional lathe, and because the multipass operation is completely automatic, the lathe does not have to be constantly tended.

One manufacturer of NC lathes synchronizes spindle and tool movement by equipping the spindle with a rotary encoder and using the encoder's 5-V square-wave output to control tool movement. The ratio of tool to spindle speed is made adjustable with an oscillator, as illustrated in Figure 11-29. The oscillator output is a multiple of the input received from the encoder. The encoder's zero reference signal is used to ensure that the tool engages the workpiece at the same circumferential position at the beginning of each pass.

Photogrammetry

Measurement and control of *xy* motion encompasses many encoder applications. Among these is photogrammetry, the precise measurement and interpretation

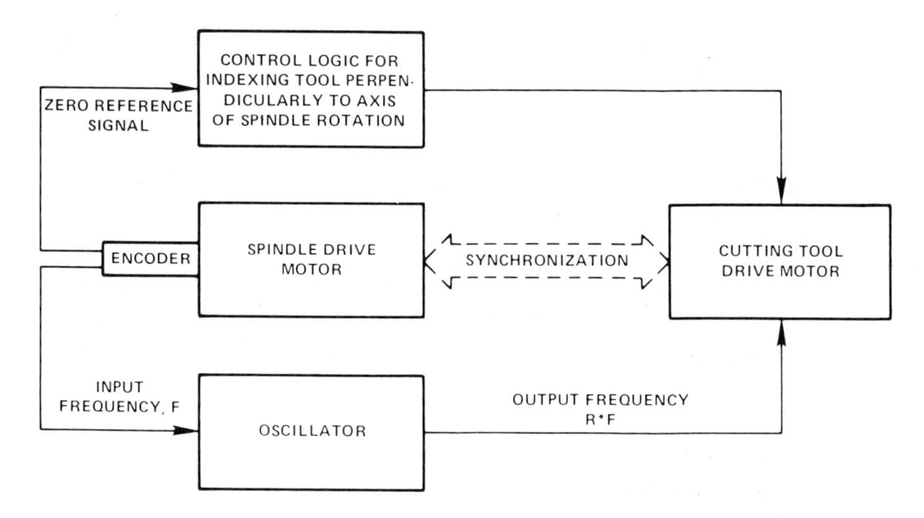

Figure 11-29 Synchronization of Tool and Spindle Drives *(Courtesy, Dynamics Research Corp.)*

of photographic images. This tool is widely used by government agencies, the military, and many businesses. Typical applications that use aerial photographs are the following:

1. Production of topographical maps
2. Land surveying
3. Determining size and location of structures
4. Agricultural control
5. Tax assessment
6. Land utilization studies

The fundamental task is accurate photographic measurement to deduce the desired distances, elevations, or structural dimensions.

Accuracy, linearity of transducer output over the full displacement, and ease of interfacing with digital systems are important transducer characteristics in these applications. Transducer error must typically be held within ± 25 μm in the least demanding applications and within ± 1 μm in the most demanding. Because of the considerable data-processing requirements, photogrammetry systems are typically designed as either computer peripherals or stand-alone units with magnetic tape or disk storage.

Both rotary and linear encoders are used in these applications. The 5-V square-wave output is coupled to a scaled up–down counter. When greater accuracy is required, the ± 1-μm linear encoder is used.

The encoder is immune to the problems of drift and nonlinearity, and the counter's BCD output is easily interfaced with computers and with stand-alone data-storage devices.

REFERENCES

The information and illustrations in this chapter are reprinted from application notes and specification sheets prepared by Dynamics Research Corporation (DRC), Wilmington, Massachusetts.

Index